Building and Training Generative AI Models

A Practical Guide to Generative AI Development and Scaling

Irena Cronin

Apress®

Building and Training Generative AI Models: A Practical Guide to Generative AI Development and Scaling

Irena Cronin
Savannah, GA, USA

ISBN-13 (pbk): 979-8-8688-2331-2 ISBN-13 (electronic): 979-8-8688-2332-9
https://doi.org/10.1007/979-8-8688-2332-9

Managing Director, Apress Media LLC: Welmoed Spahr
Acquisitions Editor: Celestin Suresh John
Editorial Assistant: Gryffin Winkler

Cover designed by eStudioCalamar

Cover image designed by Pixabay

Distributed to the book trade worldwide by Springer Science+Business Media New York, 1 New York Plaza, New York, NY 10004. Phone 1-800-SPRINGER, fax (201) 348-4505, e-mail orders-ny@springer-sbm.com, or visit www.springeronline.com. Apress Media, LLC is a Delaware LLC and the sole member (owner) is Springer Science + Business Media Finance Inc (SSBM Finance Inc). SSBM Finance Inc is a **Delaware** corporation.

For information on translations, please e-mail booktranslations@springernature.com; for reprint, paperback, or audio rights, please e-mail bookpermissions@springernature.com.

Apress titles may be purchased in bulk for academic, corporate, or promotional use. eBook versions and licenses are also available for most titles. For more information, reference our Print and eBook Bulk Sales web page at http://www.apress.com/bulk-sales.

Any source code or other supplementary material referenced by the author in this book is available to readers on GitHub. For more detailed information, please visit https://www.apress.com/gp/services/source-code.

If disposing of this product, please recycle the paper

For the memory of Daniel Cronin and his love of tech.

Table of Contents

About the Author xxix

About the Technical Reviewer xxxi

Acknowledgments xxxiii

Introduction xxxv

Chapter 1: Introduction to Generative AI Systems 1

Overview of Generative Modeling 1

1. Defining Generative Modeling 2
2. Types of Generative Models 4
3. Probabilistic Foundations 6
4. Training Methodologies and Loss Functions 8
5. Evaluation of Generative Models 9
6. Applications of Generative Models 12
7. Challenges in Generative Modeling 13
8. From Generative Models to Generative Systems 15

Applications Across Text, Vision, Audio, and Multimodal Domains 16

1. Text Generation 16
2. Vision and Image Generation 17
3. Audio and Speech Generation 18
4. Multimodal Applications 20
5. Industry Use Cases by Domain 21
6. Challenges in Cross-Modality Generation 21
7. Future Trajectories 22

Comparison with Discriminative Models....22
1. Fundamental Objective Differences....23
2. Example Use Cases by Model Type....24
3. Architecture and Training Differences....25
4. Data Requirements and Representational Learning....25
5. Evaluation Metrics....26
6. Strengths and Weaknesses of Each Model Type....27
7. Interoperability and Hybrid Architectures....28
8. Role in Human-AI Collaboration....28
9. Evolutionary Trajectories....31
More on Architecture Families (Transformers, GANs, VAEs, Diffusion)....33
1. Transformers....34
2. GANs....36
3. VAEs....38
4. Diffusion Models....39
Chapter 2: Choosing the Right Architecture....43
Transformer-Based Models for Sequence Generation....43
Origins and Core Principles....44
Configurations for Sequence Generation....45
Advantages of Transformer-Based Sequence Generation....46
Training Paradigms and Enhancements....52
Challenges and Limitations....52
Evaluation Considerations....53
Real-World Applications....54
Comparative Perspective with Other Architectures....55
Future Directions....55

GANs for Adversarial Training and Synthesis 56
Strengths of GANs 58
Challenges in GAN Training....................58
Conditional GANs 59
Architectural Innovations....................60
GANs Beyond Images 61
Comparison with Other Generative Architectures 61
Ethical and Societal Considerations 62
The Evolving Role of GANs....................62
VAEs for Probabilistic Latent Modeling 63
Core Architecture and Probabilistic Foundation 63
The Reparameterization Trick....................64
Advantages of Probabilistic Latent Modeling 64
Limitations and Challenges 65
Variants and Extensions 66
Applications of VAEs 67
Comparison with Other Generative Architectures 67
Role in Multimodal and Hybrid Architectures 68
Ethical Considerations....................68
Future Directions 69
Diffusion Models and Denoising Strategies 69
Conceptual Foundation....................70
Intuition Behind the Process....................71
Variants of Diffusion Models....................71
Denoising Strategies 72
Strengths of Diffusion Models 73
Limitations and Challenges 79

Applications........84
Comparison with Other Generative Architectures........85
Hybrid and Future Directions........90
Ethical and Societal Considerations........91

Chapter 3: Data Collection and Preparation........93

Data Types: Structured vs. Unstructured........96
The Nature of Structured Data........96
The Emergence of Unstructured Data........97
Semistructured Data: The Hybrid Zone........99
Differences in Storage and Retrieval........100
Preprocessing: Bringing Order to Disorder........101
Annotation and Labeling........103
Implications for Model Choice........104
Summary........105
Cleaning and Normalization Pipelines........107
Understanding the Purpose of Cleaning........107
Normalization As a Structural Prerequisite........109
Building a Cleaning and Normalization Pipeline........111
Handling Missing Data........113
Encoding Categorical and Text Variables........114
Detecting and Handling Outliers........115
Automating Data Cleaning........116
Monitoring for Drift........117
Case Study: Cleaning Clinical Trial Data........118
Final Thoughts........119
Synthetic Data Generation........119
What Is Synthetic Data?........120
Why Use Synthetic Data?........121

Techniques for Generating Synthetic Data 123
Evaluating Synthetic Data Quality 127
Synthetic Data in Practice 128
Challenges and Considerations 129
Future Directions 129
Dataset Bias Detection and Mitigation 130
Defining Dataset Bias 131
Forms of Dataset Bias 131
Why Dataset Bias Matters 133
Detecting Dataset Bias 133
Evaluating Bias in Model Performance 135
Mitigating Dataset Bias 135
Advanced Approaches 138
Practical Workflow 138
Case Study: Bias in Facial Recognition Systems 139
Organizational and Legal Considerations 140
Looking Ahead 141
Chapter 4: Training Fundamentals and Self-Supervised Learning ...143
Self-Supervised Learning Principles 144
From Supervised to Self-Supervised 144
Core Idea of SSL 146
Pretext Tasks 148
Real-World Applications 149
Advantages and Challenges 151
Loss Function Design 153
Why Loss Functions Matter 153
Reconstruction Losses 155
Adversarial Losses 157

Contrastive Losses 158
Hybrid Objectives....... 160
Reflections on Loss Function Design....... 161
Overfitting and Generalization....... 162
Defining Overfitting....... 163
Causes of Overfitting 164
Strategies to Improve Generalization....... 167
Regularization....... 167
Data Augmentation 167
Cross-Validation 168
Architectural Choices 168
Transfer Learning and Pretraining....... 168
Ensemble Methods....... 169
Monitoring and Validation....... 169
Modern Perspectives 169
Double Descent 170
Role of Pretraining....... 170
Implicit Regularization....... 170
Generalization in Large Language Models 171
Distribution Shifts....... 171
Ethical and Societal Dimensions 171
Monitoring Convergence and Training Progress 173
Training Dynamics....... 173
Training Diagnostics....... 175
Convergence Criteria 182
Loss Stabilization 182
Validation Metrics....... 183
Early Stopping 183

Gradient Norms 183

Representation Stability 184

Practical Monitoring Tools 184

TensorBoard 184

Weights & Biases 185

Custom Logging 185

Checkpointing 185

Distributed Monitoring 186

Pitfalls in Monitoring 186

Overfitting to Validation 186

Noisy Loss Curves 187

Metric Myopia 187

Resource Blindness 187

Emerging Techniques 188

Sharpness-Aware Monitoring 188

Representation Drift Tracking 188

Automated Alerts and Intervention 188

Evaluation Beyond Training Loss 189

Large-Scale Experiment Tracking 189

Conclusion 190

From Supervised to Self-Supervised 190

The Core Idea of Self-Supervised Learning 191

Loss Functions As Guides to Learning 192

Overfitting and Generalization 192

Monitoring Convergence and Training Progress 193

The Advantages and Challenges of Self-Supervised Learning 193

Interconnectedness of Training Concepts 194

Broader Reflections 195

Chapter 5: Optimization and Learning Strategies............................197

Why Optimization Matters..198

Historical Roots..198

Core Challenges..199

Themes of the Chapter...199

Broader Importance..200

Backpropagation and Gradient Descent..201

Historical Context of Backpropagation...202

Core Mechanics of Backpropagation...203

Gradient Descent As the Optimization Backbone 204

Practical Challenges in Gradient Descent..205

Variants of Gradient Descent..207

Broader Perspectives 209

Adaptive Optimizers (Adam, RMSProp, LAMB).......................................210

Motivation for Adaptive Methods..210

RMSProp...211

Adam 213

LAMB and Large-Batch Training..214

Comparative Analysis 217

Broader Perspectives and Future Directions.....................................220

Gradient Clipping and Normalization 221

The Problem of Exploding Gradients 222

Gradient Clipping Strategies...223

Gradient Normalization Methods..226

Interactions Between Clipping, Normalization, and Optimizers..........229

Extended Perspectives on Gradient Management.............................230

Learning Rate Schedules...231

Why Learning Rate Matters...232

Fixed vs. Dynamic Schedules ... 233
Classical Decay Methods ... 234
Modern Schedules ... 235
Large-Scale Training Considerations ... 237
Comparative Reflections and Future Trends) ... 239
Conclusion ... 240
Backpropagation and Gradient Descent: The Foundations ... 240
Adaptive Optimizers: Refinements of Gradient Descent ... 241
Gradient Clipping and Normalization: Stabilizing Training ... 242
Learning Rate Schedules: Orchestrating Progress ... 242
Interdependence of Strategies ... 243
Broader Reflections on Optimization ... 243
Looking Forward ... 244
Chapter 6: Scaling Training with Infrastructure and Distributed Systems ... 247
GPU/TPU Usage and Resource Provisioning ... 248
Evolution from CPUs to GPUs and TPUs ... 249
Architectural Characteristics of GPUs ... 250
Architectural Characteristics of TPUs ... 252
Provisioning Strategies ... 253
Utilization Challenges ... 254
Broader Implications of Hardware Scaling ... 255
Data and Model Parallelism ... 256
Conceptual Foundations of Parallelism ... 257
Data Parallelism ... 258
Model Parallelism ... 259
Pipeline Parallelism ... 260
Tensor Parallelism ... 261

Hybrid Strategies.....262
Bottlenecks and Challenges.....264
Checkpointing and Fault Tolerance.....265
The Need for Checkpointing in Large-Scale Training.....266
Historical Context of Checkpointing.....266
Principles of Fault Tolerance.....267
Models of Failures.....268
Checkpointing Strategies.....269
Trade-Offs in Frequency and Granularity.....270
Elastic Recovery in Distributed Training.....271
Conceptual Challenges and Bottlenecks.....272
Broader Reflections and Future Directions.....272
Distributed Training Frameworks.....274
The Role of Frameworks in Distributed Training.....274
Communication Primitives and Collective Operations.....275
Gradient Synchronization Strategies.....276
Fault-Aware and Elastic Framework Designs.....277
Abstractions for Parallelism (Data, Model, Pipeline, and Tensor).....278
Scalability Challenges in Framework Design.....278
Conceptual Comparison of Framework Approaches.....279
Future Directions in Distributed Training Frameworks.....281
Conclusion.....282
Infrastructure As the Third Pillar.....282
Hardware As the Foundation.....283
Parallelism As the Core Strategy.....283
Resilience Through Checkpointing and Fault Tolerance.....284
Frameworks As the Operational Backbone.....285

Interdependence of Components 285
Broader Reflections 286

Chapter 7: Fine-Tuning and Domain Adaptation 289

Transfer Learning and Pre-trained Model Utilization 290
Conceptual Foundations of Transfer Learning 291
Historical Evolution: From Feature Extraction to Pre-trained Transformers 292
Strategies for Transfer Learning 293
Benefits of Transfer Learning 295
Risks and Challenges 295
Conceptual Nuances in Pre-trained Model Utilization 297
The Broader Significance of Transfer Learning 298
Domain-Specific Dataset Curation 299
Pipeline 299
Why Domain-Specific Datasets Matter 302
Historical Context of Dataset Curation 302
Principles of Dataset Curation 303
Challenges in Collecting Domain Data 304
Annotation and Labeling Concerns 305
Engineering Workflows for Dataset Curation 306
Balancing Scale and Relevance 306
Synthetic Data Generation and Augmentation 307
Ethical and Privacy Considerations 308
Broader Implications of Domain Curation 309
Few-Shot and Zero-Shot Learning 311
Conceptual Foundations of Few-Shot and Zero-Shot Learning 311
Historical Context: From Meta-Learning to In-Context Learning 312
Few-Shot Learning Strategies 313

Zero-Shot Learning Strategies 314
Mathematical Intuition Behind Few-Shot and Zero-Shot 316
Engineering Workflows for Few-Shot and Zero-Shot Evaluation 317
Benefits and Applications 318
Challenges and Risks 319
Explainability and Uncertainty 320
Case-Style Illustrations Across Domains 321
Future Trajectories 322
Use of Adapters and LoRA 323
Motivation for Parameter-Efficient Fine-Tuning 324
Early Approaches to Modular Adaptation 325
Conceptual Foundations of Adapters 326
Engineering Practices in Adapter-Based Fine-Tuning 327
Conceptual Foundations of LoRA 328
LoRA in Practice: Efficiency and Flexibility 329
Comparative Reflections: Adapters vs. LoRA 330
Case-Style Applications Across Domains 331
Challenges and Limitations 332
Future Directions in Parameter-Efficient Adaptation 333
Conclusion 334
The Centrality of Adaptation 335
Chapter 8: Reinforcement Learning with Human Feedback (RLHF) 343
Human Annotation Pipelines 344
The Role of Human Preferences in RLHF 345
Annotation Formats: Rankings and Ratings 346
Recruiting and Training Annotators 347
Annotation Interfaces and Workflow Design 348
Quality Control and Inter-Annotator Agreement 349

Scaling Annotation Efforts 349
Case Illustration: InstructGPT 350
Challenges and Trade-Offs in Annotation Pipelines 351
Emerging Alternatives to Traditional Pipelines 351
Reward Modeling 352
Why Reward Modeling Matters in RLHF 353
Historical Roots of Reward Modeling 354
From Human Judgments to Reward Functions 355
Pairwise Comparisons and the Bradley–Terry Framework 355
Training Reward Models at Scale 356
Overfitting and Reward Hacking 357
Reward Model Generalization Across Domains 359
Case Illustrations: Summarization, Dialogue, and Instruction Following 359
Iterative Refinement of Reward Models 360
Challenges: Noise, Bias, and Misalignment 361
Alternatives and Extensions: Preference Distillation, AI-Assisted Feedback, Constitutional Signals 361
Future Directions in Reward Modeling 362
Proximal Policy Optimization (PPO) 364
Why PPO Matters in RLHF 364
Historical Context of PPO 365
Conceptual Principles of PPO 366
Integration with Pre-trained Policies 367
Engineering Practices in PPO for Language Models 367
PPO in Summarization and Dialogue Alignment 368
PPO Compared to Alternative Algorithms 369
Challenges of PPO in RLHF 369
Future Directions Beyond PPO 370

Broader Conceptual Significance of PPO....371

Challenges in Feedback Collection....372

The Cost of Feedback Collection....372

Annotation Quality and Consistency....373

Bias in Feedback Collection....374

Cultural and Linguistic Diversity....374

Annotation Noise and Reliability....375

The Problem of Over-optimization....376

Scaling Feedback Pipelines....376

Ethical Concerns in Feedback Collection....377

Emerging Alternatives to Human Feedback....377

The Fundamental Tension in Feedback Collection....378

Conclusion....379

Human Annotation Pipelines As the Foundation....380

Reward Modeling As the Surrogate for Human Judgment....380

PPO As the Engine of Alignment....381

Challenges in Feedback Collection As the Bottleneck....381

The Conceptual Significance of RLHF....382

Limitations and Open Questions....382

Chapter 9: Model Compression and Inference Optimization....385

Quantization and Weight Pruning....387

Motivation for Quantization and Pruning....388

Conceptual Foundations of Quantization....389

Types of Quantization....389

Hardware Support for Quantization....391

Conceptual Foundations of Pruning....393

Types of Pruning....394

Historical Development of Pruning....395

Trade-Offs in Quantization and Pruning 395
Engineering Practices for Quantization and Pruning 396
Case Illustrations 397
Broader Implications 397
Knowledge Distillation 399
Motivation for Knowledge Distillation 399
Historical Roots of Distillation 400
Theoretical Principles of Distillation 400
Types of Knowledge Distillation 401
Engineering Practices in Distillation 403
Applications in Natural Language Processing 403
Applications in Computer Vision and Speech 404
Challenges in Knowledge Distillation 404
Extensions and Innovations 405
Future Directions of Knowledge Distillation 405
Efficient Architectures 406
Historical Context of Efficient Architectures 407
Principles of Efficient Design 408
MobileBERT and TinyBERT 409
ALBERT and Parameter Sharing 410
TinyML and Edge Deployment 410
Sparse Attention Mechanisms 411
Trade-Offs in Efficient Architectures 412
Engineering Practices for Efficient Architectures 412
Future Directions of Efficient Architectures 413
Serving at Scale with Optimized Runtimes 415
The Challenge of Serving at Scale 417
Runtimes and Inference Engines 417

Batching Strategies for Efficiency 418
Memory Management in Large-Scale Serving 419
Operator Fusion and Graph Optimization 419
Hardware Specialization and Acceleration 420
Frameworks As Conceptual Exemplars 421
Elastic Scaling for Fluctuating Demand 421
Fault Tolerance and Reliability 422
Trade-Offs in Optimized Runtimes 422
Future Directions in Serving and Runtimes 423
Conclusion 424
Quantization and Pruning As First-Line Tools 424
Knowledge Distillation As Transfer of Learning 425
Efficient Architectures As Design for Deployment 425
Serving at Scale As the Final Bottleneck 426
Cross-Cutting Themes and Trade-Offs 426
Broader Implications of Compression and Optimization 427
Limitations and Open Questions 427

Chapter 10: Addressing Bias, Hallucinations, and Failure Modes 429

Detection and Mitigation of Hallucinated Outputs 431
Understanding Hallucinations in Generative Systems 431
Why Hallucinations Occur 432
Categories of Hallucinations 433
Risks and Implications of Hallucinations 433
Detection Strategies for Hallucinations 434
Mitigation Through Retrieval-Augmented Generation 435
Mitigation Through Decoding Constraints 435
RLHF for Hallucination Reduction 436
Hybrid Mitigation Pipelines 436

Case Illustrations of Hallucinations 437
Broader Implications 437
Managing Toxic and Biased Generations 438
Sources of Toxicity and Bias..439
Categories of Bias in Generative Systems..440
Historical Roots of Bias Awareness in AI 441
Risks and Implications of Toxic Outputs..442
Detection Strategies for Toxicity and Bias 443
Dataset Curation for Bias Mitigation..443
Algorithmic Debiasing Techniques 444
Reinforcement Learning with Human Feedback for Safety..445
Cultural and Contextual Adaptivity 445
Hybrid Mitigation Pipelines..446
Case Illustrations of Toxic and Biased Generations 447
Regulatory and Governance Frameworks 448
Open Questions and Future Directions..448
Model Safety Checks and Red Teaming 449
Rationale for Safety Checks in Generative Systems..451
Categories of Model Safety Checks..452
Automated vs. Human-in-the-Loop Safety..453
Red Teaming As a Conceptual Framework 455
Techniques of Red Teaming Generative Models 456
Lessons from Cybersecurity Red Teaming 457
Organizational Structures for Red Teaming..458
Continuous Monitoring and Adaptive Safety..459
Case Illustrations of Safety and Red Teaming 459
Trade-Offs in Safety and Red Teaming 462
Broader Implications of Safety and Red Teaming..463

Future Directions for Safety and Red Teaming 464
Historical Analogies in Safety Testing 464
Cultural and Ethical Dimensions of Safety 465
Multi-agent Red Teaming Ecosystems 466
Feedback Loops and Model Updates 467
Introduction to Feedback Loops in Generative Systems 468
Types of Feedback Loops (Positive, Negative, Reinforcing, Corrective) 468
Historical Parallels in Feedback Loops (Economics, Ecology, Control Theory) 469
Risks of Feedback in Deployed AI 470
Data Drift and Model Decay 471
Amplification of Bias Through Feedback 472
Feedback Loops in Recommender Systems As Precursors 472
Real-World Case Illustrations of Feedback Loops in Generative AI 473
Strategies for Monitoring Feedback Loops 474
Continuous Fine-Tuning and Iterative Model Updates 474
Governance and Oversight of Updates 475
Trade-Offs in Updating Models (Stability vs. Responsiveness) 476
User Feedback As a Resource and a Risk 476
Institutional Feedback Integration (Red Teams, Auditors, Regulators) 477
Technical Strategies for Safe Updating 477
Feedback Loops in Multimodal and Multi-agent Systems 478
Cultural and Ethical Dimensions of Feedback and Updates 479
Future Directions in Feedback and Updating 479
Historical Lessons in Feedback Management 480
Psychological and Behavioral Feedback Effects 480
Infrastructure Challenges in Frequent Updating 481
Long-Term Societal Risks of Runaway Feedback Loops 481
Conclusion 482

Chapter 11: Evaluation and Benchmarking of Generative Models....487

Automatic and Human Evaluation Methods....489

Historical Context of Evaluation in Generative Modeling....490

Core Principles of Evaluation....490

Automatic Evaluation Methods: Foundations....491

Human Evaluation: The Enduring Gold Standard....492

Case Study: BLEU in Machine Translation....493

Case Study: FID in Image Generation....493

Hybrid Approaches....494

Challenges of Human Evaluation....495

Toward Multidimensional Evaluation....495

Implications for Research and Deployment....496

Reference-Free Evaluation Paradigms....496

Cross-Modal Evaluation in Multimodal Systems....497

Evaluator Psychology and Cognitive Bias....498

Evaluation in High Stakes Domains....498

Cultural and Ethical Dimensions of Evaluation....499

Future Horizons in Evaluation....499

BLEU, FID, Perplexity, Diversity Scores....500

The Role of Metrics in Generative AI....501

BLEU: The Cornerstone of Text Evaluation....502

Beyond BLEU: ROUGE, METEOR, CIDEr, BERTScore....503

Perplexity: Measuring Language Model Fluency....505

Diversity Scores: Capturing Variety in Outputs....506

Benchmarking Image Generation....507

Statistical Properties and Interpretability of Metrics....508

Metric Gaming and Goodhart's Law....509

Metrics in Industry vs. Academia....509

Cultural and Linguistic Variability in Metrics 510
The Political Economy of Benchmarks 510
Metrics in Live Systems 511
Longitudinal Evolution of Metrics 511
Open Problems in Metric Design 512
Toward Meta-Metrics and Ensemble Evaluation 513
The Future of Metrics 514
A/B Testing Frameworks 516
Historical Origins of A/B Testing 517
Statistical Foundations of A/B Testing 518
Why A/B Testing Matters for Generative AI 519
Designing A/B Tests for Generative Systems 520
Metrics Within A/B Testing 520
Case Studies in Text Generation 521
Case Studies in Vision and Multimodal Systems 521
Online vs. Offline A/B Testing 522
Ethical Considerations in A/B Testing 522
Statistical Pitfalls in A/B Testing 523
Feedback Loops and Longitudinal Testing 524
Scaling A/B Testing in Industry 524
Beyond A/B: Multivariate and Bandit Testing 525
Cultural and Psychological Dimensions 525
Governance and Regulation of A/B Testing 526
Sequential and Adaptive Experimentation 526
A/B Testing for Safety and Alignment 527
Infrastructure Requirements for Large-Scale A/B Testing 527
Psychological and Sociological Effects of Testing 528

Cultural Variability in Experiment Results....528
The Future of Evaluation Ecosystems....529
Task-Based Performance Assessments....530
Conceptual Foundations of Task-Based Evaluation....531
Intrinsic vs. Extrinsic Assessment....531
Historical Roots of Task-Based Assessments....532
Domain-Specific Examples....532
Designing Authentic Tasks....533
Measuring Outcomes....533
Human–AI Collaboration in Tasks....534
Intrinsic Benchmarks vs. Real-World Integration....534
Case Study: Code Generation....535
Case Study: Education....535
Challenges of Task-Based Evaluation....535
Automating Task Evaluation....536
Human-in-the-Loop Task Evaluation....536
Formative vs. Summative Assessment....536
Safety and Fairness in Task Evaluation....537
Scaling Task-Based Evaluation....537
Theories of Assessment from Education and Psychology....537
Task Complexity and Cognitive Load....538
Adaptive and Personalized Task Evaluation....538
Infrastructure for Continuous Task-Based Monitoring....538
Risks of Task Reductionism....539
Governance of Task-Based Evaluation....539
Ethical and Cultural Dimensions....539
Future Directions....540
Conclusion....541

Chapter 12: Future Directions and Deployment in Production 545

CI/CD Pipelines for Model Updates 547

Historical Origins of CI/CD in AI 548

Conceptual Foundations of CI/CD for Generative Models 548

Core Components of CI/CD Pipelines for Generative AI 551

Data Management in CI/CD Pipelines 553

Training Pipelines for Model Updates 554

Model Registries and Versioning 555

Deployment Pipelines 556

Monitoring and Observability 556

Rollback and Recovery 557

Scaling CI/CD for Generative AI 557

Challenges and Open Problems 558

Future Directions in CI/CD Pipelines 558

Integration with DataOps 559

Cross-Organizational Collaboration 559

CI/CD and Explainability 560

Security in Model Pipelines 560

Environmental and Cost Considerations 560

Sociotechnical Dimensions 561

CI/CD for Multimodal and Agentic Systems 561

Model Monitoring and Observability 562

Historical Roots of Observability 563

Generative AI Monitoring Challenges 564

Performance and Efficiency Monitoring 565

Quality and Alignment Monitoring 565

Safety and Compliance Monitoring 566

Data Drift and Concept Drift Detection 567

User Feedback Integration 568

Infrastructure for Observability 568

Automated vs. Human-in-the-Loop Monitoring 569

Ethical and Legal Dimensions of Monitoring 569

Security Considerations 570

Multimodal Monitoring 571

Longitudinal and Life Cycle Monitoring 571

Organizational and Sociotechnical Considerations 572

Future Trends in Monitoring 572

Legal, Ethical, and Compliance Challenges 573

Historical Evolution of Tech Regulation 574

Intellectual Property and Generative Outputs 575

Privacy and Data Protection 576

Liability and Accountability 576

Bias, Fairness, and Social Equity 577

Transparency, Explainability, and Trust 578

Compliance Frameworks and Standards 578

Sector-Specific Compliance 579

Cross-Border Legal Complexities 580

Ethical Governance in Organizations 580

Human Oversight and Autonomy 581

Long-Term Ethical Risks 581

Case Studies and Precedents 582

Future Directions in Law and Policy 582

Trade Secrets and Proprietary Models 583

Antitrust and Competition Law in AI 583

Synthetic Data As a Compliance Strategy 584

Dual-Use and Misuse Dilemmas 584

Cultural and Epistemic Justice in AI Outputs 584
Comparative Global Regulatory Landscapes 585
The Economics of Compliance 585
Multimodal Fusion, Agentic Models, and AI Safety 586
Multimodal Fusion, Agentic Models, and AI Safety: Foundations and Implications 587
Technical Dimensions of Multimodal Fusion 588
Applications of Multimodal Systems 589
Ethical and Social Implications of Multimodal Fusion 590
Rise of Agentic Models 591
Technical Foundations of Agentic Models 592
Ethical and Governance Implications of Agentic Models 593
AI Safety As a Core Trend 595
Intersections of Trends 595
Case Studies Across Trends 596
Future Directions 597
Conclusion 598

Index 603

About the Author

Irena Cronin is SVP of Product for DADOS Technology, which has created an intelligent retrieval layer built to solve the performance and cost issues faced in Agentic AI implementations. She is also the CEO of Infinite Retina, which provides research to help companies develop and implement AI, AR, and other new technologies for their businesses. Prior to this, she worked for several years as an equity research analyst and gained extensive experience in evaluating both public and private companies.

Irena has a joint MBA/MA from the University of Southern California and an MS with Distinction in Management and Systems from New York University. She graduated with a BA from the University of Pennsylvania with a major in Economics (summa cum laude). Irena Cronin is the author of the book *Understanding Generative AI Business Applications* published by Apress.

About the Technical Reviewer

Sanathraj Narayan is a senior data scientist with over a decade of experience in building AI and machine learning solutions, as well as scalable data systems on AWS and Azure. He has held roles at Ericsson, Mindtree, KPMG India, and Cognizant, where he led data-driven projects across the retail, telecom, and consulting sectors. Sanathraj's expertise spans predictive modeling, recommender systems, and the deployment of end-to-end machine learning pipelines. He is also a regular conference speaker, presenting on AI, machine learning, and generative AI topics.

Acknowledgments

I am indebted to my best friend, Carol Cox, and her encouragement to make this book possible.

Introduction

Generative artificial intelligence has become one of the most transformative developments in modern computing. It has reshaped how organizations design software, how researchers analyze the world, and how creators build new forms of media. Models capable of producing text, images, audio, code, and multimodal outputs now sit at the center of AI innovation. This book provides a thorough, technically grounded, and application-oriented guide to building these systems from the ground up.

The purpose of this book is to give AI engineers, machine learning practitioners, and advanced developers a complete understanding of how to architect, train, and deploy generative AI models in real environments. It covers the full development life cycle, beginning with model selection and data preparation and moving through optimization techniques, distributed training, model compression, safety considerations, and long-term maintenance. The book explains the four foundational families of generative models used today: transformers for sequence generation, GANs for adversarial synthesis, VAEs for probabilistic latent modeling, and diffusion models for state-of-the-art image and multimodal generation. Each architecture is explored from conceptual foundations to applied use cases.

Readers will learn how to work with large, complex datasets and how to apply self-supervised learning techniques that power modern AI systems. The book offers a step-by-step explanation of optimization strategies such as backpropagation, gradient descent, adaptive learning rate schedules, and regularization. It also covers the computational infrastructure required for large-scale model training, including GPUs, TPUs, model parallelism, data parallelism, distributed systems, and performance tuning for multi-node environments. Practical guidance is provided for scaling training workloads while maintaining stability, reproducibility, and efficiency.

Fine tuning and domain adaptation are treated as essential stages of generative model development. The book shows how to use transfer learning, domain-specific dataset design, parameter-efficient tuning methods, and reinforcement learning with human feedback to adapt models to specialized tasks. Techniques such as LoRA, adapters, and reward modeling are explained in detail, along with the workflows required to collect feedback, evaluate outputs, and maintain model reliability.

In addition to training and fine-tuning, the book explains how to make models faster and more efficient for deployment. Readers will gain experience with quantization, pruning, knowledge distillation, and other compression techniques that significantly reduce inference cost. These approaches allow organizations to serve large generative models on edge devices, consumer hardware, or cloud infrastructures without sacrificing quality.

A major focus of the book is addressing failure modes and risks inherent in generative AI, including hallucinations, bias, unsafe content, and system instability. Readers will learn practical approaches for bias detection, safety filtering, hallucination reduction, and secure deployment based on red teaming, evaluation benchmarks, and feedback loops. Case studies demonstrate how these issues emerge in real applications and how to handle them systematically.

Evaluation and benchmarking are explored through both human and automated methods, covering widely used metrics such as BLEU, FID, perplexity, diversity scores, and task-specific evaluations. Readers will learn how to design robust testing pipelines, conduct A/B experiments, and measure performance in use case-specific scenarios.

The book concludes with a forward-looking perspective on deploying generative models in production environments. This includes model monitoring, observability, CI/CD pipelines for continuous updates, compliance frameworks, and future trends such as multimodal fusion, agentic systems, synthetic data pipelines, and advanced AI safety techniques.

By the end of the book, readers will have a comprehensive understanding of how to architect, train, optimize, evaluate, and deploy generative AI systems at scale. The goal is to provide a rigorous yet practical framework that helps practitioners build models that are powerful, efficient, responsible, and ready for real-world impact.

Chapter Breakdown

Chapter 1: Introduction to Generative AI Systems

Foundational overview of generative modeling and modern use cases. Covers text, vision, audio, and multimodal applications. Explains how generative models differ from discriminative models and introduces transformers, GANs, VAEs, and diffusion models.

Chapter 2: Choosing the Right Architecture

Detailed comparison of generative architectures, including strengths, limitations, and ideal application contexts. Covers transformer-based models, GAN variants, VAE formulations, and diffusion model pipelines.

Chapter 3: Data Collection and Preparation

Process of sourcing, validating, and cleaning data. Discusses structured versus unstructured data, normalization pipelines, synthetic data generation, and techniques for detecting and mitigating dataset bias.

Chapter 4: Training Fundamentals and Self-Supervised Learning

Explains the logic behind self-supervision, loss functions, overfitting, and generalization. Provides tools for tracking convergence, evaluating training progress, and stabilizing early training phases.

Chapter 5: Optimization and Learning Strategies

In-depth look at gradient descent, backpropagation, adaptive optimizers, gradient clipping, normalization techniques, and the design of learning rate schedules.

Chapter 6: Scaling Training with Infrastructure and Distributed Systems

Shows how to use GPUs and TPUs efficiently, implement parallel training, manage checkpoints, and apply distributed training frameworks such as Horovod and DeepSpeed.

Chapter 7: Fine-Tuning and Domain Adaptation

Covers transfer learning, domain specific dataset strategies, few shot and zero shot learning, and methods such as LoRA and adapter layers for efficient fine-tuning.

Chapter 8: Reinforcement Learning with Human Feedback (RLHF)

Explains how RLHF improves alignment and output quality. Provides detail on human annotation workflows, reward modeling, PPO, and practical challenges.

Chapter 9: Model Compression and Inference Optimization

Techniques for reducing model size and inference time, including quantization, pruning, knowledge distillation, mobile optimized architectures, and runtime acceleration.

Chapter 10: Addressing Bias, Hallucinations, and Failure Modes

Frameworks for detecting hallucinations, mitigating biased outputs, improving safety, and handling feedback loops. Includes case studies and best practices for red teaming.

Chapter 11: Evaluation and Benchmarking of Generative Models

Covers human and automatic evaluation methods, key metrics across modalities, A/B testing infrastructure, and task-based performance assessments.

Chapter 12: Future Directions and Deployment in Production

Discusses CI/CD automation for generative models, observability, compliance, and long-term maintenance. Explores future trends such as agentic systems, multimodal fusion, and emerging AI safety methods.

CHAPTER 1

Introduction to Generative AI Systems

Generative AI has emerged as one of the most transformative areas in AI, enabling machines to not only analyze existing data but to produce entirely new content across a wide range of formats. This chapter introduces the foundational principles of generative modeling, explains how these models differ from traditional discriminative approaches, and explores their applications in text, vision, audio, and multimodal systems. It also provides an overview of the primary architectural families driving this field, including transformers, generative adversarial networks, variational autoencoders, and diffusion models. By the end of this chapter, readers will have a clear basic understanding of what generative AI is, how it works, and where it is being deployed in real-world environments.

Overview of Generative Modeling

Generative modeling forms the intellectual and technical foundation of contemporary generative AI. It defines a category of machine learning systems specifically designed to uncover and replicate the structure

I. Cronin, *Building and Training Generative AI Models*,
https://doi.org/10.1007/979-8-8688-2332-9_1

of data distributions, enabling them to produce entirely new data that mimics the statistical properties of the original. This is a sharp departure from the more familiar discriminative models, which are trained to label, categorize, or predict based on inputs. Generative models, in contrast, seek to understand how data is formed in the first place. They aim not to react to input with a classification or a decision, but to generate the input itself.

This capability makes generative modeling not only one of the most complex but also one of the most versatile approaches in AI. These models can produce realistic images of faces that do not exist, create text that reads as if it were written by a human, simulate the acoustic properties of natural voices or music, design molecular structures that could lead to new drugs, and render coherent 3D spaces from mere textual descriptions. Their scope spans every major sensory and symbolic modality—text, vision, audio, video, and structured knowledge—making them core to many of the most visible and transformative AI applications.

1. Defining Generative Modeling

Generative modeling is the central principle that underlies generative AI systems. It refers to the process through which a machine learning model learns the structure and statistical properties of a dataset with the goal of generating new data that is similar to the original. Unlike systems that are trained to recognize or classify input data, generative models are capable of creating new content that was not previously encountered during training.

Technically, generative modeling involves estimating a probability distribution over the observed data. This statistical representation allows the model to learn how likely particular features, structures, or patterns are within the data space. Once this distribution is learned, the model can sample from it to generate new instances that share the same essential properties as the training data.

There are two main types of generative modeling frameworks. In supervised generative modeling, the model attempts to learn the joint probability distribution P(x, y), where x represents the input and y the output. This formulation allows the model to understand both how inputs and outputs behave independently and how they relate to one another. It can then generate new (x, y) pairs consistent with the training distribution. In contrast, unsupervised generative modeling focuses on learning the distribution P(x), which is the probability of the data itself. The model is trained without explicit output labels and learns to generate new input-like data points that are statistically consistent with the dataset.

The generative model's output is not a duplicate of what it has seen during training. Instead, the goal is generalization. A well-trained generative model can produce new samples that are novel but still plausible within the context of the data it was trained on. For instance, a model trained on facial images can create faces that do not belong to any real person but that exhibit realistic structure, expression, and features. A model trained on classical piano music can generate new pieces that sound like they belong in the same musical tradition without replicating any known composition.

This ability to synthesize new content, rather than merely recognize or label it, is what fundamentally distinguishes generative models from discriminative ones. Discriminative models aim to draw boundaries between classes or predict outputs from given inputs. Generative models aim to understand and replicate the underlying process by which data arises, which allows them to create new examples that were not explicitly observed.

Several criteria are used to assess the performance of generative models. One important criterion is novelty. The model must be able to generate outputs that differ meaningfully from the training data. Another is coherence. This refers to the internal logical consistency of the generated sample. For text, coherence might mean grammatical correctness, clear structure, and consistent meaning. For images, it might involve consistent lighting, perspective, and object relationships. Diversity is also a key metric. A strong generative model should be able to produce a wide range of outputs, avoiding repetition or overfitting. Fidelity is yet another criterion, which refers to how realistic or authentic the outputs appear relative to the domain.

For example, a generative model that produces city street scenes should create images that include realistic arrangements of buildings, streets, vehicles, and pedestrians, all rendered with consistency in scale, perspective, and detail. Similarly, a generative dialogue model should be able to produce responses that are appropriate to a given context, carry a consistent tone, and respect conversational norms.

In short, generative modeling is the process of learning the underlying rules of a dataset in such a way that the system can produce new and meaningful content. This process enables a model to extrapolate beyond what it has seen and contribute to new expressions of text, visuals, audio, or other forms of data. It represents a shift in the role of machine learning from analysis to synthesis, from classification to creation. As generative modeling continues to evolve, it plays an increasingly vital role in AI systems that simulate, generate, and augment human work across a wide range of disciplines.

2. Types of Generative Models

Over the past decade, generative modeling has undergone rapid evolution, with several architectures emerging as especially influential. Each of these families of models is based on a different intuition about how data should be represented, generated, and optimized.

a. Variational Autoencoders (VAEs)

VAEs approach the problem of generation by learning to encode data into a lower-dimensional latent space and then decode it back to the original format. What makes VAEs unique is their probabilistic nature: rather than encoding an input into a fixed point, they encode it into a distribution. During generation, samples are drawn from this distribution and passed through the decoder to produce novel outputs. This design enforces a smooth latent space, where similar points yield similar outputs, making it easier to interpolate between examples. VAEs are mathematically elegant, interpretable, and useful in many applications, particularly where representation learning is as important as generation itself.

b. Generative Adversarial Networks (GANs)

GANs take a very different approach. They use two networks: a generator and a discriminator. The generator creates synthetic data, while the discriminator evaluates whether the data is real or fake. The two networks are trained simultaneously in a zero-sum game, where the generator improves by trying to fool the discriminator. This adversarial setup has led to models capable of generating highly realistic images, to the point where human observers often cannot distinguish between real and generated samples. However, GANs are notoriously difficult to train and are prone to issues such as mode collapse, where the model produces limited variation in its outputs.

c. Autoregressive Models

Autoregressive models take a sequential approach to generation. They model the probability of each element in a sequence conditioned on the previous ones. In text generation, this means predicting the next word based on all the prior words in a sentence. These models have achieved enormous success due to their simplicity and effectiveness, especially in natural language processing. Transformer-based models such as GPT exemplify this category and have demonstrated high fluency and long-range coherence. They are also dominant in domains like music generation and protein sequence modeling.

d. Diffusion Models

Diffusion models are among the most recent innovations in generative modeling. Inspired by physical processes, these models learn to reverse a process that gradually adds noise to data until it becomes indistinguishable from pure noise. The model is trained to denoise step-by-step, gradually recovering the original data distribution. This method has shown exceptional performance in image generation, rivaling and often surpassing GANs in terms of quality and diversity. Diffusion models are also more stable to train and offer more controllability via conditioning techniques.

3. Probabilistic Foundations

At the heart of all generative modeling lies the mathematical framework of probability theory. Whether the model is producing images, text, audio, or video, it is fundamentally tasked with learning the underlying probability distribution of a dataset. This probabilistic view allows the model to generate new samples that are not arbitrary but rather consistent with the statistical patterns found in the original data.

Generative models can be broadly categorized based on how they approach this task. Some models represent the data distribution **explicitly**, while others do so **implicitly**. This distinction shapes nearly every aspect of how these models are built, trained, and evaluated.

In the case of **explicit generative models**, the model defines a parametric form of the probability distribution it seeks to learn. That means it specifies a functional form, typically with many learnable parameters, that directly approximates the data distribution. Training such models involves maximizing the likelihood of observed data under this parameterized distribution. This category includes architectures such as variational autoencoders (VAEs), autoregressive models like PixelCNN or

GPT, and more recently, diffusion models. These models often provide a well-defined probability density function and use it during both training and sampling.

Explicit models allow the use of standard statistical tools during optimization. For example, they can be trained using maximum likelihood estimation, which seeks to adjust model parameters to make the training data appear as probable as possible. They also allow the use of divergence measures such as Kullback-Leibler divergence to quantify how far the learned distribution deviates from the true data distribution. In models like VAEs, the training objective often involves maximizing the evidence lower bound (ELBO), a proxy objective that approximates the intractable log-likelihood while enforcing regularization through variational inference.

In contrast, **implicit generative models** do not specify a closed-form probability distribution. Instead, they are trained to produce samples that, when viewed statistically or perceptually, resemble those from the target data distribution. These models sidestep the need to compute or define likelihoods explicitly. The most well-known class of implicit models is the generative adversarial network (GAN). In a GAN, a generator network learns to produce data samples, while a separate discriminator (or critic) learns to distinguish between real and generated samples. The generator improves by learning to fool the discriminator, without ever needing to compute an explicit probability distribution.

Because implicit models lack tractable likelihoods, they require alternative strategies for training. These include adversarial loss functions, which pit the generator against the discriminator in a two-player game, as well as kernel-based methods that compare distributions in a high-dimensional feature space. In many cases, training involves approximations of statistical divergences, such as Jensen-Shannon divergence or Wasserstein distance, which can be used to align the generated and real data distributions even when exact likelihoods are unavailable.

This foundational distinction between explicit and implicit modeling affects not just training dynamics but also how models are evaluated. Explicit models offer log-likelihood estimates or ELBO values that can be directly tracked over time, providing a quantitative signal of learning progress. Implicit models lack this direct measure and often rely on surrogate metrics, such as Inception Score, Fréchet Inception Distance, or visual inspection of generated outputs, to gauge quality and diversity.

Moreover, the choice between explicit and implicit approaches has trade-offs in practice. Explicit models tend to be more stable and interpretable, but they may struggle with sharpness and detail in image generation. Implicit models, especially GANs, can produce stunningly realistic outputs but are often more sensitive to training instability, mode collapse, and evaluation challenges.

In summary, probability theory provides the foundation for generative modeling, but the way this theory is operationalized differs substantially depending on whether the model is explicit or implicit. These differences influence the architecture, the training objectives, the metrics used for evaluation, and ultimately the kinds of generative tasks the model is best suited to handle. Understanding this probabilistic foundation is essential for grasping how generative models work and how they continue to evolve.

4. Training Methodologies and Loss Functions

The training of generative models involves optimizing loss functions that often reflect complex objectives. These objectives vary depending on the model class:

- VAEs optimize a combination of reconstruction loss and a regularization term that penalizes divergence from a prior distribution.
- GANs use a min-max adversarial loss that pits the generator and discriminator against each other.

- Autoregressive models minimize the negative log-likelihood or cross-entropy of predicting the next token in a sequence.
- Diffusion models optimize a denoising loss that encourages the model to reconstruct clean data from noisy inputs.

Each loss function introduces trade-offs. For instance, reconstruction-based losses may emphasize fidelity but can lead to blurry outputs. Adversarial losses prioritize realism but may sacrifice diversity. Denoising objectives can yield sharp images but demand long training schedules and careful noise scheduling.

5. Evaluation of Generative Models

Evaluating generative models poses a fundamentally different challenge compared to discriminative models. In classification or regression tasks, the output space is well-defined, and model performance can be directly measured against ground-truth labels using accuracy, precision, recall, or mean squared error. In contrast, generative tasks, such as producing natural language, synthesizing images, or generating music, often have no single correct output. The space of plausible responses is vast, and many different outputs may be equally valid or even preferred depending on the context.

Because of this inherent subjectivity, the evaluation of generative models must account for multiple dimensions of performance. These typically include statistical fidelity to the training distribution, perceptual or aesthetic quality, semantic coherence, and task relevance. Metrics are domain-specific, and no single score can fully capture the effectiveness of a generative system.

For **image generation**, two of the most widely used evaluation metrics are

- **Inception Score (IS)**: Measures how recognizable generated images are by a pre-trained image classifier. It rewards images that are both classifiable and diverse. However, IS has been criticized for being overly dependent on the biases of the Inception model used in its computation.
- **Fréchet Inception Distance (FID)**: Compares the statistical distance between real and generated image distributions in a high-level feature space. Lower FID scores indicate that generated images are more similar to real ones. FID has largely replaced IS as the preferred standard due to its greater sensitivity to mode collapse and quality degradation.

For **text generation**, evaluation becomes even more nuanced. Commonly used metrics include

- **BLEU (Bilingual Evaluation Understudy)**: Calculates the overlap of n-grams between the generated text and one or more reference texts. Originally developed for machine translation, BLEU is simple but often fails to capture semantic meaning.
- **ROUGE (Recall-Oriented Understudy for Gisting Evaluation)**: Focuses on recall of overlapping units, particularly useful for evaluating summarization models.
- **METEOR**: Improves upon BLEU by incorporating synonym matching, stemming, and more flexible alignment strategies. It correlates more closely with human judgments but is computationally more intensive.

- **Perplexity**: Measures how well a language model predicts a sample of text. Lower perplexity generally indicates that the model is more confident and accurate in its word predictions, but it is not always aligned with human perception of quality.

In practice, these automated metrics often fail to fully capture the subtle qualities of generative output, especially in domains that require creativity, emotional intelligence, or cultural nuance. For these reasons, **human evaluation remains indispensable**. Human judges are often asked to assess generative outputs based on criteria such as fluency, relevance, coherence, diversity, novelty, and even ethical appropriateness.

- In creative writing and storytelling, human reviewers assess whether the generated narrative is engaging, believable, and consistent with tone and style.
- In visual design or generative art, humans evaluate compositional quality, stylistic fidelity, and emotional impact—factors that are difficult to measure algorithmically.
- In dialogue systems or voice-based agents, human interaction tests are used to assess how natural, responsive, and socially appropriate the AI behaves.

Human-in-the-loop evaluation is particularly critical in high-stakes domains, such as legal writing, medical assistance, psychological counseling, and autonomous decision-making, where subtle context can make a profound difference.

Recent advances are also exploring **learned evaluation models**, where neural networks are trained to act as critics of generative output. These models, such as reward models in reinforcement learning with human feedback (RLHF), attempt to simulate human preferences and scale evaluation to massive data volumes. However, such learned evaluators inherit the biases and limitations of their training data and are not yet reliable replacements for human judgment in sensitive applications.

Ultimately, evaluation in generative AI is an evolving area of research. No single metric is sufficient, and the best practices often involve a combination of automated metrics, task-specific criteria, and carefully structured human assessments. As generative models continue to improve in capability and complexity, the demand for more robust, fair, and context-aware evaluation methods will only grow.

6. Applications of Generative Models

Generative modeling supports a broad range of applications, many of which are reshaping creative, scientific, and industrial workflows:

- **Natural Language Generation**: Systems like GPT-5 generate coherent text for chatbots, content creation, summarization, and translation.
- **Image Synthesis**: Tools such as Midjourney and DALL·E produce high-quality visual content from textual prompts.
- **Drug Discovery**: Models generate novel chemical compounds with specific properties for pharmaceutical development.
- **3D Design and Simulation**: Generative models assist in producing 3D assets for AR, VR, gaming, and robotics simulation.
- **Music and Audio Synthesis**: From lifelike voice generation to entirely new musical compositions, generative audio models are advancing rapidly.

- **Code Generation**: AI models now write software code from natural language descriptions or partial programs.

7. Challenges in Generative Modeling

Although generative modeling has advanced significantly in recent years, it still faces a variety of persistent challenges. These challenges span technical obstacles, performance limitations, social risks, and ethical concerns. Addressing them is essential for improving the reliability, safety, and accountability of generative AI systems.

One of the most persistent technical issues is **mode collapse**, especially in generative adversarial networks. This occurs when the model fails to explore the full diversity of the data distribution and instead produces a narrow subset of repetitive or overly similar outputs. The lack of variety reduces the model's usefulness in creative or exploratory applications and suggests that it has not fully learned the complexity of the training data.

Bias amplification is a widely recognized problem across both language and vision models. Generative systems trained on large-scale datasets often absorb and reproduce the societal, cultural, or political biases embedded in those data sources. In some cases, these biases are not just reproduced but magnified, resulting in distorted or exclusionary outputs. For example, a text model might over-represent certain professions with specific genders or ethnicities, while an image model may fail to produce outputs that reflect demographic diversity.

Hallucination is particularly common in large-scale language models. These systems may produce text that is fluent and grammatically correct, yet factually incorrect or entirely fabricated. This becomes dangerous when the generated content is used in high-stakes settings such as education, healthcare, legal reasoning, or journalism.

- A model might generate citations to research papers that do not exist.

- It could confidently produce medical advice that sounds plausible but is unsafe.
- In customer service applications, hallucinated responses may lead to miscommunication or reputational harm.

The **computational cost** of training and deploying generative models continues to be a significant concern. Models like GPT, Claude, and Gemini require vast amounts of processing power and memory. This makes experimentation expensive and access to these systems highly centralized among large technology companies and research labs.

- Training a state-of-the-art model can cost millions of dollars in hardware, electricity, and labor.
- Inference costs can also be high, especially for applications requiring real-time generation at scale.
- These resource demands raise important questions about sustainability, scalability, and equitable access to AI technologies.

Evaluation presents another core challenge. Generative models lack clear-cut correctness, which makes it difficult to assess performance using standard benchmarks. Metrics such as BLEU, ROUGE, FID, or perplexity provide only partial insights and often fail to reflect user experience or application-specific needs. Human evaluations, while informative, are expensive and difficult to scale.

- A model may score well on automated metrics but still produce irrelevant or incoherent output in practice.
- Evaluation practices often overlook edge cases, contextual relevance, or ethical implications of generated content.

The **ethical risks** surrounding generative modeling are becoming more pronounced. The ability to generate convincing but synthetic content introduces serious concerns regarding trust, authenticity, and malicious use. These risks include but are not limited to

- Deepfake generation used for impersonation, harassment, or political manipulation
- Automated creation of fake news, conspiracy theories, or misinformation campaigns
- Unauthorized replication or modification of copyrighted content, including art, music, and literature

In summary, generative modeling continues to grapple with a range of complex challenges. From technical weaknesses such as instability and hallucination to broader societal issues like bias and ethical misuse, these obstacles require sustained attention. Solving them will demand interdisciplinary collaboration across AI research, policy, law, and ethics. Without such coordinated efforts, the risks posed by generative technologies could overshadow their transformative potential.

8. From Generative Models to Generative Systems

A single model, no matter how powerful, is often insufficient for real-world deployment. Generative systems wrap models in scaffolding that includes retrieval modules, user interfaces, task-specific tuning, alignment techniques like RLHF, and guardrails for safe operation. This systems-level approach transforms a generative model from a raw capability into a usable product or service, capable of contributing meaningfully to workflows in design, medicine, software engineering, and beyond.

Applications Across Text, Vision, Audio, and Multimodal Domains

Generative AI's influence is most visible through the applications it enables in real-world domains. These domains, text, vision, audio, and multimodal, represent the primary modalities through which humans experience and interpret information. Generative modeling has made it possible to produce new data in each of these modalities that is not only high quality but also tailored, interactive, and adaptive to context.

This section outlines how generative models have evolved into powerful tools across these domains, offering creative capabilities, efficiency improvements, and in many cases, entirely new forms of expression and functionality.

1. Text Generation

Text has been the primary modality for early large-scale adoption of generative AI, due to the wide availability of high-quality training data, the natural alignment with transformer architectures, and the clear utility across many industries.

a. Conversational Agents

Systems like ChatGPT, Claude, Gemini, and other conversational AI tools are built on top of large language models (LLMs). They generate fluent, context-aware text that supports applications, including customer service, education, writing assistance, and technical support. These agents often integrate tools such as retrieval and external APIs to augment their responses, making them not only generative but interactive and action-capable.

b. Content Generation

From blog posts and email drafts to marketing copy and screenplays, LLMs are now widely used for ideation, drafting, and polishing written content. The ability to generate multiple stylistic variations or adapt tone for specific audiences adds a level of personalization that was once costly and time-consuming.

c. Code Synthesis

Code generation has become a major focus area. Tools like GitHub Copilot, CodeWhisperer, and Replit's AI Assistant generate entire functions or even full programs based on natural language instructions. These systems reduce boilerplate work, aid in debugging, and can even suggest design patterns. They blur the line between developer and co-developer.

d. Summarization, Translation, and Editing

Generative models are used to produce concise summaries, translate text across languages, rewrite in different styles, and simplify complex material. These capabilities are valuable in journalism, education, legal, and scientific settings, where clarity and precision are essential.

2. Vision and Image Generation

Generative modeling in vision has advanced at a rapid pace, enabling machines to produce images that are stylistically diverse, photorealistic, or entirely surreal.

a. Image Synthesis

Diffusion models and GANs now generate stunning visual content based on textual prompts. Tools like DALL·E, Midjourney, and Stable Diffusion allow users to generate original artwork, product designs, architectural concepts, and more. The fidelity and coherence of these images are high enough to be used in commercial and professional settings.

b. Style Transfer and Inpainting

Generative models can blend styles from one image into another or fill in missing regions using learned context. This has significant applications in restoration, photo editing, and creative media production.

c. Medical Imaging and Scientific Visualization

In healthcare, generative models are used to simulate medical scans or enhance imaging through resolution upscaling and noise removal. These tools aid diagnostics, training, and simulation, especially in areas where labeled data is scarce.

d. 3D Model and Scene Generation

Recent advances have extended generation into three-dimensional space. Models can now create 3D assets from single 2D images or generate entire virtual scenes from descriptive text. This capability is essential for AR/VR content creation, robotics simulation, and digital twin environments.

3. Audio and Speech Generation

Audio is inherently temporal and expressive. Generative AI models are enabling synthetic audio to be produced with emotion, clarity, and control.

a. Speech Synthesis

Text-to-speech (TTS) systems powered by generative models produce natural-sounding voices with inflection, pacing, and emotional variance. These systems are now customizable, allowing users to generate speech in specific voices, accents, or characters, and have become essential tools in accessibility, gaming, voiceovers, and smart assistants.

b. Voice Cloning and Personalization

Generative audio models allow cloning of human voices with only a few seconds of audio input. This capability is both powerful and controversial. It supports personalization in applications like AI podcasting and synthetic customer service, but also introduces ethical concerns around deepfakes and consent.

c. Music Generation

Music models like Suno or Elevenlabs generate full-length compositions in specific genres, styles, or moods based on short textual prompts or seed melodies. These models can also accompany other generative systems, such as auto-generated background music for videos or games.

d. Audio Enhancement and Restoration

Generative AI is also applied to audio cleaning and reconstruction. It can remove background noise, repair corrupted audio, enhance spatial quality, or separate instruments and voices in a recording. These tools have found use in both amateur and professional production environments.

4. Multimodal Applications

The most dynamic area of generative AI lies in systems that operate across modalities. Multimodal models can process and generate combinations of text, images, audio, and even video.

a. Image-to-Text and Text-to-Image

Systems like Midjourney, Imagen, and DALL·E operate between vision and language. They can generate textual descriptions of images or generate images from textual prompts. This bidirectionality enables applications in accessibility (e.g., automatic alt text), design iteration, and education.

b. Video Generation and Editing

Generative video is still developing, but rapid progress is being made. Tools can generate short animated clips from text or still images, interpolate between video frames, and apply edits using natural language instructions. Eventually, it may be possible to generate full cinematic sequences from a paragraph of description.

c. Interactive Avatars and Digital Humans

Combining generated text, speech, and video makes it possible to create lifelike avatars. These can be used in customer service, education, therapy, and virtual influencers. The avatar responds in real time with generated voice, expressions, and gestures.

d. Robotics and Embodied AI

Multimodal models help robots interpret natural language instructions and translate them into sensorimotor actions. For example, a robot in a warehouse can receive a text or voice command, perceive its environment through vision, and then act accordingly. This requires a synthesis of language, spatial reasoning, and physical control.

e. Personalized Assistants and Cognitive Agents

Generative models that combine different modalities can power more holistic AI agents that understand context, remember past interactions, adapt responses across media, and operate tools. These systems go beyond static responses to function as collaborators, simulators, or interactive tutors.

5. Industry Use Cases by Domain

Each modality supports a broad range of domain-specific use cases:

- **Text**: Legal document drafting, tutoring, financial analysis, customer onboarding
- **Vision**: Fashion design, game asset creation, product prototyping, medical diagnostics
- **Audio**: Interactive audiobooks, real-time voice translation, accessibility for the visually impaired
- **Multimodal**: AR/VR world building, surveillance analysis, synthetic education modules

6. Challenges in Cross-Modality Generation

While the opportunities are substantial, multimodal and cross-modal systems face several technical and societal challenges:

- **Alignment across modalities** is difficult, especially in maintaining consistency (e.g., lip sync between video and audio, or relevance between image and caption).
- **Computational load** increases drastically when combining high-resolution inputs and outputs across domains.

- **Data scarcity** for multimodal supervision leads to reliance on synthetic or weakly-labeled data.
- **Bias compounding** is more likely when outputs from one domain become training inputs for another.
- **Intellectual property and attribution** become complex in generative pipelines that mix training data from multiple media types.

7. Future Trajectories

As generative models mature, modality boundaries will continue to blur. We are moving toward foundation models that are natively multimodal and capable of fluidly generating content across domains. These models will understand context from speech, respond in image, reason through video, and explain in text, without needing discrete pipelines.

The development of standardized interfaces and APIs for generative systems will enable more composable architectures, where specialized models in different modalities can cooperate. Moreover, user-facing tools will become increasingly unified, offering creators and professionals an integrated interface to generate across domains simultaneously.

Comparison with Discriminative Models

To fully understand generative AI systems, it is essential to distinguish them from their closest conceptual counterpart, discriminative models. While both belong to the broader category of machine learning models, and often share architectural similarities such as neural networks or transformers, their purposes, training objectives, data assumptions, and downstream capabilities are fundamentally different.

The comparison between generative and discriminative models provides a framework for evaluating where generative AI is most suitable, how it fits into hybrid architectures, and why it represents such a paradigm shift in the application of machine learning across industries.

1. Fundamental Objective Differences

The primary distinction lies in the types of probability distributions these models are designed to learn.

- Discriminative models aim to learn the conditional probability P(y|x), the probability of an output y given an input x. They are focused on tasks like classification or regression, where the goal is to make the most accurate prediction for y from known inputs x.
- Generative models aim to learn the joint probability P(x, y), or more generally the data distribution P(x). Once this distribution is learned, the model can not only infer y from x (via Bayes' rule) but also generate plausible new x values, simulate entire data distributions, and uncover latent structure in the data.
- This difference is not merely theoretical—it determines what the models can do in practice.

Discriminative models are inherently optimized for tasks like object detection, sentiment classification, fraud detection, or disease diagnosis. Generative models, on the other hand, can synthesize new text, generate photorealistic images, simulate physical systems, and emulate human-like interactions.

2. Example Use Cases by Model Type

Discriminative Model Use Cases:

- Image classification (e.g., identifying cats vs. dogs in a picture)
- Spam detection (e.g., classifying emails as spam or not)
- Credit scoring (e.g., predicting loan default risk)
- Speech recognition (e.g., mapping audio input to text)
- Fraud detection (e.g., predicting whether a transaction is suspicious)

These models are prized for their speed, precision, and relatively straightforward evaluation criteria. In many domains, discriminative models have been deployed at scale for more than a decade.

Generative Model Use Cases:

- Image generation (e.g., creating original artwork based on style prompts)
- Text generation (e.g., writing articles, stories, or code)
- Video simulation (e.g., generating sports replays or architectural walkthroughs)
- Drug molecule design (e.g., generating novel chemical structures)
- Audio synthesis (e.g., generating music or mimicking human voices)

These models go beyond understanding the world—they create new representations of it.

3. Architecture and Training Differences

While discriminative and generative models often share underlying architectures such as convolutional neural networks (CNNs), recurrent neural networks (RNNs), or transformers, the training objectives and data pipelines differ significantly.

- **Discriminative models** minimize classification error. This is typically implemented using loss functions like cross-entropy or mean squared error. These models focus on learning a boundary that best separates categories.
- **Generative models** minimize reconstruction loss, adversarial loss, denoising objectives, or likelihood approximation. For example, VAEs minimize the difference between original and reconstructed inputs while regularizing the latent space. GANs optimize a min-max loss between a generator and a discriminator. Autoregressive models minimize the negative log-likelihood of the next element in a sequence.

The training of generative models is usually more computationally intensive, more sensitive to hyperparameter tuning, and often more difficult to stabilize.

4. Data Requirements and Representational Learning

Discriminative models require labeled datasets, often in large quantities, to perform effectively. Their ability to generalize is tied directly to the quantity and quality of labels available.

Generative models can operate in both supervised and unsupervised settings. Many of the most powerful generative models, including GPT and DALL·E, are pre-trained on vast datasets without explicit labels. This allows them to uncover hidden patterns, relationships, and structures within the data. As a result, they are often better suited for tasks that involve representation learning, transfer learning, or domain adaptation.

Moreover, generative models can be used to augment training data for discriminative models, especially in low-resource scenarios. For instance, synthetic images generated by GANs can be used to train object detection models in medical imaging or autonomous driving.

5. Evaluation Metrics

Evaluating discriminative models is typically straightforward. Accuracy, precision, recall, F1-score, ROC-AUC, and similar metrics provide a clear picture of performance.

Evaluating generative models is more nuanced. Since there is no single "correct" output in many generative tasks, metrics like BLEU or ROUGE (for text), Fréchet Inception Distance (FID) for images, or human evaluation are commonly used. These metrics focus on fluency, realism, diversity, and relevance—qualities that are inherently more subjective.

This distinction reflects a philosophical divide: discriminative models are evaluated on being *right*, while generative models are evaluated on being *plausible*.

6. Strengths and Weaknesses of Each Model Type

Aspect	Discriminative Models	Generative Models
Objective	Predict output given input, learning P(y|x).	Model the full data distribution, learning P(x, y) or P(x).
Output	Label, category, regression value	Novel data instances such as text, image, audio
Training data	Requires labeled data	Can use unlabeled or weakly labeled data
Evaluation	Objective and metric based	Subjective and context based, often human preference based
Applications	Classification, detection, forecasting	Generation, simulation, data augmentation
Interpretability	Often more interpretable	Latent space is harder to interpret
Computational cost	Lower during training and inference	High during training and sometimes inference
Stability	Generally stable to train	Can be unstable, for example, GAN collapse or VAE blur
Diversity of outputs	Typically one prediction per input	Can produce diverse and multiple valid outputs
Data efficiency	Less data efficient without labels	More data efficient with unsupervised pretraining

7. Interoperability and Hybrid Architectures

Increasingly, systems are being designed to combine discriminative and generative elements. This hybridization allows each model type to compensate for the other's limitations.

Examples include

- **Generative pretraining followed by discriminative fine-tuning** (e.g., BERT pre-trained on language modeling but fine-tuned for classification tasks).
- **Generative augmentation for discriminative learning**, such as using synthetic faces to improve facial recognition classifiers.
- **Generative retrieval + discriminative ranking**, where a model generates multiple candidate answers and then a second model selects the best one.
- **Contrastive learning**, where self-supervised learning extracts rich representations (often generative in nature), which are then used for discriminative tasks.

This convergence hints at a larger trend in machine learning: the best systems do not fit neatly into one category but leverage strengths across both paradigms.

8. Role in Human-AI Collaboration

Discriminative and generative models contribute to human-AI collaboration in fundamentally different ways. Their distinct operational principles lead to varied roles in real-world workflows, from decision support to content creation.

Discriminative models are most often used in supportive or advisory capacities. These models are designed to predict specific outputs from input data, such as classifying an image, labeling sentiment in a review, or forecasting a numerical value. Because of their high interpretability and speed, they are well-suited for applications where humans need quick and precise answers to guide decisions.

For example, in healthcare, a discriminative model might analyze radiological images to highlight regions of concern. A human radiologist then interprets these findings within the broader clinical context. In financial systems, these models help detect fraudulent transactions, flagging anomalies for human investigators. In such scenarios, the model does not act independently. Instead, it supports human judgment with a focused analysis of existing data.

- Discriminative models tend to work best in environments where outcomes are clearly defined and the margin for error is small.
- They are typically evaluated using objective metrics, which helps ensure consistency and accountability in their use.

Generative models, in contrast, take on more expansive and collaborative roles. These models are not just identifying or labeling existing patterns; they are capable of generating entirely new data instances. This includes producing text, synthesizing images, composing music, simulating environments, and much more. As a result, generative models are often integrated into workflows that value creativity, novelty, and ideation.

Instead of simply supporting human analysis, generative models augment human imagination. In product design, they may suggest visual concepts based on descriptive prompts. In marketing, they can generate ad copy or social media posts. In software development, they may draft entire code functions or data schemas. These models act less like advisors and more like co-creators, offering raw materials that humans refine, adapt, or reframe.

- Generative models are particularly effective in the early stages of projects where exploration is prioritized over precision.
- The output of these models is often evaluated subjectively, based on coherence, usefulness, and aesthetic value rather than strict correctness.

This shift toward generative collaboration reflects a broader change in how AI is used in professional environments. Discriminative models are more likely to be embedded in systems that monitor, classify, or assess, while generative models are embedded in tools that ideate, compose, and simulate. As a result, the human role changes depending on which type of model is in use.

With discriminative models, humans remain firmly in control of the interpretive process. They query the system, receive predictions, and make final judgments. With generative models, humans become more like partners in a shared creative activity. They guide the model with prompts, evaluate outputs, and iterate in response to new ideas the system provides.

- Discriminative models encourage humans to act as decision-makers and validators.
- Generative models encourage humans to act as editors, curators, and collaborators.

In both cases, collaboration is central, but the dynamics of that collaboration are very different. Discriminative models offer clarity and certainty in structured domains. Generative models introduce possibility and variation in open-ended ones. Together, they contribute to a more comprehensive vision of what human-AI collaboration can achieve.

9. Evolutionary Trajectories

Over the past decade, discriminative models held a dominant position in machine learning. Their appeal stemmed from their ability to solve clearly defined tasks such as classification, regression, and segmentation with high precision and efficiency. These models were computationally lighter, faster to train, and easier to deploy, which made them attractive for real-world applications, especially in domains such as finance, healthcare, logistics, and content moderation.

During this period, the standard workflow revolved around supervised learning, where models learned to map inputs to outputs using labeled datasets. Tools such as logistic regression, support vector machines, random forests, and later, discriminative neural networks, became widespread. The widespread availability of labeled benchmarks and the simplicity of performance evaluation through metrics like accuracy, precision, and recall only reinforced the widespread reliance on discriminative architectures.

However, this paradigm began to shift dramatically with the rise of large-scale generative models. These systems, such as transformer-based architectures trained on vast corpora of unlabeled data, introduced a fundamentally different approach. Rather than learning to distinguish between classes, generative models learn the underlying data distribution itself. This enables them not only to generate coherent and often novel outputs, but also to perform downstream discriminative tasks through fine-tuning or prompting.

The key innovations that enabled this transition include

- **Pretraining on unlabeled data**: Generative models like GPT and LLaMA are trained on enormous datasets without requiring manual labeling. This allows them to internalize linguistic, visual, or multimodal patterns at a scale previously unattainable by traditional discriminative models.

- **Scaling laws**: Empirical findings show that model performance improves predictably with increases in compute, data, and parameter count. This insight has encouraged the development of increasingly large and capable generative architectures.
- **Foundation models**: These are pre-trained models that serve as general-purpose engines for a wide range of tasks. They can be fine-tuned for classification, summarization, translation, or question answering, often outperforming specialized discriminative models that were trained from scratch.

This convergence has led to a notable reversal in how tasks are approached. In many modern workflows, generative models now serve as the foundation, while traditional discriminative objectives are treated as special cases or fine-tuning targets. For example, a single generative language model can be prompted to act as a classifier, an extractor, a translator, or even a programmer, depending on the context of use.

- Generative models can now achieve state-of-the-art results on benchmark classification tasks without task-specific retraining.
- Techniques such as in-context learning, few-shot prompting, and zero-shot generalization have allowed generative models to bypass the need for large, labeled datasets in many cases.

This shift also raises complex architectural and operational questions. Discriminative models remain more efficient in constrained environments, especially when speed, interpretability, or strict reproducibility is needed. They can be deployed at the edge or in real-time systems with limited computational resources. On the other hand, generative models, due to their size and generality, often require high-end infrastructure and significant energy resources to operate effectively.

As generative modeling becomes increasingly dominant, new research questions have emerged:

- What roles will discriminative models retain in a generative-first ecosystem?
- How can we make large generative models more efficient, secure, and explainable for enterprise or embedded applications?
- Should future AI systems be designed primarily as generative engines with modular discriminative components, or will hybrid systems emerge that blend the best of both paradigms?
- How will this trajectory affect AI education, tooling, and workforce preparation in both technical and creative industries?

This ongoing evolution is not merely a technical transition. It represents a change in how AI is conceived, built, and integrated across domains. Discriminative models are no longer the default endpoint in machine learning pipelines. Instead, generative models increasingly serve as the universal foundation upon which a wide variety of tasks and applications are constructed.

More on Architecture Families (Transformers, GANs, VAEs, Diffusion)

The power of generative AI arises not just from high-level conceptual design but from the underlying architectures that make learning and generation possible. While generative models vary widely in their outputs, objectives, and use cases, most belong to a few foundational architecture families. These include Transformers, GANs, VAEs, and Diffusion Models.

Each of these architecture families represents a different approach to learning and generating from data. They are designed with different assumptions, loss functions, and optimization strategies, and each has unique strengths, limitations, and practical domains of use. Understanding these families is key to selecting the right approach for a given problem, interpreting model behavior, and innovating new applications.

1. Transformers

a. Origins and Design

Transformers were introduced in 2017 through the seminal paper "Attention Is All You Need" by Vaswani et al. Originally designed for sequence-to-sequence tasks in machine translation, transformers quickly displaced previous recurrent architectures due to their scalability, parallelism, and ability to model long-range dependencies.

The core innovation in the transformer is the self-attention mechanism. Instead of processing sequences sequentially, as RNNs and long short-term memory (LSTMs) do, transformers process all tokens simultaneously and allow each token to attend to every other token in the sequence. This allows for powerful contextual modeling and parallel computation.

b. Architecture Components

Self-attention layers compute the influence of all other tokens on each token, enabling contextual understanding. Multi-head attention captures multiple relationships in parallel, allowing richer representations. Positional encoding injects information about token order into the architecture since the attention mechanism itself is order-agnostic. Feedforward layers apply nonlinear transformations independently to each token. Layer normalization and residual connections stabilize training and allow deeper networks.

c. Generative Use

In generative settings, transformers are used in an autoregressive fashion. A decoder-only transformer predicts the next token based on previous tokens, generating sequences one step at a time. This is the architecture behind GPT, LLaMA, and other LLMs. For multimodal models like Midjourney or Stable Diffusion, transformers are adapted to handle image tokens, audio tokens, or cross-modal fusion.

d. Enterprise Applications

Transformers have become foundational in industries where language, reasoning, or contextual understanding are central. In legal technology, they summarize and extract clauses from long-form contracts. In financial services, they generate reports, interpret market data, and answer regulatory compliance queries. In customer service, transformer-based chatbots resolve queries across multiple languages and channels. In marketing, they automate personalized email and social media content.

e. Hybrid Use

Many diffusion models use transformers as their core denoising networks. Transformer layers are also integrated into VAEs for richer latent space modeling. Multimodal transformer architectures such as Flamingo and Kosmos use transformer backbones with modality-specific encoders. This enables flexible fusion of vision, language, and audio for applications such as virtual tutors or digital assistants.

f. Advantages

Transformers are excellent at modeling long sequences, scalable to large parameter counts, and capable of using unlabeled data for general-purpose language and multimodal modeling. They are flexible across tasks with minimal architectural changes.

g. Challenges

They require large compute resources for training and inference. They are sensitive to data and scale for achieving strong generalization and difficult to interpret internally.

2. GANs

a. Origins and Concept

Introduced in 2014 by Ian Goodfellow and colleagues, GANs represent a fundamentally different approach to generative modeling. Rather than explicitly modeling the data distribution, GANs use a game-theoretic setup involving two networks: a generator and a discriminator.

The generator produces data samples from noise, while the discriminator learns to distinguish between real samples and fake samples. The two networks are trained simultaneously, each improving in response to the other, until the generator learns to produce data that the discriminator cannot reliably distinguish from real data.

b. Architecture Components

The generator takes a random noise vector as input and maps it to a synthetic data instance. The discriminator receives a data instance and outputs a probability indicating whether it is real or generated. The adversarial loss drives the generator to fool the discriminator while the discriminator learns to detect fakes.

c. Use in Generation

GANs have been especially successful in high-quality image generation. They are widely used for producing art, faces, synthetic datasets, and domain transfer. Advanced variants include StyleGAN, BigGAN, and CycleGAN, each designed to improve stability, diversity, or realism.

d. Enterprise Applications

In entertainment and media, GANs generate character concepts and game environments. In fashion and retail, they design synthetic garments and perform virtual try-on simulations. In automotive, they produce simulated sensor data for testing autonomous systems. In surveillance, GANs improve resolution in low-light video.

e. Hybrid Use

GANs have been combined with VAEs to form VAE-GANs, which improve output sharpness while maintaining structured latent spaces. Some diffusion workflows also use a GAN as a final refinement stage after coarse generation. In these hybrid designs, GANs act as detail enhancers rather than primary generators.

f. Advantages

GANs are capable of generating sharp, photorealistic images, offer efficient inference once trained, and operate with an intuitive conceptual framework of learning through competition.

g. Challenges

Training is notoriously unstable due to the adversarial setup. Mode collapse is a frequent issue, and evaluation is more difficult because GANs do not learn an explicit probability distribution.

3. VAEs

a. Origins and Objective

VAEs were introduced as a way to combine the strengths of probabilistic graphical models and neural networks. They are designed to learn a latent space that captures the underlying factors of variation in data, enabling sampling, interpolation, and structured generation.

Unlike traditional autoencoders, VAEs treat the latent representations as random variables. They learn to approximate the true data distribution using variational inference, a technique that makes the intractable posterior estimation problem manageable through optimization.

b. Architecture Components

The encoder maps input data to a distribution over latent variables. The latent space is a continuous, smooth, structured space from which new samples can be drawn. The decoder reconstructs data from samples drawn from the latent space. The loss function combines reconstruction loss with a regularization term that keeps the latent space close to a standard normal distribution.

c. Use in Generation

VAEs are used in domains where smooth interpolation or understanding of latent structure is important. Applications include anomaly detection, molecule generation, handwriting synthesis, and representation learning.

d. Enterprise Applications

In manufacturing and aerospace, VAEs detect anomalies in sensor logs. In healthcare, they assist in detecting rare diseases from medical images. In biotech, VAEs explore latent structures of proteins or RNA sequences. In recommendation systems, VAEs learn user preferences for personalized product suggestions.

e. Hybrid Use

As mentioned, VAE-GANs offer structured latent control with realistic output fidelity. Transformer-VAEs are also used in text generation tasks that require control over attributes like tone, style, or topic. Latent diffusion models incorporate VAE encoding before denoising in the latent space rather than pixel space.

f. Advantages

VAEs provide an interpretable, structured latent space, are capable of continuous interpolation between samples, and can be trained in a stable and probabilistically grounded manner.

g. Challenges

Outputs tend to be blurrier than GANs. Balancing reconstruction and regularization is difficult. VAEs struggle with high-fidelity generation of complex data.

4. Diffusion Models

a. Origins and Breakthroughs

Diffusion models emerged from the study of stochastic processes and gained prominence in deep learning through papers like DDPM and were later improved by models such as Stable Diffusion and Imagen.

Diffusion models learn to generate data by reversing a process of gradual corruption. During training, clean data is slowly corrupted by Gaussian noise over several steps. The model then learns to reverse this process, reconstructing data from noise.

b. Architecture Components

The forward process adds noise to data over many steps, creating a trajectory toward complete randomness. The reverse process involves a neural network that predicts and removes noise at each step. Starting from pure noise, the model denoises the data iteratively until a coherent sample emerges.

c. Use in Generation

Diffusion models are state of the art in image generation, with applications in design, branding, fashion, and industrial modeling. They are expanding into video, audio, and 3D domains. Text-to-image diffusion models, powered by transformer-based conditioning, have become powerful tools for visual ideation and asset creation.

d. Enterprise Applications

In branding, diffusion models create custom logos and packaging. In e-commerce, they generate product shots for SKUs without needing photography. In publishing, they produce illustrations from editorial text. In real estate and construction, they render interiors from blueprint descriptions.

e. Hybrid Use

Diffusion models frequently integrate transformers for conditioning. They are often preceded by VAE encoding to operate in latent space. Some implementations use GANs to refine the final output. Diffusion models are also being adapted for controllable generation, combining classifier guidance and prompt embeddings.

f. Advantages

They produce highly realistic and diverse samples, avoid mode collapse by modeling the full data distribution, and support flexible conditioning.

g. Challenges

Sampling is slow due to the iterative process. Training is computationally expensive. They are still maturing for long-form video or narrative text.

In our next chapter, we tackle choosing the right architecture.

CHAPTER 2

Choosing the Right Architecture

Generative AI encompasses a range of model architectures, each with its own design philosophy, strengths, and limitations. Choosing the right architecture is a critical step in building systems that meet the specific needs of a given application. This chapter examines four of the most influential approaches: transformer-based models for handling complex sequences, generative adversarial networks for high-fidelity synthesis through adversarial training, variational autoencoders for learning structured probabilistic representations, and diffusion models for gradual denoising and high-quality sample generation. By exploring the principles, advantages, and trade-offs of each, this chapter provides the knowledge needed to align architectural choice with project objectives, performance requirements, and resource constraints.

Transformer-Based Models for Sequence Generation

Transformer-based models have fundamentally reshaped the landscape of sequence generation in artificial intelligence. They are now the default choice for a broad range of applications in natural language processing, computer vision, audio generation, and multimodal AI systems. The rapid

I. Cronin, *Building and Training Generative AI Models*,
https://doi.org/10.1007/979-8-8688-2332-9_2

and widespread adoption of transformers stems from three central factors: their architectural innovations, their scalability to unprecedented model sizes, and their unmatched ability to capture long-range dependencies in sequential data more effectively than older recurrent or convolutional approaches.

Before the introduction of transformers, sequence modeling was dominated by RNNs and their more advanced variants such as LSTMs and gated recurrent units (GRUs). These architectures processed data sequentially, maintaining hidden states that evolved with each step. While powerful for short sequences, their inherently sequential computation made them slow to train and prone to difficulties in capturing long-range dependencies. Convolutional approaches, while faster, also faced limitations in representing long contextual relationships without very deep networks. The arrival of transformers disrupted this paradigm entirely.

Origins and Core Principles

The transformer architecture replaced recurrence and convolution with **self-attention**, a mechanism that allows the model to compute relationships between every pair of elements in a sequence in parallel. This was a major departure from the step-by-step processing of RNNs, enabling transformers to scale much more efficiently and to better capture global dependencies in data.

The self-attention mechanism works by representing each token in three ways: as a query, a key, and a value. The model computes similarity scores between the query of one token and the keys of all tokens in the sequence, normalizing these scores to determine how much weight to assign to each value. This process allows the model to focus dynamically on the most relevant parts of the input for a given task.

The architecture uses **multi-head attention**, in which several attention layers operate in parallel, each learning to capture different types of relationships, such as syntactic structure, semantic meaning, or positional patterns, within the same sequence. These parallel attention heads are then combined and passed through feed-forward layers and residual connections, ensuring stable gradient flow and better representational capacity.

Configurations for Sequence Generation

Transformer-based models for sequence generation typically follow one of two main configurations:

- **Encoder-decoder architecture**: This design is common in machine translation and summarization. The encoder processes the input sequence into a rich contextual representation. The decoder then uses this representation, combined with previously generated tokens, to produce the output sequence one step at a time. Models such as T5 and BART are prime examples.
- **Decoder-only architecture**: This approach models the probability of the next token given all previous tokens in an autoregressive fashion. It is the architecture behind GPT-style models, which excel at free-form text generation, long-form narrative, and interactive dialogue.

Both configurations benefit from positional encodings that help the model track the order of tokens, since self-attention itself is inherently order-agnostic.

Advantages of Transformer-Based Sequence Generation

Transformer-based models have become the dominant architecture for sequence generation because they combine theoretical elegance with practical performance benefits across a wide range of domains. Their success is not the result of a single innovation but rather a combination of architectural design, training efficiency, scalability, and adaptability. The following sections expand on the key advantages that explain their widespread adoption.

Long-Range Dependency Modeling

One of the most significant advantages of transformers is their ability to model long-range dependencies in sequential data. Unlike RNNs and LSTMs, which must propagate information step-by-step through a hidden state, transformers employ a self-attention mechanism that allows any token in a sequence to directly attend to any other token. This means that relationships between distant elements can be modeled in a single computation step rather than gradually over many steps.

This property is particularly valuable in tasks where global context is essential. In document summarization, for example, important information may appear at both the beginning and end of a text, and the model must be able to connect these distant parts to produce a coherent summary. In code generation, a function defined hundreds of lines earlier may need to be referenced in later sections. In multi-turn dialogue, transformers can maintain awareness of earlier conversation turns to produce consistent and contextually relevant responses.

From a theoretical standpoint, the ability to capture arbitrary pairwise interactions in a single layer means that transformers can represent dependencies that might otherwise require multiple recurrent steps to approximate. This advantage grows with sequence length, making transformers particularly effective for long documents, extended audio sequences, or multi-stage reasoning chains.

Parallelizable Training

Training efficiency is another area where transformers outperform traditional sequential models. RNNs and LSTMs process sequences token by token during training, making it impossible to parallelize operations across time steps. Transformers, on the other hand, process the entire sequence simultaneously thanks to their non-recurrent architecture. The self-attention mechanism computes dependencies between tokens in parallel, enabling full exploitation of modern hardware accelerators such as GPUs and TPUs.

This parallelism drastically shortens training times and makes it feasible to train models on massive datasets, a critical factor in the success of large-scale generative AI systems. Models like GPT-5 and LLaMA owe much of their training feasibility to this architectural property. For large organizations, the reduction in training time translates into significant cost savings. For researchers, it opens the door to more experiments in less time, accelerating innovation.

Parallelizable training also facilitates batch processing of sequences of varying lengths, allowing efficient use of computational resources. Combined with optimized attention implementations, this makes transformers capable of handling workloads that would be prohibitive for recurrent models.

Cross-Domain Adaptability

Transformers are not limited to natural language processing. Their core design—a mechanism for computing relationships between elements in a set—can be applied to any data that can be represented as a sequence or a set of tokens. This has led to their adoption in diverse fields:

- **Text**: In natural language generation, transformers have set state-of-the-art benchmarks for tasks like machine translation, summarization, and conversational AI.
- **Vision**: Vision transformers (ViTs) represent images as sequences of patches, enabling transformers to excel in image classification, object detection, and even image generation when combined with generative training objectives.
- **Audio**: By treating audio spectrograms as sequences, transformers have been used for speech recognition, text-to-speech synthesis, and music generation.
- **Biological data**: Transformers have been applied to protein sequence modeling, RNA structure prediction, and drug discovery, where the data consists of symbolic sequences with complex dependencies.
- **Multimodal systems**: Multimodal systems integrate vision and language into unified transformer-based architectures for tasks that require reasoning across modalities.

The adaptability of transformers to multiple domains stems from their modular design: the same attention and feed-forward layers can operate on a wide variety of token types, with only minimal preprocessing needed to adapt to a new modality.

Scalability

Another defining advantage is the predictable improvement in performance as model size and training data increase. This property, often referred to as the "scaling law" for transformers, has been documented extensively in empirical studies. When more parameters and data are added, transformers tend to exhibit smooth, consistent gains in accuracy, generalization, and generation quality.

Scalability has been central to the rise of large language models like GPT-5, Anthropic's Claude, and Google's PaLM. These systems leverage billions or trillions of parameters trained on massive text corpora, enabling them to handle complex reasoning, follow nuanced instructions, and produce highly coherent long-form content.

The scaling advantage is not limited to text. In image generation, scaling up transformer backbones in models like DALL·E and Parti has led to better compositional abilities and finer detail reproduction. In protein modeling, scaling transformer models has improved the accuracy of structure prediction and functional annotation.

Importantly, scalability is not just about raw size; it also encompasses scaling in terms of computational budget, dataset diversity, and context length. Transformers have been extended to handle longer sequences through techniques like sparse attention, memory-augmented layers, and retrieval-augmented architectures, further enhancing their utility.

Flexible Conditioning

Transformers excel at integrating diverse forms of conditioning information into the generation process. Conditioning refers to providing the model with additional signals—beyond the primary sequence—to guide its output. Because attention mechanisms naturally allow the mixing of different input streams, transformers can incorporate conditioning data in a seamless way.

Forms of conditioning include

- **Prompts**: Natural language instructions or examples that guide output generation in models like ChatGPT.
- **Images**: For image captioning or text-to-image generation, an image can be encoded and fed into the transformer alongside text tokens.

- **Structured data**: Tables, graphs, or JSON structures can be converted into sequences and used to condition outputs.
- **Contextual embeddings**: Outputs from other models, such as embeddings from a vision encoder, can be integrated as part of the attention input.
- **Control signals**: In creative applications, sliders or discrete codes can control attributes like style, tone, or layout.

Flexible conditioning makes transformers ideal for controlled generation tasks, where fidelity to the conditioning signal is as important as fluency in generation. This is critical in applications like data-to-text generation, personalized content creation, and multi-turn conversational agents that must maintain topic coherence.

Synergy of Advantages

The real power of transformers in sequence generation lies in how these advantages reinforce one another. Long-range dependency modeling enables rich contextual awareness, while parallelizable training makes it practical to train on enormous datasets. Cross-domain adaptability ensures that the same architecture can be reused for text, vision, audio, and more, making it a universal backbone for generative AI. Scalability guarantees that investments in larger models and more data yield predictable returns. Flexible conditioning allows fine control over outputs, making transformers useful in real-world applications where precision and reliability are required.

For example, consider a multimodal storytelling system that generates illustrated narratives from a short prompt. A transformer-based architecture can

1. Interpret the prompt and establish a storyline (long-range dependency modeling).
2. Generate text and illustrations concurrently using separate but aligned transformer modules (cross-domain adaptability).
3. Train efficiently on paired text-image datasets using parallelizable attention layers (parallelizable training).
4. Scale to handle long narratives and high-resolution images as hardware budgets allow (scalability).
5. Accept user feedback mid-story to adjust plot direction or visual style (flexible conditioning).

This type of integrated workflow is possible precisely because transformers embody all of these advantages.

For researchers, the advantages of transformers mean shorter iteration cycles, fewer architecture-specific constraints, and a clear pathway for scaling experiments. For businesses, transformers offer a single architecture that can be adapted to multiple product lines—chatbots, summarization tools, content generators, and multimodal assistants—reducing the need to maintain entirely separate systems for each modality. For end users, the benefits translate into more capable, responsive, and customizable AI systems.

The dominance of transformers in sequence generation is the result of architectural properties that align exceptionally well with the needs of modern AI: the ability to model long-range dependencies, efficient parallel training, adaptability across domains, scalability with data and compute, and flexibility in conditioning. These strengths are not isolated—they work together to create models that are both powerful and practical. As research continues, innovations in efficient attention, retrieval-augmented generation, and multimodal integration will likely amplify these advantages, ensuring that transformers remain at the forefront of sequence generation for years to come.

Training Paradigms and Enhancements

Training transformer-based sequence generators usually involves two stages:

1. **Large-scale pretraining**: Models are trained on massive datasets, often in the hundreds of billions of tokens, to learn general-purpose sequence representations. Pretraining objectives include next-token prediction (causal language modeling), masked token prediction (masked language modeling), or sequence-to-sequence prediction.
2. **Fine-tuning**: The pretrained model is adapted to specific downstream tasks using smaller labeled datasets. Fine-tuning can be supervised or semi-supervised, and may use reinforcement learning to optimize for human feedback (RLHF).

Recent advances include **instruction tuning**, in which models are trained to follow natural language instructions across a wide variety of tasks, and **parameter-efficient tuning** techniques such as adapters and LoRA (Low-Rank Adaptation), which make fine-tuning large models more computationally affordable.

Challenges and Limitations

Despite their strengths, transformers are not without drawbacks:

- **Quadratic attention cost**: The standard self-attention mechanism scales quadratically with sequence length in both computation and memory, making very long sequences costly to process.
- **Resource requirements**: Training large transformers requires enormous datasets, high-end hardware, and significant energy consumption, raising accessibility and sustainability concerns.
- **Output quality control**: Without careful fine-tuning and alignment, transformers may produce factually incorrect, biased, or contextually inappropriate content.
- **Inference latency**: While training is parallelizable, autoregressive generation at inference time still proceeds step-by-step, which can limit throughput in real-time applications.

To address these issues, researchers have developed efficient transformer variants such as Longformer, Performer, and Reformer, which modify the attention mechanism to reduce computational costs for long sequences.

Evaluation Considerations

Evaluating transformer-based sequence generation models depends heavily on the application domain:

- **Language generation**: Metrics such as BLEU, ROUGE, METEOR, and perplexity are common, but often need to be supplemented with human evaluation for qualities like coherence, creativity, and adherence to instructions.

- **Code generation**: Automated evaluation can include compilation checks, unit tests, and functional correctness benchmarks.
- **Image-to-text generation**: Evaluation considers relevance, accuracy, and faithfulness to the input content.

In high-stakes domains, human evaluation remains indispensable for ensuring quality and safety.

Real-World Applications

Transformer-based sequence generation has achieved state-of-the-art results in numerous applications:

- **Machine translation**: Models like Marian, M2M-100, and mBART outperform older neural and statistical approaches across many languages.
- **Conversational AI**: Systems like ChatGPT and LaMDA use transformers to maintain context across multiple dialogue turns while producing coherent and contextually relevant responses.
- **Code synthesis**: Models such as Codex and AlphaCode generate production-grade code from natural language descriptions.
- **Content creation**: Transformers assist in writing articles, creating marketing copy, and producing interactive storytelling experiences.
- **Music generation**: Models trained on symbolic music representations or audio spectrograms compose new pieces in a range of styles.

- **Multimodal reasoning**: Models like Flamingo and Kosmos integrate text, images, and other modalities to perform complex reasoning tasks.

Comparative Perspective with Other Architectures

In the broader context of generative modeling

- **Compared to GANs**, transformers are far better suited to long-form structured generation, though GANs may still produce sharper images in some settings.
- **Compared to VAEs**, transformers generally achieve higher output quality in sequence tasks but lack the explicit probabilistic latent space that VAEs provide for interpolation and sampling control.
- **Compared to diffusion models**, transformers are faster at inference for sequence generation because diffusion models require many iterative denoising steps. However, diffusion models have surpassed transformers in some benchmarks for image fidelity.

Future Directions

Research on transformer-based sequence generation is exploring several promising directions:

- **Efficient attention mechanisms** that scale linearly or sublinearly with sequence length, enabling transformers to handle documents or audio recordings many times longer than current limits.

- **Multimodal integration**, where a single transformer processes and generates sequences across text, vision, audio, and other modalities without separate model components.
- **Grounded generation**, combining transformers with retrieval systems or external knowledge bases to improve factual accuracy and reduce hallucination.
- **Personalization and controllability**, allowing models to adapt to specific user preferences or follow stylistic constraints in real time.

Given their adaptability and scalability, transformers are expected to remain the foundation of sequence generation for the foreseeable future, while evolving to become more efficient, controllable, and trustworthy.

GANs for Adversarial Training and Synthesis

GANs represent one of the most influential developments in generative modeling over the past decade. Introduced by Ian Goodfellow and colleagues in 2014, GANs provided a new framework for training generative models through an adversarial process. Rather than explicitly modeling a data distribution or optimizing a likelihood function, GANs train two competing neural networks in a game-theoretic setting, leading to highly realistic sample generation in many domains.

At their core, GANs consist of two components: a generator and a discriminator. The generator's job is to produce synthetic samples that resemble the real data as closely as possible. The discriminator's job is to distinguish between real samples drawn from the training set and fake samples produced by the generator. Both models are trained simultaneously in a zero-sum game: the generator improves by trying to fool the discriminator, while the discriminator improves by getting better at telling real from fake.

This adversarial process can be thought of as a competition between a counterfeiter (the generator) and a police inspector (the discriminator). The counterfeiter tries to create increasingly convincing fake currency, while the inspector works to detect the fakes. As training progresses, the counterfeiter learns to create notes so realistic that they are nearly indistinguishable from real currency, and the inspector must continually adapt to detect even subtle flaws.

The training of a GAN involves the following steps in each iteration:

- The generator samples a set of random latent vectors, often drawn from a Gaussian or uniform distribution, and maps them into synthetic data samples.
- The discriminator is given a batch containing both real samples from the dataset and fake samples from the generator. It predicts whether each sample is real or fake.
- The discriminator's parameters are updated to improve its classification accuracy.
- The generator is then updated based on the discriminator's feedback, using a gradient signal that indicates how convincingly it fooled the discriminator.

The key innovation of GANs lies in their ability to learn to generate data without explicitly defining or calculating the probability density of the data distribution. Instead, the discriminator's judgments provide an implicit measure of how closely the generator's outputs match the true data distribution.

Strengths of GANs

GANs have several strengths that have made them a central architecture in generative modeling:

- They can produce samples of exceptionally high visual quality, often sharper and more detailed than those generated by earlier methods such as Variational Autoencoders.
- They do not require explicit likelihood computation, which can be difficult or intractable for complex data distributions.
- Their adversarial training objective directly optimizes for realism in the output space, making them well-suited for tasks where perceptual quality is critical.
- GANs can be conditioned on labels, text, images, or other modalities, allowing controlled synthesis of highly specific outputs.

One of the earliest and most striking demonstrations of GAN capabilities was in generating photorealistic human faces, as seen in architectures such as Progressive Growing of GANs and StyleGAN. These models could produce faces that were not only indistinguishable from real photographs but also highly diverse in features such as age, gender, hairstyle, and lighting.

Challenges in GAN Training

While GANs are powerful, they are also notoriously difficult to train. Some of the main challenges include

- **Mode collapse**: The generator may learn to produce a limited set of outputs that successfully fool the discriminator, ignoring large portions of the data distribution. This reduces diversity and makes the model less useful for creative or exploratory tasks.
- **Training instability**: Because GAN training is framed as a minimax optimization problem, it can be sensitive to hyperparameter choices and prone to oscillations, divergence, or failure to converge.
- **Evaluation difficulty**: Measuring the quality of GAN-generated outputs is challenging. Metrics like Inception Score (IS) and Fréchet Inception Distance (FID) provide approximate measures of realism and diversity but do not always correlate perfectly with human judgment.

Numerous improvements have been proposed to address these issues, such as Wasserstein GANs (WGANs), which replace the original loss function with an approximation of the Wasserstein distance, providing more stable gradients and reducing mode collapse. Spectral normalization, gradient penalties, and architectural innovations have also contributed to more reliable training.

Conditional GANs

A significant extension of the original GAN framework is the Conditional GAN (cGAN). In a cGAN, both the generator and discriminator are given additional information alongside the noise input and generated sample. This conditioning information can be anything from class labels to full text descriptions or images.

For example, in image-to-image translation tasks such as turning a sketch into a photorealistic image, the conditioning input is the sketch itself. The generator learns to map this structured input to the target domain, while the discriminator ensures that the output matches the desired style and realism. Notable models such as Pix2Pix and CycleGAN have shown how cGANs can enable style transfer, domain adaptation, and high-fidelity image translation without paired datasets in some cases.

Architectural Innovations

Over the years, several architectural enhancements have made GANs more powerful and flexible:

- **Deep Convolutional GANs (DCGANs)** introduced the use of convolutional layers in both generator and discriminator, greatly improving image quality and stability.
- **Progressive Growing of GANs** gradually increases the resolution of generated images during training, allowing the model to first learn large-scale structure before focusing on fine details.
- **StyleGAN** introduced style-based modulation, allowing unprecedented control over features at different levels of detail and enabling fine-grained manipulation of generated outputs.
- **BigGAN** scaled up GAN architectures to very large parameter counts, achieving state-of-the-art performance on class-conditional image synthesis.

GANs Beyond Images

Although GANs are most famous for image generation, they have been applied to a wide range of domains:

- **Text-to-image synthesis**, where models like DALL·E combine transformer-based architectures with adversarial losses to improve visual realism
- **Speech synthesis and voice conversion**, generating realistic human voices from text or transforming one speaker's voice into another's
- **Music generation**, producing melodies, harmonies, and instrument timbres
- **Data augmentation**, creating synthetic but realistic training data for improving model performance in low-data scenarios
- **3D object and scene generation**, enabling applications in gaming, simulation, and augmented or virtual reality

Comparison with Other Generative Architectures

In the broader context of generative modeling, GANs have specific strengths and trade-offs compared to other approaches:

- Compared to transformer-based generators, GANs excel in producing high-resolution, realistic imagery, but they are less suited to long-form, structured sequence generation such as natural language.
- Compared to VAEs, GANs generally produce sharper and more visually appealing outputs but lack a well-defined probabilistic latent space, making controlled sampling and interpolation more difficult.

- Compared to diffusion models, GANs are typically faster at inference since they generate outputs in a single forward pass, while diffusion models require multiple denoising steps. However, diffusion models have recently surpassed GANs in some benchmarks for photorealism and diversity.

Ethical and Societal Considerations

GANs also raise important ethical questions, particularly because of their ability to create highly convincing synthetic media. Deepfake technology, which uses GAN-like architectures to replace faces or manipulate videos, has been used for entertainment, satire, and accessibility applications but also for misinformation, political manipulation, and harassment.

The capacity to generate synthetic identities, voices, and imagery has implications for security, privacy, and trust. This has led to calls for watermarking, detection tools, and regulatory oversight to mitigate potential harms.

The Evolving Role of GANs

Although GANs are no longer the sole leader in generative modeling due to the rise of transformers and diffusion models, they remain highly relevant. Their ability to produce high-quality outputs in a single step makes them attractive for real-time applications, and ongoing research continues to address their training challenges. Hybrid models that combine adversarial losses with other generative objectives have emerged as a way to leverage the strengths of GANs while mitigating weaknesses.

Recent trends include combining GANs with diffusion processes to enhance both speed and fidelity, integrating attention mechanisms to improve feature learning, and using GANs for tasks where conditional control and realism are both critical.

VAEs for Probabilistic Latent Modeling

VAEs are a class of generative models that combine the representational power of deep neural networks with the mathematical rigor of probabilistic modeling. Introduced by Kingma and Welling in 2013, VAEs provided a scalable and efficient way to perform variational inference in deep latent variable models, enabling them to learn complex data distributions while also providing interpretable latent representations.

The defining characteristic of VAEs is their explicit probabilistic formulation. Rather than learning to directly map inputs to outputs, as in traditional autoencoders, VAEs model the joint probability distribution of inputs and latent variables. This allows them to generate new data samples by first sampling from the latent distribution and then decoding those latent variables into data space.

Core Architecture and Probabilistic Foundation

At their core, VAEs consist of two neural networks: an encoder (also called the recognition network or inference model) and a decoder (also called the generative model).

- The **encoder** maps input data xx to a probability distribution over latent variables zz. Instead of outputting a single point in latent space, the encoder outputs parameters of a distribution, typically the mean and variance of a Gaussian. This distribution represents the model's uncertainty about the underlying latent representation of the input.
- The **decoder** maps samples from the latent distribution back into the data space, producing reconstructions of the input or entirely new samples if the latent vector is drawn independently.

The probabilistic foundation of VAEs lies in their objective function, which maximizes the **evidence lower bound (ELBO)** on the marginal likelihood of the data. This involves two terms:

1. A **reconstruction term** that measures how well the decoder can reconstruct the input from the latent code.
2. A **regularization term** that measures how close the encoder's approximate posterior distribution $q(z|x)$q(z|x) is to a chosen prior $p(z)$p(z), usually a standard normal distribution.

The regularization term is computed using the **Kullback-Leibler (KL) divergence**, which encourages the latent space to follow the prior distribution and thus enables smooth sampling and interpolation.

The Reparameterization Trick

A key innovation that made VAEs trainable with backpropagation is the **reparameterization trick**. Since sampling from a probability distribution is non-differentiable, the reparameterization trick allows gradients to flow through stochastic latent variables by expressing a sample z as:

$z=\mu+\sigma\cdot\epsilon$z = \mu + \sigma \cdot \epsilon

where ϵ\epsilon is drawn from a standard normal distribution, and μ\mu and σ\sigma are outputs of the encoder network. This separates the stochasticity from the network parameters, enabling standard gradient-based optimization.

Advantages of Probabilistic Latent Modeling

VAEs offer several advantages compared to purely deterministic generative models:

- **Structured latent space**: Because the latent space is regularized to follow a known distribution, it becomes smooth and continuous. This allows for meaningful interpolation between points and controlled manipulation of latent dimensions.
- **Probabilistic reasoning**: VAEs model uncertainty explicitly, which can be useful in tasks that require confidence estimation or handling ambiguous inputs.
- **Efficient sampling**: Once trained, generating new samples is as simple as sampling from the prior distribution in latent space and decoding.
- **Flexibility**: The probabilistic framework is adaptable to many types of data, including images, text, audio, and multimodal data.

Limitations and Challenges

Despite their strengths, VAEs are not without drawbacks:

- **Blurriness in generated images**: VAEs optimize for pixel-wise reconstruction accuracy, which can lead to outputs that are overly smooth and lack fine-grained detail.
- **Trade-off between reconstruction and regularization**: Strong KL regularization can lead to underfitting, while weak regularization can cause overfitting to the training data and poor generative performance.

- **Posterior collapse**: In some settings, especially with powerful decoders, the encoder may learn to ignore the latent variables, leading to a degenerate posterior distribution.

To address these issues, researchers have proposed various improvements, such as β-VAEs (which adjust the weight of the KL term), vector-quantized VAEs (which discretize the latent space), and hierarchical VAEs (which introduce multiple layers of latent variables).

Variants and Extensions

Over the years, the basic VAE framework has been extended in multiple directions:

- **β-VAE**: Introduces a hyperparameter β\beta to scale the KL divergence term, encouraging disentangled representations in the latent space.
- **VQ-VAE**: Uses a discrete latent space with vector quantization, improving generative quality and enabling applications like neural audio codecs.
- **Conditional VAE (CVAE)**: Conditions the encoder and decoder on auxiliary information (labels, text descriptions, or other data) for controlled generation.
- **Hierarchical VAE**: Stacks multiple layers of latent variables to capture complex data distributions more effectively.
- **Semi-supervised VAE**: Combines generative modeling with classification objectives to improve performance with limited labeled data.

Applications of VAEs

VAEs have been applied across a wide range of domains:

- **Image synthesis and editing**: VAEs can generate new images, interpolate between existing ones, and manipulate attributes via latent space operations.
- **Representation learning**: The smooth latent space of VAEs is useful for tasks such as clustering, retrieval, and anomaly detection.
- **Text generation**: VAEs have been used to model sentence-level variability and generate diverse outputs.
- **Speech and audio synthesis**: VAEs can learn compact representations of speech signals for compression, denoising, and transformation tasks.
- **Drug discovery and molecular design**: VAEs generate novel chemical structures by sampling from a learned latent distribution over molecular graphs.

Comparison with Other Generative Architectures

When compared to other major generative model families, VAEs occupy a distinct niche:

- **Versus GANs**: GANs often produce sharper and more visually appealing outputs, but VAEs provide a structured latent space and explicit likelihood estimation, which GANs lack.
- **Versus transformers**: Transformers excel at sequence modeling and capturing long-range dependencies, whereas VAEs focus on compact latent representations and probabilistic reasoning.

- **Versus diffusion models**: Diffusion models have recently surpassed VAEs in image fidelity, but VAEs are faster at inference and require only a single forward pass to generate samples.

Role in Multimodal and Hybrid Architectures

VAEs are often integrated into larger systems to complement other modeling approaches. Examples include

- **VAE-GAN hybrids**, which combine the structured latent space of VAEs with the adversarial training of GANs to improve perceptual quality.
- **Variational transformers**, which incorporate VAE-style latent variables into transformer architectures to enhance diversity and handle uncertainty in sequence generation.
- **Diffusion-VAE hybrids**, where VAEs are used to initialize or compress diffusion processes for faster sampling.

Ethical Considerations

As with all powerful generative models, VAEs raise ethical questions. While their outputs are often less photorealistic than those from GANs or diffusion models, they can still be used to generate synthetic content, including faces, voices, or text. This creates potential for misuse in impersonation, misinformation, or copyright infringement. Responsible deployment of VAEs includes considering watermarking, dataset curation, and transparent disclosure of synthetic outputs.

Future Directions

Research on VAEs continues to evolve in several promising directions:

- **Improving sample fidelity**: Combining VAEs with perceptual loss functions, adversarial training, or more expressive decoders to reduce blurriness.
- **Disentangled representation learning**: Encouraging latent dimensions to correspond to interpretable factors of variation.
- **Scalable VAEs**: Developing architectures and training procedures that work effectively with very large datasets and high-resolution data.
- **Multimodal VAEs**: Building models that can jointly represent and generate across multiple data modalities, such as text and images or video and audio.

VAEs remain an important tool in the generative modeling toolkit. Their probabilistic foundation, structured latent spaces, and flexibility make them valuable both as stand-alone models and as components within more complex hybrid systems.

Diffusion Models and Denoising Strategies

Diffusion models have emerged as one of the most significant advances in generative AI, particularly for high-quality image synthesis. In recent years, they have surpassed Generative Adversarial Networks (GANs) in certain benchmarks, especially in terms of fidelity and diversity, while providing a more stable training process. Their power lies in their probabilistic formulation, iterative denoising process, and flexibility in conditioning on various types of input data. Models like Denoising Diffusion Probabilistic

Models (DDPMs), Denoising Diffusion Implicit Models (DDIMs), and Latent Diffusion Models (LDMs) have set new standards in generative modeling for images, audio, and multimodal systems.

Conceptual Foundation

At a high level, diffusion models learn to generate data by reversing a gradual noising process. The idea is rooted in nonequilibrium thermodynamics: one can simulate the forward process of adding noise to data until it becomes pure noise and then learn the reverse process to recover the data from that noise. By chaining many small denoising steps together, the model learns how to produce highly detailed and realistic outputs.

The process has two main components:

1. **Forward (diffusion) process**: The model defines a Markov chain that gradually adds Gaussian noise to the data over a fixed number of time steps. After enough steps, the data distribution becomes indistinguishable from an isotropic Gaussian distribution.

2. **Reverse (denoising) process**: The model learns the reverse of the diffusion process, predicting the noise that was added at each step. By subtracting this predicted noise iteratively, the model reconstructs data samples from pure noise.

The elegance of this formulation lies in the fact that the forward process is fixed and known, while the reverse process is learned. This separation simplifies training and provides a mathematically grounded approach to generative modeling.

Intuition Behind the Process

One way to think about diffusion models is as a series of denoising autoencoders, each trained to remove a small amount of noise from its input. By stacking many such steps, the model can learn to recover complex structures from heavily corrupted inputs. Because each denoising step is small, the learning problem at each step is relatively easy, and the overall process is stable.

This stability is a key reason why diffusion models avoid some of the pitfalls of GANs, such as mode collapse. The iterative refinement process also allows them to produce exceptionally fine-grained detail in outputs.

Variants of Diffusion Models

Several variants have been proposed to improve efficiency, quality, and flexibility:

- **DDPM (Denoising Diffusion Probabilistic Models)**: The original formulation that formalized the forward and reverse processes as a probabilistic graphical model.
- **DDIM (Denoising Diffusion Implicit Models)**: A non-Markovian variant that allows for deterministic sampling paths, enabling faster inference and the ability to trade off between speed and diversity.
- **Score-based generative models (SGMs)**: Related to diffusion models through the concept of score matching, where the model learns the gradient of the data log-density with respect to inputs.

- **Latent Diffusion Models (LDMs)**: Instead of operating in pixel space, these models run the diffusion process in a lower-dimensional latent space learned by an autoencoder. This reduces computational cost and makes high-resolution synthesis more practical. Stable Diffusion is a prominent example.
- **Guided diffusion**: Incorporates conditioning signals (such as class labels, text prompts, or images) into the denoising process, often through classifier guidance or classifier-free guidance.

Denoising Strategies

The denoising step is where the generative power of diffusion models resides. Several strategies have been developed to improve denoising efficiency and control:

- **Classifier guidance**: Uses an external classifier to guide the sampling process toward desired attributes. This can enhance fidelity to conditioning labels but may reduce diversity.
- **Classifier-free guidance**: Trains the diffusion model itself to handle both conditioned and unconditioned generation, blending the two during sampling to control the strength of conditioning.
- **Progressive denoising**: Starts with a very noisy input and gradually removes noise over many steps, refining details at each step. This mirrors human drawing processes, where broad strokes come before fine details.

- **Deterministic sampling (DDIM)**: Reduces the number of steps required for sampling, producing high-quality outputs in fewer iterations.
- **Hybrid denoising**: Combines denoising with other objectives, such as perceptual loss or adversarial loss, to improve specific aspects of output quality.

Strengths of Diffusion Models

Diffusion models have emerged as one of the most influential developments in generative AI in recent years. Their success in applications such as image synthesis, text-to-image generation, audio creation, and even scientific modeling is due to a set of core strengths that differentiate them from earlier approaches like GANs and VAEs. These strengths are rooted in their iterative denoising framework, probabilistic formulation, and flexibility in conditioning.

Stable Training

One of the most important advantages of diffusion models is their stability during training. In GANs, the training process is adversarial: the generator and discriminator are engaged in a two-player minimax game. While this setup can lead to sharp outputs, it is notoriously sensitive to hyperparameters, prone to oscillations, and often results in failure modes such as mode collapse or vanishing gradients.

Diffusion models avoid these pitfalls by using a fixed and well-defined forward process (gradually adding Gaussian noise) and a reverse process that is learned through supervised training. The training objective—predicting the noise added at each step—reduces to a simple mean squared error loss in many implementations. This makes the optimization landscape smoother and less sensitive to hyperparameter tuning.

The absence of a discriminator also removes the instability caused by competing objectives. As a result, diffusion models can often be trained with fewer tricks and more predictable progress, even on diverse and complex datasets. This stability is one reason why they have been so quickly adopted for high-quality generative tasks in industry.

High-Output Fidelity

The iterative refinement process of diffusion models is key to their ability to produce outputs with exceptional detail and realism. Unlike single-pass generators such as GANs or VAEs, diffusion models gradually construct the output over many denoising steps. Each step focuses on removing a small amount of noise, which allows the model to make fine-grained corrections to texture, lighting, and structure.

For example, in image generation, early denoising steps establish the global composition—shapes, perspective, and object placement—while later steps refine edges, add texture, and ensure consistency in small-scale details. This multi-stage process mirrors how an artist might first sketch an outline and then add layers of detail and shading.

In practice, this means that diffusion models can handle complex visual patterns such as intricate hair textures, realistic reflections, and natural-looking landscapes more effectively than many single-pass architectures. The same principle applies to other domains: in audio generation, diffusion models can capture both the broad structure of a musical piece and the subtle nuances of timbre and reverb.

Mode Coverage

Another notable strength of diffusion models is their ability to cover the full diversity of the target data distribution. GANs, while capable of producing sharp images, are prone to **mode collapse**, where the generator focuses on a small subset of possible outputs that fool the discriminator, neglecting other valid variations in the data. This can lead to low diversity in generated samples.

Diffusion models, by contrast, optimize a likelihood-based objective (or a close approximation to it), which encourages them to match the entire data distribution. The gradual noise-removal process means the model learns to map noisy samples back to many possible clean examples, rather than converging on a narrow set of outputs. As a result, diffusion models are better at producing varied samples that reflect the richness of the training data.

This mode coverage is particularly valuable in creative applications, where diversity is as important as realism. For instance, in text-to-image systems like Stable Diffusion, a single prompt can yield many plausible but visually distinct outputs, giving users multiple options to choose from.

Flexibility in Conditioning

Diffusion models are inherently flexible in how they can be conditioned. Conditioning refers to providing the model with additional input information that guides the generation process toward specific outcomes. This can take many forms:

- **Text prompts**: Text-to-image systems such as DALL·E, Imagen, and Stable Diffusion use natural language descriptions to steer the denoising process toward producing images that match the prompt.
- **Class labels**: Class-conditional diffusion models generate outputs belonging to specific categories, useful in tasks like data augmentation for classification.
- **Semantic maps**: In tasks like semantic-to-image synthesis, the model takes a high-level scene layout and generates a photorealistic image matching that layout.

- **Partial images or masks**: In inpainting tasks, the model fills in missing regions of an image while keeping the rest unchanged.
- **Multimodal conditioning**: Diffusion models can also integrate inputs from multiple sources—text, images, audio—within the same generative process.

This flexibility is possible because conditioning information can be injected into the denoising network at each step, influencing how noise is removed. Approaches like **classifier guidance** and **classifier-free guidance** provide further control, allowing users to adjust the strength of conditioning to balance fidelity to the input with creative variation.

Theoretical Grounding

Diffusion models are built on a solid theoretical foundation, drawing from concepts in statistical physics, stochastic processes, and probabilistic modeling. The forward process of gradually adding noise is mathematically analogous to diffusion processes in physics, while the reverse process relates to score-based generative modeling and denoising autoencoders.

This grounding provides several benefits:

- **Interpretability**: The connection to well-understood physical processes makes the behavior of diffusion models easier to reason about compared to some other black-box generative methods.
- **Principled design**: The probabilistic framework offers clear pathways for extending the models—for example, by modifying the noise schedule, changing the prior distribution, or incorporating different types of stochastic processes.

- **Cross-domain applicability**: The generality of the formulation means that diffusion models can be adapted to domains far beyond images, including molecular modeling, weather simulation, and audio synthesis.

Theoretical insights have also led to practical improvements, such as better noise schedules for faster convergence, alternative parameterizations for more stable training, and efficient sampling methods that reduce the number of required denoising steps.

Synergy of Strengths

The strengths of diffusion models reinforce each other in ways that make them especially powerful for modern generative AI:

- Stable training ensures that models can be scaled to large datasets and high resolutions without collapsing or requiring extensive manual tuning.
- High output fidelity makes them suitable for applications where detail and realism are critical, such as commercial design or cinematic visual effects.
- Strong mode coverage ensures variety in outputs, which is crucial for creative industries and scientific simulations.
- Flexible conditioning allows them to be deployed in interactive systems where users can steer generation with natural language or other inputs.
- A solid theoretical foundation supports ongoing innovation and adaptation to new domains.

For instance, in a text-to-image workflow, a diffusion model might first establish the scene layout according to the prompt (leveraging conditioning), then refine details iteratively (ensuring fidelity), and finally produce several diverse variations (providing mode coverage). The training process for such a model can proceed reliably without the instability that might plague an equivalently complex GAN.

Practical Implications

For researchers, the stability and theoretical grounding of diffusion models make them attractive for experimentation. They offer a flexible platform for exploring new conditioning methods, architectures, and training objectives. For practitioners, the combination of high quality, diversity, and controllability makes diffusion models well-suited for integration into creative tools, design software, and AI-assisted research pipelines.

In commercial contexts, these strengths translate into tangible value: marketing teams can use them to generate varied campaign visuals from a single prompt, film studios can rapidly prototype scenes, and scientific teams can simulate complex phenomena without expensive physical experiments.

The rise of diffusion models is a direct result of their core strengths: stable training, high output fidelity, comprehensive mode coverage, flexible conditioning, and a solid theoretical foundation. These attributes not only explain their rapid adoption but also suggest that they will remain central to generative AI for years to come. As research continues to improve their efficiency and extend their reach into new modalities, diffusion models are likely to play an even more prominent role in shaping the next generation of AI systems.

Limitations and Challenges

Diffusion models have demonstrated remarkable capabilities, but like any generative architecture, they are not without drawbacks. Understanding these limitations is important for researchers, practitioners, and organizations considering their deployment in real-world systems. The primary challenges include slow sampling speed, high computational cost, significant memory requirements, and the risk of over-smoothing in generated outputs. While many of these challenges are being actively addressed through research, they remain key factors that can influence the choice of architecture for specific applications.

Sampling Speed

One of the most cited drawbacks of diffusion models is their slow sampling speed during inference. This limitation is inherent to their design. Generating an output typically involves a sequence of denoising steps, starting from a sample of pure noise and gradually refining it toward a clean sample that matches the learned data distribution.

A standard diffusion process might require hundreds or even thousands of steps, each involving a forward pass through a large neural network. In contrast, GANs can produce an output in a single forward pass of the generator. This difference means that, for high-resolution outputs, diffusion models can be orders of magnitude slower.

For example, producing a single high-quality 1024×1024 image with a vanilla diffusion model might take several seconds on a high-end GPU, whereas a GAN could do it in milliseconds. This speed gap can be a significant disadvantage in latency-sensitive applications such as real-time video generation, interactive creative tools, or gaming environments where immediate feedback is essential.

Recent innovations like **Denoising Diffusion Implicit Models (DDIM)**, **Progressive Distillation**, and **Score-Based Generative Modeling with Fewer Steps** have reduced the number of sampling steps required. Nevertheless, even with these optimizations, diffusion models still lag behind the near-instantaneous generation capabilities of single-pass architectures.

Computational Cost

Training diffusion models is computationally demanding. Each training iteration involves simulating the forward diffusion process by adding noise and then training the reverse denoising process step by step. Since the model must learn to reverse the process at multiple noise levels, the training loop effectively runs the network many times per batch, leading to high GPU or TPU utilization.

When training on high-resolution data, such as 4K images or long-form audio, the computational burden increases sharply. Not only does each step require more floating-point operations due to larger inputs, but the number of necessary steps to achieve high fidelity can also increase. This results in substantial training times, often spanning days or weeks on clusters of powerful GPUs.

For example, training state-of-the-art text-to-image models like Imagen or DALL·E required hundreds of thousands of GPU-hours. This cost makes large-scale diffusion modeling inaccessible to most smaller organizations and independent researchers, limiting experimentation and innovation to well-funded institutions or collaborative open-source efforts.

While techniques like **Latent Diffusion Models (LDMs)** reduce training costs by operating in a compressed latent space rather than pixel space, they still require substantial computational resources to reach the quality levels demanded in modern generative AI applications.

Memory Footprint

Another practical challenge is the memory footprint of diffusion models. High-resolution data is inherently high-dimensional, and training or inference in these spaces requires storing activations, intermediate representations, and gradients for multiple denoising steps. This can quickly exceed the memory capacity of a single GPU, especially when training with large batch sizes.

For example, a 512×512 RGB image has over 786,000 pixels, and representing these at multiple noise levels for a batch of samples can consume tens of gigabytes of memory. This makes it difficult to train large models without specialized hardware setups or advanced memory optimization techniques such as gradient checkpointing or activation offloading.

Latent diffusion addresses part of this problem by performing the denoising process in a lower-dimensional latent space obtained via an autoencoder. This can reduce memory usage by an order of magnitude, enabling training of high-resolution models on more modest hardware. However, for tasks where working directly in the original data space is essential—such as medical imaging or scientific simulations—memory constraints remain a challenge.

Over-smoothing Risk

While diffusion models are celebrated for producing realistic and detailed outputs, there is an inherent risk of over-smoothing, particularly if the model is not tuned correctly or if the noise schedule is poorly designed. Over-smoothing occurs when the denoising process averages out fine details in an attempt to remove noise, resulting in outputs that look slightly blurred or lack high-frequency texture.

This effect is more pronounced in earlier or less-optimized diffusion architectures. In images, over-smoothing can manifest as skin textures that appear too soft, fabric patterns that lose their weave detail, or landscapes that look as if they have been lightly airbrushed. In audio generation, over-smoothing can remove the natural crispness of percussive sounds or the subtle timbre variations in instruments.

State-of-the-art implementations mitigate this problem through improved training objectives, better noise schedules, and hybrid approaches that incorporate adversarial losses to encourage sharper details. Nonetheless, in scenarios where maximum sharpness is critical, such as fine art reproduction or high-precision texture generation, GAN-based methods may still have a slight edge.

Trade-Offs and Application-Specific Considerations

These limitations interact in ways that require careful consideration depending on the intended use case:

- **Real-time applications**: For scenarios requiring immediate feedback, the slow sampling speed of diffusion models can be prohibitive. Hybrid models or alternative architectures might be more suitable.
- **Large-scale deployment**: The high computational and memory costs can limit the practicality of deploying diffusion models at scale, especially for consumer-facing services with millions of daily queries.
- **High-precision needs**: While over-smoothing is less of an issue in modern designs, it can still be relevant in niche applications where ultra-fine detail is non-negotiable.

These trade-offs mean that diffusion models are often best suited for offline or batch generation tasks, such as creating assets for films, games, or design projects, where quality is prioritized over latency.

Ongoing Research Toward Mitigation

The rapid progress in diffusion model research is addressing these challenges:

- **Speed improvements**: Methods like **accelerated sampling schedules**, **deterministic samplers**, and **score distillation** aim to reduce the number of steps without compromising quality.
- **Efficiency optimizations**: Latent diffusion and mixed-precision training have significantly cut down memory and compute requirements.
- **Hybrid approaches**: Combining diffusion models with GAN-style discriminators can preserve sharpness while retaining stability.
- **Adaptive noise schedules**: Optimizing the progression of noise levels can reduce over-smoothing and improve both speed and fidelity.

These developments suggest that many current limitations may be less severe in the coming years, potentially opening the door for real-time and low-resource deployments of diffusion-based generative systems.

While diffusion models have revolutionized generative AI with their stability, flexibility, and fidelity, their limitations in sampling speed, computational cost, memory footprint, and potential over-smoothing are important factors that must be considered when selecting an architecture for a given application. These challenges reflect the trade-offs inherent in any design and highlight the importance of aligning technical capabilities with the practical constraints and goals of the target use case.

Ongoing research is actively working to overcome these issues, and the trajectory suggests that many of today's bottlenecks will be mitigated or resolved in the near future. Until then, practitioners must weigh the strengths and weaknesses carefully, often blending diffusion models with other techniques to achieve the best balance between performance, efficiency, and output quality.

Applications

Diffusion models have proven effective in a wide range of applications:

- **Image synthesis**: Generating photorealistic images from noise or from conditioning inputs such as text prompts.
- **Image editing**: Modifying specific regions of an image while preserving the rest, often using inpainting techniques.
- **Super-resolution**: Increasing the resolution of images while maintaining detail and texture.
- **Style transfer**: Applying the visual style of one image to the content of another.
- **Text-to-image generation**: Models like DALL·E, Imagen, and Stable Diffusion have demonstrated the ability to create detailed images from textual descriptions.
- **Audio and speech synthesis**: Diffusion models have been adapted to generate waveforms or spectrograms, producing high-quality audio.
- **Video generation**: Early work is extending diffusion-based generation to the temporal domain.
- **Scientific data generation**: Applications include molecular structure prediction, climate modeling, and medical imaging.

Comparison with Other Generative Architectures

Diffusion models have rapidly become one of the four dominant paradigms in generative modeling, alongside Generative Adversarial Networks (GANs), Variational Autoencoders (VAEs), and transformer-based generative models. Each of these approaches embodies distinct design philosophies, training dynamics, and areas of application. Understanding the comparative strengths and weaknesses of these architectures is essential for making informed decisions about which to use in a given context.

Diffusion Models vs. GANs

GANs, introduced in 2014, have long been the benchmark for generating high-fidelity images and other media. They operate through an adversarial framework in which two neural networks, the generator and the discriminator, compete. The generator attempts to produce samples indistinguishable from real data, while the discriminator tries to tell them apart. This adversarial setup has proven capable of producing visually striking results, particularly in the realm of image synthesis.

However, GANs are notoriously unstable to train. The adversarial loss can oscillate or collapse entirely if the balance between generator and discriminator is not maintained. Mode collapse is a persistent problem.

Diffusion models sidestep many of these stability issues because their training objective is based on denoising rather than adversarial competition. They do not require a discriminator network, eliminating one source of instability. Instead, diffusion models are trained to reverse a fixed, known noise process, which leads to more predictable convergence.

In terms of output diversity, diffusion models generally achieve better **mode coverage** than GANs, meaning they more faithfully represent the full variety of the training distribution. This is particularly important in applications where coverage matters as much as fidelity, such as scientific simulations or medical image generation, where missing rare but critical patterns can have serious consequences.

The trade-off is speed. GANs can generate a complete output in a single forward pass of the generator, making them well suited for real-time applications. Diffusion models, by contrast, require many iterative denoising steps during inference, leading to slower generation times. This latency difference can be orders of magnitude, especially when comparing a standard GAN with a vanilla diffusion model.

In summary:

- **Diffusion strengths over GANs**: Stability, diversity of outputs, and better coverage of the data distribution
- **GAN strengths over diffusion**: Extremely fast inference, often sharper details in certain settings, and lower computational cost at inference

Diffusion Models vs. VAEs

VAEs, introduced in 2013, were among the first deep learning models to bring probabilistic latent variable modeling to large-scale datasets. VAEs encode input data into a continuous latent space, then decode from that space back into data space. The model learns both the mapping to and from the latent representation, and training optimizes a combination of reconstruction loss and a regularization term that encourages the latent space to follow a known prior distribution, often Gaussian.

One of the primary advantages of VAEs is the explicit structure of their latent space. Because the latent variables follow a well-defined distribution, it is straightforward to interpolate between points, generate new samples, and manipulate high-level attributes of generated outputs.

This makes VAEs particularly attractive for applications requiring controlled and interpretable generation, such as in scientific modeling, representation learning, and structured data synthesis.

Diffusion models, on the other hand, do not typically learn such an explicit latent space in their original formulation. While they implicitly capture a representation of the data distribution through the denoising process, they lack the same direct, low-dimensional latent interface that VAEs provide. This makes certain forms of controlled generation more challenging unless additional conditioning mechanisms are added.

In terms of output fidelity, diffusion models generally outperform VAEs. VAEs often produce outputs that are overly smooth or blurry due to the assumptions made in their likelihood-based training objective, which can limit high-frequency detail. Diffusion models, with their iterative refinement process, are better at capturing fine-grained details and textures.

Hybrid approaches, such as **latent diffusion models**, have sought to combine the strengths of both by using an autoencoder-like component to compress data into a latent space before applying the diffusion process. This enables higher efficiency while still retaining much of the high-fidelity advantage of diffusion models.

In summary:

- **Diffusion strengths over VAEs**: Higher fidelity outputs, finer detail preservation, and better perceptual quality.
- **VAE strengths over diffusion**: Explicit, structured latent space for interpretable and controlled generation, faster inference, and lower computational demands for certain tasks.

Diffusion Models vs. Transformers

Transformers, first popularized for sequence modeling in natural language processing, have since been adapted for a wide range of generative tasks. At their core, transformers rely on self-attention mechanisms, which allow each element in an input sequence to directly attend to every other element. This makes them highly effective for capturing long-range dependencies in data.

When used for generation, transformers can take several forms. In language modeling, **decoder-only** transformers such as GPT produce text in an autoregressive fashion, predicting the next token given all previous tokens. In vision, **Vision Transformers (ViTs)** process images as sequences of patches, enabling them to model complex spatial relationships. There are also multimodal transformers that integrate text, images, and audio into a single generative framework.

The key strength of transformers lies in **sequence modeling**. For data that is naturally sequential—text, audio waveforms, symbolic music, protein sequences—transformers can excel. They can generate coherent, contextually rich sequences that span thousands of elements. Their training and inference can also be highly parallelized during the forward pass, though autoregressive decoding still introduces sequential dependencies.

Diffusion models, by contrast, excel in **iterative refinement of spatial or continuous data**. For tasks like high-resolution image synthesis, super-resolution, and inpainting, the denoising process can gradually add structure and detail in a way that is particularly well suited to spatial patterns. While transformers can be adapted for image generation (for example, by predicting pixels or patches autoregressively), these approaches are often less efficient and less competitive in perceptual quality compared to state-of-the-art diffusion-based systems.

Interestingly, the two paradigms are increasingly being combined. Diffusion models have been implemented within transformer backbones to leverage the attention mechanism's strength in modeling global context. Conversely, transformers have been used to parameterize parts of the diffusion process, particularly in large-scale text-to-image models like Imagen and Stable Diffusion.

In summary:

- **Diffusion strengths over transformers**: Superior iterative refinement for high-dimensional spatial data, better texture and detail synthesis in images, robustness in training for generative tasks without requiring autoregressive steps.
- **Transformer strengths over diffusion**: Exceptional performance on long-range sequence modeling, versatility across domains including text and multimodal tasks, and strong scaling behavior with large datasets and model sizes.

Choosing the Right Architecture

The decision between diffusion models, GANs, VAEs, and transformers depends on the application's priorities:

- If **training stability** and **output diversity** are critical, diffusion models are often the best choice.
- If **speed** is paramount, GANs may be preferable.
- If **interpretable latent spaces** are important, VAEs offer a clear advantage.
- If the task involves **long sequences or multimodal integration**, transformers are a natural fit.

In practice, teams often experiment with multiple approaches or combine them, especially as open-source implementations and pretrained models become more accessible.

Hybrid and Future Directions

Research is increasingly exploring hybrid models that combine diffusion with other approaches:

- **Diffusion-GAN hybrids**: Incorporating adversarial loss to speed up sampling and sharpen details
- **Diffusion-transformer hybrids**: Using transformers within the denoising network to better capture global context
- **Latent diffusion in multimodal settings**: Extending latent diffusion to text, video, and 3D generation

Future directions for diffusion research include

- **Accelerated sampling**: Reducing the number of denoising steps needed without sacrificing quality
- **Better conditioning mechanisms**: Improving controllability in text-to-image and other conditional generation tasks
- **Unified multimodal models**: Integrating diffusion into systems that handle multiple modalities in a coherent way
- **Energy efficiency**: Optimizing architectures and training procedures to reduce the carbon footprint of large-scale diffusion models

Ethical and Societal Considerations

As with all powerful generative technologies, diffusion models raise ethical concerns:

- **Deepfakes and misinformation**: The ability to produce convincing synthetic media can be misused for deception.
- **Copyright and ownership**: Generated content may infringe on intellectual property rights, especially when models are trained on copyrighted data.
- **Bias and fairness**: Models trained on biased datasets can reproduce or amplify those biases in generated content.

Addressing these issues will require a combination of technical solutions (such as watermarking and bias mitigation), policy development, and public education.

Diffusion models, with their probabilistic foundations and iterative refinement, have reshaped the field of generative modeling. Their ability to generate high-fidelity, diverse outputs in a stable and theoretically grounded framework makes them a cornerstone of modern AI. As research continues to address their speed and efficiency limitations, diffusion-based generation is likely to expand its dominance across more domains and modalities.

The next chapter focuses on data collection and preparation.

CHAPTER 3

Data Collection and Preparation

The success of any generative AI system is inseparable from the quality, structure, and integrity of the data that feeds it. Whether the objective is to generate text, images, video, audio, simulations, or multimodal outputs, every capability originates from data. Data is not only the material generative models are trained on; it is also the medium through which these models express their learned understanding of the world. In generative AI, the data defines the edges of what is possible—what can be learned, what can be synthesized, and ultimately, what can be trusted.

Despite this central role, data is often treated as a secondary concern relative to model architecture or training infrastructure. This misalignment can be catastrophic. No matter how advanced a transformer model or diffusion network may be, if the underlying data is corrupted, poorly structured, incomplete, or biased, the results will reflect those flaws. In practice, the most impactful gains in model reliability and generalization often come not from increasing model size or tweaking learning rates, but from improvements in how the data is prepared, processed, and curated.

This chapter explores the foundational components of data management and processing for generative AI systems, breaking down the steps that transform raw, real-world inputs into high-value, learning-ready formats. The chapter is organized into four interlinked subtopics, each addressing a different layer of the data pipeline.

I. Cronin, *Building and Training Generative AI Models*,
https://doi.org/10.1007/979-8-8688-2332-9_3

We begin with a systematic overview of **data types—structured, unstructured, and semistructured**. Understanding these categories is critical, as each one requires distinct preprocessing methods, storage solutions, and modeling strategies. Structured data, typically clean, tabular, and machine-readable, lends itself to classical analytics and straightforward machine learning applications. Unstructured data, such as text, images, video, and audio, requires sophisticated techniques for parsing, labeling, and interpretation. Semistructured data, including JSON, XML, and log formats, lies between these two, blending machine readability with flexibility. Navigating across these data types is essential for building multi-modal generative systems and for operating in complex, dynamic environments.

The next subtopic focuses on **cleaning and normalization pipelines**, which serve as the foundation of any trustworthy data pipeline. Raw data is noisy, inconsistent, and full of implicit ambiguity. Duplicate entries, missing values, inconsistent formats, mislabeled categories, and semantic contradictions are the norm rather than the exception. Cleaning transforms chaotic raw inputs into coherent, well-structured signals. Normalization standardizes scales, formats, and encodings, making data compatible with the numerical operations required by machine learning and deep learning frameworks. Without robust cleaning and normalization, even the most sophisticated generative models will produce unstable, biased, or meaningless outputs. These pipelines not only protect model performance—they are the operational backbone of AI reliability.

We then turn to **synthetic data generation**, a rapidly growing discipline that addresses some of the most pressing limitations of real-world data: scarcity, privacy concerns, cost of labeling, and class imbalance. In many applications, autonomous driving, healthcare diagnostics, rare event modeling, there simply isn't enough diverse, annotated real-world data to train high-quality models. Synthetic data offers an alternative, enabling developers to simulate complex environments, balance skewed distributions, generate edge cases, and protect user identities through privacy-preserving practices. The ability to generate data that complements or augments real-world corpora is becoming a core competency in modern AI development. Moreover, generative models can now be trained on synthetic data created by other generative models, forming self-improving loops that challenge the traditional reliance on real-world input.

Finally, we explore one of the most consequential and politically sensitive aspects of data preparation: **dataset bias detection and mitigation**. All datasets carry implicit biases, reflecting not only gaps in data collection but also the social, historical, and economic inequalities embedded in the systems that produce the data. In generative AI, these biases are not just statistical anomalies. They manifest as discriminatory outputs, exclusion of marginalized perspectives, reinforcement of stereotypes, and even safety risks when deployed in user-facing products. Detecting and mitigating these biases is a technical, ethical, and operational necessity. This section explores how bias arises, how it can be systematically detected through audits and metrics, and what techniques can be used to reduce its influence—from reweighting and rebalancing to adversarial debiasing and fairness-aware training. It also emphasizes that bias mitigation must be proactive, continuous, and embedded throughout the data life cycle, not just a last-minute fix before deployment.

Collectively, these four subtopics define the end-to-end data journey in a generative AI system. They move from understanding what data is, to how it is cleaned and normalized, how it can be extended synthetically, and how it must be interrogated and corrected for bias. Treating these activities as core, not peripheral, is essential for the development of systems that are robust, generalizable, interpretable, and aligned with both user intent and societal values.

In short, data is not just fuel for generative models. It is infrastructure. It is design. It is governance. And it is one of the most powerful levers we have for shaping the behavior, quality, and impact of artificial intelligence in the real world.

Data Types: Structured vs. Unstructured

Understanding the different types of data is the starting point for any serious engagement with AI and data science systems. At the most fundamental level, data can be categorized into two broad types: structured and unstructured. These categories are not merely convenient taxonomies; they reflect fundamentally different properties that affect how data is collected, stored, processed, analyzed, and utilized in real-world applications.

Structured data tends to be machine-readable and organized according to predefined schemas, making it ideal for conventional analytics. Unstructured data, in contrast, is inherently more ambiguous, context-dependent, and varied in form. It is harder to parse but often richer in meaning. These two broad data types shape everything from storage architecture and preprocessing pipelines to model design and deployment strategy. The ability to navigate between them is critical for building modern AI systems capable of solving real-world problems.

The Nature of Structured Data

Structured data is what most people traditionally envision when thinking about data: cleanly arranged numeric or textual entries presented in rows and columns. This type of data adheres to a fixed schema, where every element has a defined type, structure, and location. Examples include relational databases, spreadsheets, tab-separated values, transactional logs, and time-series sensor outputs.

Common domains where structured data dominates:

- Financial transactions (e.g., bank statements, stock trades)
- Customer databases (e.g., CRM systems)

- Inventory management systems
- Web analytics platforms
- Sensor telemetry in manufacturing

Structured data is particularly well-suited to traditional statistical analysis and classical machine learning algorithms. Since the format is predictable, data can be readily ingested by models that require specific input dimensions and types. Models such as logistic regression, decision trees, random forests, and support vector machines operate efficiently on structured datasets where each row represents a well-defined observation and each column represents a measurable feature.

Structured data facilitates

- Efficient querying and filtering via SQL
- Rapid data aggregation and summarization
- Easy integration with dashboards and reporting tools
- Straightforward application of transformation pipelines such as normalization, imputation, and encoding

However, structured data also imposes constraints. It cannot easily express relationships that are implicit, contextual, or dynamic. Textual nuances, spatial-temporal complexity, and multimodal information do not compress well into tabular formats. This limitation becomes especially evident in domains like healthcare, media, education, or customer engagement, where the richness of meaning is often embedded in language, images, or sequences rather than numeric values.

The Emergence of Unstructured Data

Unstructured data represents information that does not conform to a rigid schema. It lacks a predefined model and typically requires parsing, interpretation, or transformation before it becomes computationally useful. This form of data includes natural language text, audio recordings, images, video files, documents, web content, and real-time user interactions.

Examples include

- Text: emails, chat logs, social media posts, legal contracts
- Audio: customer service calls, voice memos, podcast recordings
- Images: photographs, scanned documents, satellite imagery
- Video: security footage, YouTube uploads, cinematic content

Unstructured data accounts for the vast majority of digital information. It is estimated that over 80% of enterprise data is unstructured. While historically underutilized due to its complexity, the rise of deep learning has made it increasingly accessible to machine processing.

Advances in fields such as

- **Natural Language Processing (NLP)** allows for parsing and understanding of written and spoken language.
- **Computer Vision (CV)** enables automated interpretation of images and videos.
- **Audio Signal Processing** converts waveform data into spectral or phonemic representations.
- **Multimodal Fusion** allows the integration of text, image, audio, and metadata into unified machine-readable formats.

This shift reflects the changing nature of information in the digital age. Businesses, governments, and individuals now communicate, interact, and record experiences in free-form formats that do not fit into traditional databases. Extracting insights from this flood of unstructured content is both a technical challenge and a strategic advantage.

Semistructured Data: The Hybrid Zone

Semistructured data bridges the gap between structured and unstructured data. It contains organizational properties that make it easier to parse than pure unstructured data, but it lacks the rigidity of fully structured tables. This format typically relies on metadata, tags, or schema-on-read approaches to impose just enough order to make the data useful.

Common examples of semistructured formats:

- JSON (JavaScript Object Notation)
- XML (eXtensible Markup Language)
- YAML (Yet Another Markup Language)
- HTML documents
- Log files with labeled events
- Configuration files with nested attributes

Semistructured data is widespread in modern web services, APIs, and data interchange systems. When a user requests information from an API, the response is often returned in a JSON format that includes nested user profiles, status codes, geolocations, and metadata. Though not tabular, this structure allows for parsing and extraction.

In AI workflows, semistructured data often requires partial transformation:

- Parsing JSON to extract structured features
- Dropping or embedding irregular entries

- Flattening hierarchical data into tabular form
- Retaining complex segments for downstream NLP or CV models

This dual nature makes semistructured data both powerful and complex. It provides rich context but demands flexible tooling and robust preprocessing routines.

Differences in Storage and Retrieval

The choice of storage mechanism is directly influenced by the type of data being handled.

Structured data is typically stored in

- Relational databases (e.g., PostgreSQL, MySQL, Oracle)
- Data warehouses (e.g., Snowflake, Amazon Redshift)
- Online transaction processing (OLTP) systems

These systems enforce schemas, ensure data consistency, and support query languages such as SQL. They are optimized for fast reads and writes, indexing, and join operations.

Unstructured data is stored in

- File systems (e.g., local drives, Hadoop Distributed File System)
- Object stores (e.g., Amazon S3, Google Cloud Storage)
- NoSQL databases (e.g., MongoDB, Cassandra)
- Blob storage (binary large objects)

These platforms focus on scalability, replication, and distributed access. Since they do not enforce a schema, the onus is on downstream applications to interpret the data correctly.

Semistructured data often resides in

- Document-oriented databases (e.g., MongoDB, Couchbase)
- Time-series databases (e.g., InfluxDB)
- Columnar databases (e.g., Apache Parquet, Delta Lake)

These offer more flexibility than relational systems and allow schema evolution over time, which is crucial for real-time applications and streaming data pipelines.

Preprocessing: Bringing Order to Disorder

Preprocessing is the set of transformations applied to raw data to make it suitable for machine learning models. The complexity and requirements of preprocessing vary significantly based on data type.

Structured data preprocessing typically includes

- Data type casting
- Missing value imputation (mean, median, interpolation)
- Normalization and standardization
- Categorical variable encoding (e.g., one-hot, ordinal, frequency encoding)
- Feature scaling (min–max, z-score)

Because structured data follows a schema, many preprocessing steps can be automated and reused across datasets.

Unstructured data preprocessing is more diverse and domain-specific:

- **Text**:
    - Tokenization
    - Stopword removal
    - Lemmatization or stemming
    - Embedding (Word2Vec, GloVe, BERT)
    - Named Entity Recognition (NER)
- **Images**:
    - Resizing
    - Cropping or padding
    - Normalization
    - Data augmentation (flipping, rotation, brightness adjustment)
    - Conversion to grayscale or channel separation
- **Audio**:
    - Noise filtering
    - Fourier transformation
    - Spectrogram generation
    - Segmentation and voice activity detection

These steps often require libraries such as OpenCV, spaCy, NLTK, librosa, or Hugging Face Transformers, and must be validated carefully to avoid introducing artifacts or noise.

- **Key preprocessing steps by modality**:
    - Text: tokenization, embedding, normalization
    - Image: resizing, color alignment, augmentation
    - Audio: denoising, frequency transformation, segmentation

Annotation and Labeling

Labels are essential for supervised learning. In structured data, labels often exist as dedicated fields—binary classifications, numerical targets, or categorical classes. The process of assigning them is typically rule-based or extracted from historical records.

In contrast, unstructured data often lacks explicit labels. For example:

- A product review does not inherently state whether it is positive or negative.
- A video clip must be watched and manually tagged for action recognition.
- A voice recording may contain multiple speakers and emotional tones.

Labeling unstructured data usually requires

- Manual annotation by human experts
- Crowdsourcing platforms (e.g., Amazon Mechanical Turk)
- Assisted labeling via heuristics or model suggestions
- Annotation tools with bounding boxes, transcriptions, or classification tags

This introduces challenges such as

- Inconsistency in subjective judgments
- Annotator bias or fatigue
- Time and cost constraints
- Need for domain expertise (e.g., in medical imaging)

To scale this process, organizations often turn to

- **Self-supervised learning**: Derive labels from internal data structure
- **Weak supervision**: Use noisy or proxy labels (e.g., hashtags, clicks)
- **Active learning**: Focus labeling on ambiguous or informative samples

Implications for Model Choice

The structure of input data influences which models are suitable.

For structured data, classical models still offer strong performance:

- Linear models (logistic, ridge, lasso)
- Decision trees and ensemble methods (random forest, XGBoost, LightGBM)
- Shallow neural networks for tabular learning

These models are interpretable and computationally efficient and require less data.

For unstructured data, deep learning architectures dominate

- CNNs for image and video
- RNNs, LSTMs, and GRUs for sequential data
- Transformers (e.g., BERT, GPT, ViT) for text, vision, audio, and multimodal tasks

These models can learn hierarchical representations and contextual relationships but require large training sets and significant compute.

Foundation models have emerged as a unifying paradigm, capable of processing structured prompts and unstructured content. With sufficient scale and fine-tuning, these models

- Generalize across modalities
- Reduce the need for manual labeling
- Enable few-shot and zero-shot learning
- Serve as flexible backbones for task-specific adaptation

The choice between classical and neural models depends on data type, size, availability of compute, and the interpretability needs of the application.

Summary

Structured, unstructured, and semistructured data represent the foundational formats in which digital information exists. Each type offers different trade-offs in terms of accessibility, richness, interpretability, and readiness for machine learning.

- Structured data provides high precision, speed, and usability for traditional analytics and models but lacks expressive power for human-like content.

- Unstructured data holds tremendous potential for insights and personalization but requires sophisticated algorithms, annotations, and infrastructure to unlock.
- Semistructured data offers the flexibility of unstructured content with some of the structure necessary for parsing and transformation, making it an increasingly common format in real-time and scalable systems.

For example, Netflix uses all three types together in a single recommendation flow: your account details, subscription plan, and watch history live in structured tables; streaming logs and device metadata arrive as semistructured JSON events; and the movies themselves come with unstructured subtitles, plot summaries, thumbnails, and trailers. The recommendation system fuses these signals so that when you open the app, it can quickly match your structured profile with patterns in semistructured logs and the meaning extracted from unstructured media, then surface a personalized row of titles that you are likely to click and enjoy.

Mastering the nuances of these data types is essential for building robust, fair, and intelligent AI systems. Whether constructing preprocessing pipelines, selecting storage architectures, or choosing modeling strategies, the nature of the data remains the most critical determinant of success.

As AI systems expand into evermore complex environments, the ability to navigate between structured tables, JSON logs, image archives, and conversational transcripts is no longer a niche skill. It is a foundational requirement for modern data science and AI engineering.

Cleaning and Normalization Pipelines

Once data has been collected, whether from sensors, databases, APIs, web scraping, surveys, or human input, it is rarely suitable for immediate use in AI systems. In its raw form, data is often noisy, incomplete, inconsistent, and unstructured. This messiness can manifest in dozens of ways: missing values, inconsistent formats, duplicate records, semantic ambiguity, labeling errors, anomalies, or shifts in underlying distributions. If not addressed through methodical cleaning and normalization, these issues can compromise even the most sophisticated AI architectures, leading to unreliable predictions, unstable behavior, and flawed conclusions.

Cleaning and normalization pipelines are the essential intermediaries that bridge the chaotic real world and the highly structured requirements of machine learning. These processes define how raw information is transformed into analytically useful, statistically sound, and ethically responsible inputs for AI systems. The quality of a model's output is directly proportional to the quality of its input. Thus, investing in robust preprocessing is not an optional task—it is a core requirement.

Understanding the Purpose of Cleaning

At its most basic level, data cleaning involves detecting and correcting, or removing, errors and inconsistencies. These can occur at multiple levels: syntactic (e.g., mismatched formats), semantic (e.g., logically invalid values), or relational (e.g., conflicting records between linked tables). Cleaning operations differ depending on the type of data, its source, its intended use, and the modeling task.

Cleaning tasks may seem mundane, but they are deeply consequential. For instance, one extra space in a string label (" New York" vs. "New York") can create a new class during one-hot encoding, leading to model fragmentation. An improperly parsed date can break time-series predictions. A missing value in a critical field can trigger misclassification in a medical diagnosis algorithm.

Common cleaning steps include

- **Removing duplicates**: Duplicated records can inflate certain patterns and result in overfitting or false correlations.
- **Handling missing values**: Entries may be absent due to failed data transmission, sensor malfunction, or user non-response. These gaps must be evaluated for imputation, flagging, or deletion.
- **Fixing typos and inconsistencies**: Especially in categorical variables. "yes," "Yes," and "YES" must be harmonized.
- **Filtering outliers**: Data points far from the norm may be true anomalies, measurement errors, or artifacts. Context determines whether to retain or remove them.
- **Standardizing formats**: Dates, currencies, measurement units, and encodings must be normalized for consistency.
- **Types of data quality issues**:
    - Null or blank entries in mandatory fields
    - Categorical misalignment due to case sensitivity or spelling errors
    - Misformatted timestamps, corrupted encodings, inconsistent delimiters

- Redundant or duplicated IDs
- Logical inconsistencies (e.g., negative ages, future birth dates)

A key principle in cleaning is context awareness. Data cannot be cleaned in a vacuum. Understanding the domain, source systems, and business rules is essential for making intelligent decisions about what constitutes valid or invalid data. For example, in healthcare data, a heart rate of 500 should almost certainly be flagged as an error, while a value of 40 might be acceptable for an athlete; in retail transactions, a negative quantity could mean a product return rather than a bad record; in a subscription business, a missing cancellation date might be intentional for active customers but a serious issue for customers marked as churned. These domain-specific expectations guide how you treat outliers, missing values, and inconsistencies, and they often make the difference between overaggressive cleaning that destroys signal and targeted cleaning that improves data quality without losing important information.

Normalization As a Structural Prerequisite

Once cleaned, data must often be normalized—transformed into standardized scales or representations to ensure numerical compatibility and algorithmic stability. Without normalization, differences in measurement scale can distort model behavior. For example, a variable representing income (ranging from 10,000 to 1,000,000) may dominate another variable like age (ranging from 18 to 100), even if both are equally important to the prediction task.

Normalization is especially critical in algorithms sensitive to feature magnitude:

- **Gradient descent-based models** (e.g., neural networks, logistic regression): large-scale features may produce erratic gradients, hindering convergence.
- **Distance-based models** (e.g., KNN, clustering): unnormalized features can bias similarity computations.
- **Kernel methods** (e.g., SVM): performance may degrade if variables are not standardized.

Common normalization techniques include

- **Min–max scaling**: Rescales feature values to a fixed interval, typically [0, 1]. It preserves the shape of the original distribution but changes the absolute scale, which is useful when the minimum and maximum bounds are meaningful in the model or downstream system.
- **Z-score standardization**: Transforms values so that the feature has mean 0 and variance 1. This is often called "standardization" rather than "normalization" and is especially helpful for methods that assume features are centered and have comparable spread.
- **Clarification**: In strict technical usage, *scaling* refers to adjusting a variable's range or variance, while *normalization* refers to transforming data so it follows a particular distribution such as a standard normal. Many practitioners use "normalization" loosely for any rescaling step, so it is helpful to be explicit about which operation you mean.

- **Robust scaling**: Uses median and IQR to scale data. Ideal when outliers are present.
- **Log transformation**: Useful for right-skewed distributions, reducing the impact of extreme values.
- **Quantile transformation**: Maps data to a uniform or normal distribution.
- **L2 normalization**: Common in text/vector data. Ensures that the Euclidean norm is 1.

Normalization can also involve encoding categorical variables:

- **One-hot encoding**: Converts categories into binary vectors. Suitable for nominal categories with few unique values
- **Ordinal encoding**: Assigns integer values to ordered categories
- **Binary encoding**: Efficient for high-cardinality fields
- **Embeddings**: Dense, learned representations for categories with semantic meaning (e.g., user IDs, item types)

Feature normalization is often paired with dimensionality reduction (e.g., PCA, t-SNE) to prepare the dataset for modeling. The success of these techniques depends on consistent preprocessing.

Building a Cleaning and Normalization Pipeline

Modern AI workflows rely on structured, repeatable, and scalable pipelines. Rather than applying cleaning and normalization as ad hoc steps in a notebook, robust systems encapsulate them as reusable modules with validation, versioning, and automation built in.

Typical pipeline stages include

1. **Ingestion**: Read raw data from APIs, files, message queues, or databases.
2. **Schema validation**: Ensure fields conform to expected data types, units, and constraints.
3. **Profiling**: Generate summaries of value distributions, missingness, cardinality, and skewness.
4. **Cleaning**: Apply domain-specific logic to filter, replace, or correct problematic entries.
5. **Feature engineering**: Derive new features or transform existing ones.
6. **Normalization**: Apply scaling and encoding techniques.
7. **Validation**: Test for drift, completeness, and model compatibility.
8. **Audit logging**: Record all transformations for reproducibility.
9. **Export**: Save outputs to a feature store, model input, or analytics platform.

Tools that support these pipelines include

- **pandas**: Flexible data wrangling for tabular data
- **PySpark**: Distributed data processing for large-scale workloads
- **Scikit-learn Pipelines**: Modular preprocessing and modeling steps

- **TensorFlow Transform (TFT)**: Preprocessing for TensorFlow-based models
- **Apache Beam**: Streaming and batch data transformation
- **Airflow/Kubeflow**: Orchestration of data pipelines across environments

Handling Missing Data

Missing data is unavoidable. What matters is how it is understood and handled. There are three canonical types of missingness:

- **MCAR (Missing Completely At Random)**: No relationship between missingness and data values. Deletion won't bias the analysis.
- **MAR (Missing At Random)**: Missingness depends on observed variables. Imputation can reduce bias if dependencies are modeled.
- **MNAR (Missing Not At Random)**: Missingness is related to the missing value itself. Requires modeling the missingness mechanism explicitly.

Methods for handling missing data:

- **Listwise deletion**: Remove rows with missing values. Acceptable for MCAR but can reduce sample size.
- **Simple imputation**: Replace with mean, median, or mode. Easy but may distort distributions.
- **Regression imputation**: Predict missing values using other features.

- **Multiple imputation**: Create several plausible imputed datasets and combine results to capture uncertainty.
- **KNN imputation**: Use nearest neighbors to estimate missing values.
- **Interpolation**: Useful in time-series data.
- **Indicator flags**: Create binary variables indicating whether a field was missing.

The choice depends on the amount of missingness, its pattern, and its likely cause. Regardless of method, all imputed values should be explicitly documented.

Encoding Categorical and Text Variables

Raw categorical and textual inputs must be converted into numerical representations before being used in models. The chosen method can dramatically affect performance and interpretability.

Categorical Encoding Techniques:

- **One-hot encoding**: Expands categorical variables into binary columns. Suitable for small cardinality.
- **Ordinal encoding**: Assigns an integer per category. Should only be used when order matters.
- **Frequency encoding**: Replaces each category with its count or frequency in the dataset.
- **Target encoding**: Replaces categories with average target values. Powerful but prone to leakage.
- **Hash encoding**: Uses a hash function to reduce dimensionality.

Text Encoding Techniques:

- **Bag of Words (BoW)**: Counts of token appearances.
- **TF-IDF**: Weighted representation of term importance.
- **Word embeddings**: Dense vectors capturing semantic similarity (Word2Vec, GloVe).
- **Contextual embeddings**: Use pretrained transformers (e.g., BERT) to derive context-aware vectors.
- **Encoding best practices**:
    - Avoid ordinal encoding for nominal categories.
    - Apply target encoding within cross-validation folds to avoid leakage.
    - Clip rare categories or group into "Other" to improve stability.
    - Use pre-trained embeddings when labeled data is scarce.

Detecting and Handling Outliers

Outliers can arise from data entry errors, sensor glitches, natural variation, or true anomalies. The decision to retain or remove them must be grounded in the application context.

Detection methods:

- **Z-score**: Values beyond 3 standard deviations.
- **Interquartile Range (IQR)**: Points beyond 1.5× IQR from the quartiles.
- **Boxplots and histograms**: Visual methods for small datasets.

- **Mahalanobis distance**: Multivariate outlier detection.
- **Isolation Forests**: Tree-based anomaly detection algorithm.
- **DBSCAN and LOF**: Clustering-based anomaly detection.

Handling methods:

- **Winsorization**: Replace outliers with percentile-based thresholds.
- **Transformation**: Apply log or square root to compress skew.
- **Binning**: Aggregate into categories.
- **Flagging**: Retain but mark for downstream analysis.

In high-stakes domains like healthcare or security, outliers may be more important than central patterns. Removing them prematurely can erase rare but vital signals.

Automating Data Cleaning

Automation enables scalability, reproducibility, and monitoring. However, human judgment is still essential, especially for semantic errors or nuanced inconsistencies.

Automation tools and methods:

- **Great Expectations**: Defines tests and expectations for data integrity.
- **Datafold**: Tracks data drift and schema changes.
- **Deequ**: Declarative testing of data quality for Spark workloads.

- **HoloClean**: Uses probabilistic modeling to repair dirty data.
- **Pandera**: Enforces data types and constraints in Python.
- **Data validation with TensorFlow Data Validation (TFDV)**: Ensures input pipeline integrity.

Automated checks can catch schema violations, null thresholds, unexpected distributions, and changes in data cardinality. These guardrails are critical in production environments where data flows continuously.

Monitoring for Drift

Once a pipeline is live, data quality does not remain static. Input distributions, source systems, and business rules evolve. Drift must be detected and addressed.

Types of drift:

- **Covariate drift**: Input features change distribution
- **Label drift**: Distribution of labels shifts over time
- **Concept drift**: Relationship between features and labels changes

Detection strategies:

- Monitor summary statistics (mean, variance).
- Use statistical tests (Kolmogorov-Smirnov, Chi-squared).
- Compare distributions using divergence metrics (KL divergence, JS divergence).
- Log model performance metrics over time.

If drift is detected, pipeline adjustments may include retraining, feature recalibration, or updating normalization parameters. Monitoring tools such as EvidentlyAI, Fiddler, or Arize help automate this process.

Case Study: Cleaning Clinical Trial Data

In a multi-site clinical trial, patient data is collected from hospitals with varied formats and systems. Common issues include

- Inconsistent measurement units (e.g., weight in pounds vs. kilograms)
- Missing data in vitals due to device malfunctions
- Ambiguous date fields (DD/MM/YYYY vs. MM/DD/YYYY)
- Unstructured clinician notes alongside structured vitals

The cleaning pipeline includes

1. Converting all units to standardized metrics
2. Parsing and standardizing date formats using heuristics
3. Imputing vitals using time-series modeling and windowed medians
4. Removing notes with ambiguous language or converting to embeddings
5. Applying rule-based filters to flag out-of-range measurements

This results in a normalized, trustworthy dataset used for survival prediction modeling, drug response analysis, and early-warning alert systems.

Final Thoughts

The success of AI systems does not begin with model architecture—it begins with data. Cleaning and normalization transform raw, unruly information into the lifeblood of intelligent computation. These processes ensure consistency, accuracy, fairness, and reproducibility across all stages of the AI life cycle.

From removing nulls to encoding categorical fields, from eliminating bias in text labels to detecting systemic drift in input streams, preprocessing is both technical and epistemological. It defines what knowledge the model sees, how it interprets the world, and what decisions it will make. There is no such thing as "just preprocessing"—there is only good preprocessing and bad.

In a future increasingly driven by machine intelligence, the craftsmanship of data cleaning and the discipline of normalization will be among the most valuable skills in AI development. They are the difference between insight and illusion, between fairness and harm, between function and failure.

Synthetic Data Generation

In data-driven AI systems, the quantity, diversity, and representativeness of training data are essential to model performance. However, real-world data is often incomplete, imbalanced, privacy-sensitive, or expensive to acquire. In such cases, synthetic data generation becomes a strategic and often essential approach. Synthetic data refers to information that is artificially created rather than collected from actual events or observations. It mimics the statistical properties and structural characteristics of real data, and when done properly, it can be used for model training, testing, validation, or augmentation with substantial benefits.

Far from being a workaround or a lesser alternative, synthetic data generation is now a core component in the design of modern AI systems. It enables simulation, diversification, personalization, and ethical data stewardship. In fields ranging from autonomous driving to healthcare diagnostics, synthetic data bridges gaps left by real-world constraints and unlocks new capabilities in model development.

What Is Synthetic Data?

Synthetic data is any data that is artificially generated rather than obtained by direct measurement. It can be fully simulated, partially generated using seed data, or produced via models that learn from existing real-world datasets. The generated data retains the statistical features, relationships, and functional patterns of the source domain, but it does not correspond to real individuals, events, or transactions.

There are several key categories of synthetic data:

- **Fully synthetic**: Generated from scratch using rules, simulations, or generative models with no direct trace to real-world instances.
- **Partially synthetic**: Combines real and synthetic components. For example, a real dataset with imputed or fabricated values in selected columns.
- **Hybrid synthetic**: Augmented with transformations, noise injection, or domain-specific alterations while maintaining core real examples.

Synthetic data can take many forms: numerical tables, text, audio, video, images, or multimodal combinations. Advances in generative modeling, particularly GANs, VAEs, diffusion models, and large language models, have made synthetic generation feasible across all modalities.

Why Use Synthetic Data?

The reasons for using synthetic data are diverse and often domain-specific, but they fall into several broad categories.

1. Privacy Preservation

Data containing personal information (such as medical records, financial transactions, or user interactions) poses serious legal and ethical challenges. Regulations like GDPR, HIPAA, and CCPA restrict the use of such data for training or sharing. Synthetic data offers a privacy-preserving alternative that can be statistically similar to the original without revealing actual identities.

Privacy-safe synthetic datasets enable

- Federated training across organizations without sharing raw data
- Public release of datasets for academic or commercial research
- Safer model debugging, testing, and prototyping

2. Data Scarcity

In many scenarios, especially involving rare diseases, edge cases in autonomous systems, or high-risk industrial operations, real data is extremely limited or prohibitively expensive to collect. Synthetic data fills these gaps by simulating scenarios that are too rare, costly, or dangerous to observe directly.

3. Balanced Class Representation

Real-world data is often imbalanced. For instance, fraud detection datasets might contain 0.5% fraudulent cases, and facial recognition datasets might underrepresent certain skin tones or age groups. Synthetic data can be used to rebalance class distributions and mitigate bias.

- Generate minority class samples to improve recall.
- Simulate edge cases and corner conditions.
- Ensure equal representation across demographic groups.

4. Data Augmentation and Generalization

Synthetic data is not just for replacing missing data—it is a proactive tool for improving generalization. By creating variations of existing inputs, it exposes the model to a broader functional space, increasing its robustness.

Examples include

- Rotating and lighting transformations in image datasets
- Paraphrasing and entity substitution in NLP tasks
- Environmental simulation in reinforcement learning

5. Controlled Experimentation

Synthetic data allows for full control over variables. This is especially useful in scientific applications, benchmarking, algorithmic fairness testing, and causal inference.

For example, in causal modeling, researchers can

- Hold confounders constant while varying treatments
- Simulate interventions without real-world harm
- Create counterfactual scenarios not present in original data

Techniques for Generating Synthetic Data

The methods used to generate synthetic data vary depending on the type of data, the intended use, and the degree of fidelity required.

1. Rule-Based Simulations

This is the most traditional form of synthetic data generation. Domain experts design systems that simulate processes or environments according to known rules or physics.

Examples:

- Flight simulators that train pilots
- Physics engines for simulating collisions in robotics
- Population simulations using demographic rules for epidemiology

Advantages:

- High interpretability and control
- No dependency on real data
- Reproducible and scalable

Disadvantages:

- Limited realism and diversity
- Expensive to build complex simulations

2. Probabilistic Models

Statistical methods can be used to estimate the joint or conditional distributions of a dataset and sample from them to create new data points.

Techniques include:

- Gaussian Mixture Models
- Bayesian networks
- Copula models

These methods work well for structured tabular data and are often used in financial modeling or risk analysis. They offer controllable randomness and interpretability but struggle with high-dimensional or multimodal data.

3. GANs

GANs are among the most powerful tools for creating high-quality synthetic data. They consist of two neural networks: a generator and a discriminator, which play a competitive game. The generator tries to create realistic samples, while the discriminator attempts to distinguish real from fake.

GANs have been used to generate

- Photorealistic faces (e.g., StyleGAN)
- Fashion images and 3D renderings
- Synthetic voice and music
- Tabular datasets with correlations and rare cases

Advantages:

- High realism and visual quality
- Scalable and customizable
- Works across many modalities

Challenges:

- Training instability
- Mode collapse (lack of diversity)
- Difficulty with discrete or categorical data

4. VAEs

VAEs are probabilistic generative models that encode data into a latent space and then decode it back into the original space. Unlike GANs, VAEs optimize a reconstruction loss and a regularization term (Kullback-Leibler divergence).

Strengths of VAEs:

- Smooth and interpretable latent spaces
- Easier training than GANs
- Good for structured data, time series, and sequence modeling

Limitations:

- Lower fidelity for images
- Blurriness in generated outputs
- Less diversity in samples compared to GANs

5. Diffusion Models

A more recent class of generative models, diffusion models work by gradually transforming noise into structured data through a reverse denoising process. These models have achieved state-of-the-art results in image, video, and audio generation.

They are particularly effective for

- Complex, high-resolution imagery
- Text-to-image generation (e.g., DALL·E, Stable Diffusion)
- Generating diverse samples with controlled conditioning

Diffusion models offer better mode coverage than GANs and greater stability during training, though they tend to be slower at inference.

6. Language Models for Textual Data

In text domains, large-scale autoregressive models such as GPT, T5, and PaLM have become the standard tools for synthetic text generation. These models learn the statistical structure of language from massive corpora and can generate human-like paragraphs, documents, summaries, dialogues, and code.

Applications include

- Simulating customer conversations
- Augmenting question-answer datasets
- Creating synthetic documentation or code snippets
- Generating alternate translations and paraphrases

Text generation models offer rich controllability through prompt engineering, fine-tuning, or parameter-efficient adapters.

Evaluating Synthetic Data Quality

The effectiveness of synthetic data hinges on its quality. Poorly generated data can introduce noise, bias, or artifacts that degrade model performance. Therefore, rigorous evaluation is essential.

There are three broad criteria for assessing synthetic data:

1. Fidelity

Fidelity measures how closely the synthetic data resembles the real data. This includes matching statistical distributions, feature correlations, and domain-specific properties.

Metrics:

- Distributional similarity (KL divergence, Wasserstein distance)
- Coverage of real data feature space
- Visual or semantic realism

2. Utility

Utility reflects how well the synthetic data can substitute for or augment real data in actual modeling tasks.

Tests:

- Train-on-synthetic, test-on-real (ToSToR)
- Downstream task performance (e.g., classification accuracy, F1-score)
- Model robustness and generalization

3. Privacy

The data must not leak sensitive information. High-fidelity synthetic data can unintentionally reproduce real-world examples if the generator overfits.

Checks:

- Membership inference attacks
- Nearest-neighbor overlap
- Differential privacy guarantees

Balancing these three—fidelity, utility, and privacy—is the central challenge of synthetic data generation.

Synthetic Data in Practice

Synthetic data is already in widespread use across industries:

- **Healthcare**: Synthetic medical records for research without exposing patient data.
- **Autonomous Vehicles**: Simulation of rare driving scenarios (e.g., night, fog, accidents).
- **Finance**: Generation of synthetic transactions for fraud detection.
- **Retail and Marketing**: Simulated customer behaviors and purchase journeys.
- **Cybersecurity**: Creation of attack scenarios for defense testing.

Organizations such as NVIDIA (Omniverse Replicator), Microsoft (Counterfit), Mostly AI, Gretel.ai, and Synthetaic are offering platforms for scalable synthetic data creation. Some use simulation, others use generative modeling, and many combine both.

Challenges and Considerations

While synthetic data offers many advantages, it also comes with significant challenges:

- **Bias Amplification**: If trained on biased real data, synthetic data can reinforce those biases.
- **Overfitting and Memorization**: Generative models can inadvertently leak training samples.
- **Lack of Ground Truth**: Evaluation is difficult without a gold standard for comparison.
- **Complexity of Generation**: High-quality data generation, especially in 3D or multimodal scenarios, requires substantial computational resources and domain expertise.
- **Regulatory Ambiguity**: While synthetic data helps with privacy, its use in regulated environments still faces legal uncertainty.

Proper governance, versioning, documentation, and testing frameworks are essential for responsible deployment of synthetic data systems.

Future Directions

The field of synthetic data generation is moving rapidly, driven by improvements in generative models, simulation tools, and AI safety requirements. Promising developments include

- **Generative Foundation Models**: Multimodal models capable of generating text, images, code, and more from unified latent spaces

- **Zero-shot and Few-shot Synthetic Creation**: Reducing reliance on large-labeled datasets through advanced conditioning and prompting
- **Synthetic Benchmarks and Test Suites**: Using synthetic environments to create robust test beds for AI evaluation
- **Differentially Private Generators**: Building synthetic datasets with provable privacy guarantees
- **Sim2Real Transfer**: Training in synthetic environments and deploying in real-world systems

Synthetic data is not merely a backup plan; it is becoming a core part of AI strategy.

Dataset Bias Detection and Mitigation

Dataset bias is one of the most pervasive and damaging problems in AI development. While algorithms are often assumed to be neutral, the data they are trained on is shaped by human choices, social patterns, and institutional structures. This means that bias can enter an AI system long before the first line of model code is written. In practice, dataset bias affects every downstream stage of AI, from model behavior to user experience, fairness, and even legal compliance.

This section outlines the nature of dataset bias, techniques to detect it, and strategies for mitigating its effects during data preparation. These steps are critical for creating AI systems that are not only accurate but equitable, generalizable, and trustworthy.

Defining Dataset Bias

Dataset bias refers to systematic errors or distortions in the data that lead to unfair or skewed model behavior. It results from imbalanced representation, flawed measurement, subjective labeling, historical inequalities, or contextual oversights during data collection and preprocessing. Even when unintentional, dataset bias can compromise the ethical and functional integrity of AI systems.

Forms of Dataset Bias

Sampling Bias

Sampling bias arises when the dataset does not adequately represent the target population. For example, if a health dataset underrepresents elderly individuals, any predictive model trained on it may fail for this demographic.

- Undercoverage: Certain groups are excluded or poorly sampled.
- Overrepresentation: Some groups dominate due to ease of access or convenience sampling.
- Survivorship bias: Data reflects only successful cases, ignoring failures.

Measurement Bias

This occurs when data collection tools or processes introduce systematic distortions. It can stem from poor instrumentation, inconsistent data entry practices, or context-sensitive metrics.

- Sensor inaccuracies
- Temporal measurement inconsistencies
- Subjective human assessment

Labeling Bias

Labeling bias is introduced during annotation, especially when labels reflect societal prejudices or annotator subjectivity.

- Inconsistent or inaccurate labels across groups
- Annotator bias due to cultural or cognitive framing
- Toxicity or sentiment mislabeling in text datasets

Historical Bias

Historical bias reflects the unjust conditions or decisions in the system being modeled. Even accurate data can encode past discrimination.

- Policing data reinforcing racial profiling
- Medical datasets reflecting treatment disparities
- Employment records that reflect gender imbalance

Representation Bias

This occurs when the features used do not capture the full diversity of the domain or misrepresent certain subpopulations.

- Inappropriate feature selection
- Linguistic underrepresentation in NLP datasets
- Misleading proxies for protected characteristics

Why Dataset Bias Matters

Dataset bias can significantly harm AI outcomes:

- Reduces generalizability across subpopulations
- Embeds discrimination into automated systems
- Produces unequal error rates across groups
- Undermines user trust and system adoption
- Exposes organizations to regulatory and reputational risk

Real-world examples include face recognition systems with high error rates for dark-skinned individuals, hiring algorithms that penalize women, and credit scoring models that deny loans based on ZIP codes tied to ethnicity.

Detecting Dataset Bias

Exploratory Data Analysis (EDA)

Begin by analyzing distributions, frequency counts, and correlations, particularly for sensitive attributes such as age, gender, race, and location.

- Are some groups over- or underrepresented?
- Do missing values cluster by group?
- Are certain labels or outcomes concentrated in specific populations?

Statistical Testing

Quantify disparities using formal statistical techniques:

- Chi-square test for independence
- Kolmogorov-Smirnov test for distribution divergence
- Mutual information or correlation analysis between sensitive features and labels

Counterfactual Testing

Hold all features constant except one (e.g., gender) and observe whether the model's outcome changes. If results shift drastically, this may indicate direct bias.

Embedding and Latent Analysis

In vision and language models, embeddings may reveal group-specific clustering.

- Use t-SNE or PCA to visualize group separation.
- Examine whether representations encode stereotypes.
- Inspect whether names or terms from different cultures are treated equivalently.

Bias Audits

Perform a structured audit of dataset composition, labeling processes, and feature representations.

- Group-level parity checks
- Label balance by subgroup
- Feature completeness across demographics

Evaluating Bias in Model Performance

Detecting dataset bias is closely linked to understanding how it manifests in trained models.

- **Accuracy Parity**: Do models achieve similar accuracy across groups?
- **Equal Opportunity**: Are true positive rates consistent?
- **False Positive Rate Parity**: Are mistakes made equally often?
- **Calibration**: Do predicted probabilities match observed outcomes across subgroups?

Evaluating these metrics can help trace back performance disparities to data issues.

Mitigating Dataset Bias

Bias mitigation strategies are often classified according to when they are applied in the machine learning pipeline.

Preprocessing Techniques

Rebalancing

Adjust sampling or weights to reflect more balanced distributions.

- Undersampling overrepresented classes
- Oversampling minority groups using SMOTE or ADASYN
- Weighting examples inversely to their frequency

Data Augmentation

Use synthetic data or transformations to expand minority subgroups.

- Generate counterfactual examples
- Modify inputs via paraphrasing, translation, or image manipulation
- Simulate rare or protected attributes to balance coverage

Feature Filtering

Remove features that are strongly correlated with protected attributes but not causally related to the target.

- Drop ZIP codes that serve as race proxies.
- De-identify names or specific geographic tags.
- Apply adversarial removal of sensitive information from latent space.

In-Processing Techniques

Fairness-Constrained Learning

Incorporate fairness metrics directly into the training objective.

- Add loss penalties for demographic disparity.
- Use multi-objective optimization: accuracy + fairness.
- Train adversarial debiasing models to remove group signal.

Representation Learning

Create intermediate feature representations that minimize group separability.

- Autoencoders trained with group-invariant objectives
- Domain adaptation networks
- Disentangled latent spaces

Postprocessing Techniques

Threshold Adjustment

Set group-specific decision thresholds to equalize error rates or predictive parity.

- Raise thresholds for high false positives
- Lower thresholds for underrepresented groups

Outcome Calibration

Adjust model confidence scores to align with observed outcome probabilities.

- Platt scaling
- Isotonic regression per demographic segment

Reject Option

Flag uncertain cases near the decision boundary for manual review or intervention.

- Particularly valuable in high-risk applications like healthcare, finance, and criminal justice

Advanced Approaches

Causal Inference

Use causal models to distinguish between correlation and causation. This helps identify whether disparities are due to legitimate variables or discriminatory factors.

- Build structural causal models (SCMs).
- Block causal pathways that reflect bias.
- Simulate interventions to test counterfactual fairness.

Intersectional Fairness

Analyze groups at the intersection of multiple identities (e.g., race and gender) rather than in isolation.

- Measure error rates for Black women, not just all women or all Black individuals.
- Avoid masking compound disadvantages by aggregating.

Practical Workflow

1. **Define fairness goals**: Choose metrics based on context, domain, and stakeholders.
2. **Conduct a bias audit**: Profile dataset distributions, feature correlations, and label imbalances.
3. **Test for representation**: Analyze whether key subpopulations are missing or distorted.

4. **Implement mitigation**: Select preprocessing, in-processing, or postprocessing techniques.

5. **Evaluate iteratively**: Measure fairness metrics alongside accuracy and generalization.

6. **Document changes**: Track data modifications, rationale, and impact on performance.

7. **Consult stakeholders**: Include legal, ethical, and affected community perspectives.

Case Study: Bias in Facial Recognition Systems

A commercial facial recognition API was found to have drastically different accuracy rates across demographic groups:

- 99% accuracy for white men
- 85% for women overall
- <65% for dark-skinned women

Investigation revealed

- The training dataset was composed mostly of celebrity photos, predominantly white and male.
- The model had never seen faces under varied lighting, occlusion, or facial structures common in underrepresented populations.

Mitigation included

- Acquiring a balanced dataset of public domain faces across demographics
- Augmenting lighting conditions using synthetic transformations

- Introducing loss penalties for misclassification on underrepresented groups

Post-mitigation, the model achieved 90–95% parity across all subgroups without sacrificing global accuracy.

Organizational and Legal Considerations

Governance

- Maintain documentation for dataset sourcing, feature selection, and cleaning practices.
- Require bias assessments as part of model validation workflows.
- Establish review committees for high-stakes applications.

Regulation

Bias mitigation is increasingly mandated by law:

- **GDPR** (EU): Right to explanation and non-discrimination.
- **CCPA** (California): Transparency in automated decisions.
- **NYC Local Law 144**: Requires bias audits for automated hiring tools.
- **EU AI Act**: High-risk AI systems must demonstrate fairness and human oversight.

Tools and Standards

Several open-source and commercial tools support bias detection and mitigation:

- IBM AI Fairness 360
- Google's What-If Tool
- Fairlearn
- Aequitas
- Microsoft's Fairness Dashboard

Looking Ahead

Bias detection and mitigation is not a solved problem—it is a continuous, context-specific process. Future directions include

- Generative modeling for debiased data synthesis
- Active learning to focus labeling efforts on uncertain or biased regions
- Fair federated learning across decentralized data sources
- Greater involvement of ethicists, legal scholars, and community stakeholders
- Normative frameworks that define acceptable trade-offs across accuracy, fairness, and utility

The success of AI depends not only on how well it performs but on who it serves and how responsibly it behaves.

The next chapter covers training fundamentals and self-supervised learning.

CHAPTER 4

Training Fundamentals and Self-Supervised Learning

Training lies at the heart of AI. No matter how sophisticated an architecture may be, its performance depends on how effectively it learns from data. This process involves more than simply exposing a model to examples. It requires careful choices about objectives, the management of complexity, and continuous monitoring to ensure that learning is both efficient and generalizable.

This chapter introduces the foundations of training modern AI systems with a focus on self-supervised learning, a paradigm that has become central to contemporary advances. The discussion begins with the principles of self-supervised learning, showing how raw data can act as its own source of supervision. It then explores the design of loss functions, which serve as the guiding signals of optimization. From there, the chapter examines the challenges of overfitting and the pursuit of generalization, a balance that remains central despite increasing model scale. Finally, it turns to the monitoring of convergence and training progress, a practical but essential part of guiding large models toward useful outcomes.

I. Cronin, *Building and Training Generative AI Models*,
https://doi.org/10.1007/979-8-8688-2332-9_4

By the end of this chapter, the reader will understand the conceptual and practical elements that underlie successful training, and why these elements form the backbone of self-supervised learning systems.

Self-Supervised Learning Principles

From Supervised to Self-Supervised

The earliest breakthroughs in machine learning relied almost exclusively on supervised learning. In this paradigm, researchers and practitioners build models that are trained on pairs of inputs and outputs:

- An image of a cat matched with the label "cat."
- A sentence in English paired with its French translation.
- A financial record annotated as fraudulent or not fraudulent.

The model's task is to discover a mapping from the input space to the output space, with the goal of minimizing some loss function that measures how far the model's predictions diverge from the correct labels. This framework has proven immensely powerful, and it underlies many of the technologies that became mainstream in the first two decades of the deep learning era. Image classification systems, speech recognition models, and machine translation pipelines were all designed with supervised methods at their core.

Yet this reliance on labeled data created an immediate bottleneck. For supervised learning to work, one needs vast datasets where every sample has been carefully annotated by humans or, in some cases, by specialized experts. The celebrated ImageNet dataset, for example, required thousands of hours of human labor to label millions of images across more than a thousand categories. Building datasets of such scale is prohibitively expensive, especially in domains where expertise is required.

- Radiological scans must be annotated by trained physicians.
- Legal documents require expert review to assign categories.
- Niche languages demand rare linguistic expertise for translation datasets.

In addition, human labeling introduces subjectivity and bias, which become encoded in the dataset and subsequently in the trained models. This makes it difficult to ensure fairness or objectivity.

The problem is further compounded by scalability issues. Modern deep neural networks thrive on enormous amounts of data. As models have grown in parameter count, from millions to billions to even trillions, the appetite for data has escalated accordingly. While unlabeled data such as images, text, or audio can be collected cheaply and in abundance from the Internet and other digital repositories, labeled data does not scale in the same way.

This bottleneck led researchers to explore alternatives:

- **Semi-supervised learning:** Combine a small-labeled dataset with a large unlabeled one, yielding modest gains.
- **Unsupervised learning:** Attempt to cluster or compress data without labels, often producing weak or brittle representations.
- **Self-supervised learning:** Generate labels from the data itself, enabling scalability and richness without manual annotation.

This shift from external annotation to internal signal generation is more than a technical adjustment. It is a profound conceptual change. SSL matters because it breaks the dependence on human labeling and unleashes the potential of vast reservoirs of unlabeled data. It allows models to scale with raw data availability, not just with annotated corpora. Furthermore, it enables the construction of general-purpose representations that can be fine-tuned for a variety of downstream applications, creating an efficient pipeline from raw data to useful models.

Core Idea of SSL

At its heart, self-supervised learning is a deceptively simple idea: the data supervises itself. Instead of relying on human-provided labels, the system defines a task where the input provides both the training signal and the evaluation criterion.

This usually involves obscuring, altering, or splitting the data in such a way that the model must

- Recover missing information
- Predict future information
- Align different representations

The model's ability to succeed in these surrogate or "pretext" tasks leads to the acquisition of representations that transfer well to real-world, human-defined tasks.

For instance, in natural language processing, a text sequence can be partially masked, with the model tasked to fill in the blanks. Consider the sentence: *"The cat sat on the ___."* The model must infer the missing word "mat" from the surrounding context. By repeatedly solving this kind of task across billions of sentences, the model develops a nuanced understanding of

- Syntax
- Semantics
- Pragmatics

In vision, the paradigm is similar but adapted to the properties of images. A model might be shown two augmented versions of the same image, such as

- A cropped view
- A color-shifted version

It is then tasked to recognize that they are derived from the same underlying content, while simultaneously distinguishing them from other images.

Another variant is predictive: given part of an audio waveform, predict the next segment. This approach encourages the model to capture

- Temporal regularities of speech
- Structural features of music
- Repeated patterns in environmental sounds

The key insight is that regardless of domain, the world itself provides signals that can be harnessed as labels. The role of the researcher is to design pretext tasks that expose these signals in ways that encourage models to learn representations that generalize.

SSL's elegance is that it resonates with how humans and animals learn. Babies learn language not by reading dictionaries but by inferring meaning from repeated patterns, context, and prediction. Likewise, SSL equips machines to bootstrap intelligence from raw sensory streams without explicit instruction.

Pretext Tasks

The design of pretext tasks is the creative engine of self-supervised learning. These are artificially constructed challenges that force the model to engage with the structure of the data, producing representations that are transferable to downstream applications.

One of the most celebrated examples is **masked language modeling (MLM)**, popularized by BERT:

- Randomly replace words in a sentence with a special mask token.
- Train the model to predict the missing words from context.
- Learn grammar, semantics, and world knowledge through repetition.

Closely related is **next-sentence prediction (NSP):**

- Provide the model with two sentences.
- Ask it to determine whether the second follows the first in the original text.
- Model discourse-level coherence beyond single-sentence semantics.

In the visual domain, contrastive learning has proven revolutionary. Two key paradigms are

- **SimCLR:** Create multiple augmented "views" of the same image (cropping, color jittering, blurring), pull them together in embedding space, and push apart embeddings of different images.

- **MoCo:** Extend contrastive learning with a momentum encoder and a memory bank of negative examples, ensuring a rich and dynamic training signal.

Other visual pretext tasks include

- **Image inpainting:** Obscure a region of an image and train the model to reconstruct it, forcing global structural understanding.
- **Rotation prediction:** Rotate an image by 0°, 90°, 180°, or 270°, and train the model to predict the angle, thereby learning object orientation.

In speech and audio, **wav2vec** introduced a highly effective strategy:

- Quantize raw audio into discrete units.
- Mask portions of the sequence.
- Predict the masked units using context from surrounding frames.

Its successor, **HuBERT**, pushed this further by clustering acoustic units and training the model to predict these clusters, achieving strong results for phoneme and word-level tasks.

Each of these tasks embodies the same principle: construct a challenge that cannot be solved without learning meaningful structure. Some tasks emphasize syntax, others semantics, still others global coherence. The genius of SSL lies in how these artificial puzzles yield representations that generalize.

Real-World Applications

The true measure of self-supervised learning is its impact on real-world systems. Far from being a theoretical curiosity, SSL has become the backbone of the most powerful AI models in existence.

Large language models (LLMs) like GPT, LLaMA, and PaLM owe their success to SSL. Their central principle is deceptively simple:

- Train on massive corpora of text drawn from the Internet, books, and archives.
- Use next-token prediction as the training objective.
- Scale up to billions or trillions of parameters, letting capabilities emerge.

Through this simple objective, models acquire not only linguistic competence but also factual knowledge, reasoning ability, and emergent abilities such as in-context learning. The entire LLM revolution is essentially a testament to SSL at scale.

In vision, foundation models like CLIP and DINO illustrate the versatility of SSL:

- **CLIP:** Train on hundreds of millions of image–caption pairs using a contrastive objective, aligning visual and textual embeddings in a shared space. This enables zero-shot recognition of categories never seen in training.
- **DINO:** Use self-distillation, where a student model learns to predict the representations of a teacher across augmented versions of the same image, yielding strong features without labels.

Speech and audio have also been transformed:

- **Wav2vec 2.0:** Pretrain on raw audio to produce features transferable to speech recognition.
- **HuBERT:** Leverage cluster prediction tasks to achieve robust phonetic and semantic representations.

Beyond text, vision, and speech, SSL is spreading rapidly:

- **Biology:** Models trained on unlabeled protein sequences discover folding patterns and improve drug discovery pipelines.
- **Reinforcement learning:** Agents use SSL to predict future states or encode sensory streams before tackling reward-driven tasks.
- **Robotics:** SSL is used to interpret camera and sensor data, predicting object trajectories and enabling manipulation.

The expansion of SSL across domains shows its universality. By making raw data sufficient for learning, it lowers entry barriers and accelerates innovation across industries.

Advantages and Challenges

The advantages of SSL are compelling. Perhaps the most important is scalability: unlabeled data is abundant, and SSL makes it possible to harness this resource.

- Text data can be scraped from the Internet.
- Images are generated at staggering volumes every day.
- Audio is collected from microphones embedded in virtually every device.

Another major advantage is transfer learning. SSL models acquire general-purpose representations that can be adapted for specific downstream tasks with relatively little labeled data. This shift has already changed industrial practice: instead of assembling bespoke labeled datasets for every problem, practitioners fine-tune or adapt foundation models pretrained with SSL.

SSL also resonates with human cognition:

- Children acquire language by predicting missing meanings from context.
- Humans build mental models by filling in gaps and making predictions about the world.
- The machine analog of this predictive learning is masked prediction or contrastive objectives.

Nevertheless, challenges remain. Shortcut learning is one: models often exploit superficial cues rather than building deep understanding.

- In vision, contrastive models might learn to separate images by background textures instead of objects.
- In text, a model might rely on surface statistical regularities rather than true semantic reasoning.

Another issue is the risk of spurious correlations and bias amplification. Since SSL models are trained on web-scale data, they inherit whatever prejudices and imbalances exist in that data.

Resource intensity is another challenge:

- Training GPT-class models requires thousands of GPUs or TPUs running for weeks.
- Hyperparameter tuning in SSL is often sensitive, requiring expert intervention.
- Storage requirements for web-scale datasets are immense.

Despite these hurdles, the trajectory is clear: SSL has become the foundation of modern AI. Its ability to unlock the value of unlabeled data, scale to massive corpora, and produce versatile representations makes it indispensable.

Loss Function Design

Loss functions are the heartbeat of training in machine learning. They quantify the difference between what the model predicts and what it should ideally predict, offering a signal that drives the learning process through gradient descent and backpropagation. Choosing the right loss function is not merely a technical detail but a foundational decision that shapes what the model learns, how it generalizes, and what behaviors emerge. In the context of self-supervised learning, loss function design is especially important because there are no external labels. Instead, the loss must be crafted to capture meaningful structures in the data and to encourage representations that are transferable to downstream tasks.

This section explores several families of loss functions that have proven central to modern AI systems. These include reconstruction losses, adversarial losses, contrastive losses, and hybrid objectives that combine elements of different approaches. Each family has distinct motivations, mathematical properties, and real-world applications. Understanding these families provides insight not only into how current systems work but also into where innovations are likely to arise.

Why Loss Functions Matter

Training a model involves adjusting parameters so that predictions minimize some notion of error. The loss function formalizes this error and acts as the compass that guides optimization. If the compass is poorly chosen, the model may optimize for the wrong objective, producing representations that look correct in training but fail in deployment.

For instance, if one were to train a speech recognition model with a loss that simply counts the number of character errors, the system might learn to approximate words but ignore higher-level phonetic or semantic structure. On the other hand, if a more nuanced loss like Connectionist Temporal Classification is chosen, the model is encouraged to capture sequential alignment, leading to better generalization.

The role of the loss function is particularly pronounced in self-supervised learning. Because there are no human labels to rely on, the loss must create a learning signal directly from the data. This makes the art of designing pretext tasks inseparable from the art of designing appropriate losses.

Connection to Gradient-Based Optimization in Self-Supervised Learning

Once a loss is defined, gradient based optimizers such as stochastic gradient descent or Adam translate that loss into parameter updates. At each training step, the model computes the self-supervised loss on a batch of examples, then backpropagation calculates gradients that indicate how each parameter contributed to the error. The optimizer uses these gradients, along with a learning rate and possibly momentum or adaptive step sizes, to nudge the parameters in directions that reduce the loss on future batches.

In self supervised settings, this loop is especially important because the loss encodes the entire learning objective. If the learning rate is too high, the model may bounce around and fail to discover stable representations that satisfy the self-supervised task. If it is too low, the model may underfit and never fully exploit the signal encoded in the loss. Learning rate schedules, such as warm-up followed by decay, are often paired with self supervised losses to let the model first explore the space of representations and then gradually refine them.

In practice, designing a successful self-supervised method means choosing a loss that exposes useful structure in the data and pairing it with an optimization setup that can reliably minimize that loss at scale. The objective shapes the landscape, and gradient-based optimization is the process that walks that landscape toward representations that transfer well to downstream tasks.

Within the broad family of loss functions, several categories stand out:

- **Reconstruction losses**, which measure how well the model can reproduce data that has been obscured or compressed
- **Adversarial losses**, which emerge from the competition between a generator and a discriminator
- **Contrastive losses**, which measure the similarity between related pairs and dissimilarity between unrelated pairs
- **Hybrid losses**, which combine multiple objectives to balance competing priorities

Reconstruction Losses

Reconstruction losses are among the most intuitive. They ask the model to reconstruct data that has been corrupted, masked, or compressed. By doing so, the model is forced to learn internal representations that capture essential structure.

Common formulations include

- **Mean Squared Error (MSE):** Measure the squared difference between predicted and target values. This is widely used in autoencoders, where the input is compressed into a latent representation and then reconstructed.

- **L1 Loss (Mean Absolute Error):** Measure the absolute difference between predicted and target values, often producing sharper reconstructions compared to MSE.
- **Cross-Entropy Loss:** Particularly useful for categorical outputs, such as predicting masked tokens in language models.

In practice, reconstruction losses are used in a variety of contexts:

- **Autoencoders:** Compress inputs into a latent code and reconstruct them with minimal error, forcing the model to learn compressed but informative representations.
- **Variational Autoencoders (VAEs):** Add a probabilistic component, combining reconstruction loss with a regularization term that shapes the latent space distribution.
- **Masked language modeling:** Replace random tokens in text with a mask and train the model to predict the original words, effectively a categorical reconstruction task.
- **Image inpainting:** Obscure regions of an image and train the system to generate plausible replacements, teaching the model both texture and global structure.

Reconstruction losses are simple and stable to optimize. However, they often suffer from an averaging problem. For example, in image generation tasks, MSE encourages predictions that minimize pixel-wise difference. If multiple reconstructions are possible, the model averages them, producing blurry outputs. This limitation has spurred the development of more sophisticated loss families that encourage sharper or more realistic results.

Adversarial Losses

Adversarial losses emerged with the introduction of Generative Adversarial Networks. The key idea is to pit two models against each other:

- A **generator** tries to create synthetic data samples that resemble real data.
- A **discriminator** tries to distinguish between real and synthetic samples.

The generator is trained to maximize the probability that the discriminator misclassifies its outputs as real, while the discriminator is trained to minimize misclassification. This creates a minimax game where the generator improves by learning to fool the discriminator.

The original adversarial loss is defined as

- **Generator objective:** maximize $\log(D(G(z)))$.
- **Discriminator objective:** maximize $\log(D(x)) + \log(1 - D(G(z)))$.

Here, D represents the discriminator, G the generator, z the random noise input, and x the real data sample.

Over time, adversarial losses have evolved into more stable formulations:

- **Wasserstein loss:** Replace the original objective with an Earth-Mover distance, producing smoother gradients and improving training stability.
- **Hinge loss and least squares loss:** Alternative forms that avoid vanishing gradients and improve convergence.

- **Gradient penalty:** An additional regularization term that enforces Lipschitz continuity, reducing mode collapse.

Applications of adversarial losses are wide-ranging:

- **Image generation:** GANs produce realistic images of faces, objects, and scenes.
- **Super-resolution:** Adversarial loss encourages reconstructions that look photo-realistic rather than blurry.
- **Style transfer:** Models learn to generate images that match both the content of one domain and the style of another.
- **Speech synthesis:** GANs have been used to improve the naturalness of generated audio waveforms.

The strength of adversarial losses lies in their ability to encourage outputs that are indistinguishable from real data. Their weakness lies in training instability. Because two networks are competing, optimization can oscillate or collapse if not carefully balanced. Despite this, adversarial losses remain one of the most influential innovations in modern AI.

Contrastive Losses

Contrastive losses have become the cornerstone of many self-supervised learning methods. The principle is straightforward:

- Bring representations of similar or related pairs closer together in embedding space.
- Push representations of unrelated pairs farther apart.

The challenge lies in defining what counts as similar and what counts as dissimilar.

One of the most widely used formulations is the **InfoNCE loss**, originally introduced in the context of maximizing mutual information. In practice, it is implemented as

- Select a positive pair, such as two augmented views of the same image.
- Select multiple negative samples, such as views from other images.
- Train the model to maximize the similarity of the positive pair while minimizing similarity to negatives.

Contrastive approaches have been central to breakthroughs in vision and multimodal learning:

- **SimCLR:** Create multiple augmentations of the same image and train the model with InfoNCE to cluster views of the same image together.
- **MoCo:** Extend contrastive learning by maintaining a large memory bank of negative examples, ensuring that each batch has a rich set of comparisons.
- **CLIP:** Train on image-caption pairs using a contrastive objective, aligning visual and textual embeddings so that corresponding pairs are close and mismatched pairs are far apart.

Contrastive losses have also been applied in language, audio, and even reinforcement learning. In each case, the construction of positive and negative pairs defines the effectiveness of the method.

Challenges include

- **Sampling strategy:** Selecting informative negatives is critical, as trivial negatives provide little learning signal.
- **Batch size:** Larger batches provide more negatives, which improves learning but increases computational demand.
- **Shortcut learning:** Models may rely on superficial cues if augmentations are not carefully designed.

Despite these challenges, contrastive losses have proven to be one of the most versatile and effective self-supervised objectives. They produce embeddings that are highly transferable across domains.

Hybrid Objectives

As the field has matured, researchers have increasingly recognized that no single loss function is sufficient for all purposes. Hybrid objectives combine multiple losses to balance strengths and weaknesses.

Examples include

- **VAE with adversarial loss:** Combine reconstruction loss with adversarial loss to encourage both faithful reconstructions and realistic outputs.
- **Perceptual loss with pixel loss:** Use deep features from pretrained networks as an additional objective, ensuring that reconstructions match human perceptual similarity rather than just pixel similarity.
- **Multimodal training objectives:** Combine contrastive loss for alignment with generative loss for reconstruction, as seen in models that handle both vision and language.

The benefits of hybrid objectives are clear:

- They allow the model to satisfy multiple constraints simultaneously.
- They reduce the risk of overfitting to artifacts of a single objective.
- They often produce richer representations that generalize better.

However, hybrid objectives introduce new challenges. Balancing the weights of different losses is a nontrivial hyperparameter tuning problem. If one component dominates, the benefits of the others may be lost. Furthermore, hybrid objectives often increase computational requirements, as multiple loss terms must be calculated and backpropagated in each iteration.

Nevertheless, hybrid loss design reflects the growing recognition that real-world tasks are multifaceted. Just as humans learn by combining multiple signals (visual, auditory, linguistic, and contextual), AI systems benefit from training objectives that integrate diverse sources of supervision.

Reflections on Loss Function Design

Loss function design has evolved from simple error measures to sophisticated formulations that encode complex structures and alignments. In supervised learning, loss functions often reflect straightforward prediction errors. In self-supervised learning, they act as the scaffolding that makes representation learning possible.

- Reconstruction losses encourage models to capture internal structure by filling in missing data.
- Adversarial losses push models to generate outputs indistinguishable from real samples.

- Contrastive losses teach models to build discriminative embeddings by aligning positives and separating negatives.
- Hybrid objectives combine elements of each, reflecting the complexity of real-world tasks.

Understanding these categories equips researchers and practitioners to make informed choices when designing self-supervised systems. It also highlights opportunities for innovation. New loss functions often precede major advances in the field, as seen with the emergence of adversarial and contrastive methods. Future progress is likely to come from creative formulations that better capture the richness of human experience and the structure of real-world data.

Overfitting and Generalization

The effectiveness of a machine learning system depends not only on its ability to minimize training error but also on its capacity to generalize to new, unseen data. This balance between fitting the known and adapting to the unknown is one of the oldest and most fundamental challenges in artificial intelligence. It is particularly important in self-supervised learning, where models often train on massive corpora and must apply their acquired representations to downstream tasks with different distributions or objectives.

This section examines the concepts of overfitting and generalization in depth. It begins with a clear definition of overfitting and the historical roots of the problem, then moves to causes of overfitting in modern systems. It continues by discussing strategies for promoting generalization and concludes with modern perspectives that challenge and refine traditional wisdom.

Defining Overfitting

Overfitting occurs when a model learns to perform extremely well on its training data but fails to generalize to new data. It memorizes noise, idiosyncrasies, or spurious correlations present in the training set, instead of capturing the underlying structure that applies more broadly.

A classic example involves training a decision tree on a small dataset. If allowed to grow without constraint, the tree may create extremely specific branches for every training sample. While the training error drops to zero, the test error soars because the tree has not learned meaningful rules but rather encoded peculiarities.

In deep learning, overfitting manifests differently but follows the same principle. Large models with millions or billions of parameters have the capacity to memorize training examples. If this memorization occurs without learning general features, the model will struggle to adapt to new inputs.

The bias-variance trade-off provides a conceptual lens. A model with too much bias fails to capture the complexity of the data and underfits. A model with too much variance captures not only the signal but also the noise, resulting in overfitting. The ideal lies in a balance where the model has enough capacity to represent the signal but not so much flexibility that it memorizes irrelevant details.

Examples help clarify the concept:

- A handwriting recognition model that performs perfectly on training digits but misclassifies slightly different handwriting styles in the test set has overfit.
- A speech model that memorizes background noise in training samples and fails when tested on clean audio has overfit.

- A language model that repeats entire passages verbatim from its training corpus rather than generating new sentences shows memorization rather than generalization.

In self-supervised learning, overfitting can be subtle. Because the training objective often involves predicting masked tokens, reconstructing inputs, or aligning pairs, a model might learn shortcuts that satisfy the training loss but do not yield representations useful for real-world applications. For example, in masked language modeling, a system might over-rely on local word co-occurrence patterns without capturing deeper semantics.

Causes of Overfitting

Overfitting arises from multiple factors that interact in complex ways. Understanding these causes is critical for designing models that generalize.

Model Capacity

The most straightforward cause is excessive model capacity. Deep neural networks can represent an extraordinary range of functions. If the model is much larger than the dataset, it may memorize examples instead of learning patterns. This does not mean large models always overfit. With appropriate regularization and massive data, large models can generalize remarkably well. Still, capacity relative to data availability is a key risk factor.

Dataset Size and Quality

The size of the training dataset plays a central role.

- **Small datasets:** If the training set is too small, the model has little choice but to overfit to the few examples it sees.
- **Noisy datasets:** If labels are inconsistent or incorrect, the model may memorize errors, leading to poor generalization.
- **Unrepresentative datasets:** If the training set fails to reflect the diversity of real-world cases, the model generalizes poorly outside its narrow scope.

Data Leakage

Data leakage occurs when information from outside the training set inadvertently influences the model during training.

- A speech model trained on recordings that accidentally include metadata tags about speaker identity may learn to classify based on these tags rather than the audio itself.
- A fraud detection model trained on financial records that include fields only available after the transaction is completed may perform unrealistically well in training but fail in deployment.

Leakage leads to illusory performance gains that vanish when the model is evaluated properly.

Spurious Correlations

Machine learning models are prone to exploiting correlations that happen to exist in the training data but do not generalize.

- An image classifier may learn to associate cows with green pastures, misclassifying cows on beaches.
- A sentiment analysis model may learn that the word "not" always signals negativity, failing on sentences like "not bad."
- A medical diagnosis system may learn to associate the presence of a hospital watermark on images with certain outcomes.

These spurious cues reduce generalization because they do not reflect causal relationships.

Excessive Training

Even with good data and reasonable model capacity, training for too long can lead to overfitting. As optimization continues, the model begins to fit to noise and outliers. This is often visible in learning curves: training error continues to decrease while validation error plateaus or increases.

Lack of Regularization

Regularization methods intentionally constrain the learning process to prevent overfitting. Without them, models are free to memorize. Common regularization strategies include dropout, weight decay, and data augmentation. A lack of such techniques increases overfitting risk.

Strategies to Improve Generalization

Over the decades, researchers have developed a toolkit of strategies to mitigate overfitting and promote generalization. These range from simple heuristics to sophisticated techniques grounded in theory.

Regularization

Regularization imposes constraints on the learning process, discouraging the model from fitting to noise.

- **Weight decay (L2 regularization):** Penalize large weights to encourage simpler functions.
- **L1 regularization:** Encourage sparsity in weights, leading to more interpretable models.
- **Dropout:** Randomly deactivate neurons during training, preventing co-adaptation and forcing robustness.
- **Early stopping:** Halt training when validation error stops improving, preventing memorization of noise.

Data Augmentation

Data augmentation increases effective dataset size by creating transformed versions of training examples.

- **Vision:** Rotate, crop, flip, or adjust brightness of images.
- **Language:** Rephrase sentences, replace words with synonyms, or shuffle segments.
- **Audio:** Add noise, shift pitch, or stretch time.

These augmentations expose the model to greater variation, encouraging it to learn invariant features rather than memorizing exact patterns.

Cross-Validation

Cross-validation provides a robust estimate of generalization. By repeatedly splitting data into training and validation folds, researchers can detect overfitting early. Models that perform well across folds are less likely to have memorized spurious correlations.

Architectural Choices

Certain model architectures are inherently more robust. Convolutional networks, for example, exploit spatial locality in images, reducing the risk of overfitting to irrelevant pixel patterns. Recurrent and transformer architectures use weight sharing across positions, limiting the number of free parameters relative to input size. These design choices act as implicit regularization.

Transfer Learning and Pretraining

Pretraining on large datasets with self-supervised objectives and fine-tuning on smaller labeled sets has become a standard strategy. Pretrained models begin with generalizable features, reducing the risk of overfitting to small downstream datasets. For instance:

- A BERT model pretrained on billions of words generalizes well when fine-tuned on a few thousand sentiment analysis samples.
- A ResNet pretrained on ImageNet generalizes well to medical image classification after fine-tuning.

Ensemble Methods

Ensembles combine predictions from multiple models, reducing variance. Bagging, boosting, and stacking are classical techniques, but in deep learning, ensembles are often created by averaging checkpoints or training multiple models with different initializations. Ensembles reduce the likelihood of overfitting to peculiarities of a single model.

Monitoring and Validation

Careful monitoring of training curves and validation metrics is essential. Indicators of overfitting include

- Divergence between decreasing training error and increasing validation error
- Large gaps between training accuracy and validation accuracy
- Highly confident predictions on training data but uncertainty on new inputs

By monitoring these signs, practitioners can intervene before overfitting worsens.

Modern Perspectives

Traditional discussions of overfitting and generalization are framed around small to medium models and datasets. However, modern deep learning has introduced new phenomena that challenge classical assumptions.

Double Descent

The double descent phenomenon shows that test error does not always follow the classic U-shaped curve implied by the bias-variance trade-off. Instead, as model capacity increases, test error may first decrease, then increase due to overfitting, and then decrease again as capacity grows further.

This occurs because very large models can interpolate training data perfectly yet still generalize when trained with appropriate optimization and regularization. Double descent explains why models with billions of parameters can generalize well despite having the capacity to memorize.

Role of Pretraining

Pretraining with self-supervised learning has changed the dynamics of generalization. Models trained on vast corpora acquire broad, general-purpose features. Fine-tuning them on downstream tasks is less prone to overfitting because the pretrained representations already encode useful knowledge. This makes SSL not only a data-efficient strategy but also a generalization-enhancing one.

Implicit Regularization

Researchers have observed that certain optimization algorithms, particularly stochastic gradient descent, act as implicit regularizers. Even without explicit regularization terms, SGD tends to find flatter minima in the loss landscape, which correlate with better generalization. This has shifted focus from explicit penalties to understanding the dynamics of optimization.

Generalization in Large Language Models

Large language models challenge traditional definitions of overfitting. On the one hand, they are capable of memorizing training data verbatim, which raises concerns about privacy and originality. On the other hand, they exhibit remarkable generalization, generating coherent responses to novel prompts and transferring to tasks never explicitly seen in training. This duality suggests that memorization and generalization are not mutually exclusive but can coexist in complex ways.

Distribution Shifts

Modern generalization challenges often involve distribution shifts, where the training and test data differ in significant ways. Examples include

- **Domain shifts**: Training on formal news text and testing on informal social media posts
- **Temporal shifts**: Training on data collected in one period and testing on data from later periods
- **Geographical shifts**: Training on medical images from one region and testing on another

Generalization under distribution shifts is harder than classical generalization within a fixed dataset. Addressing this requires domain adaptation, continual learning, and robustness strategies.

Ethical and Societal Dimensions

Overfitting is not only a technical problem but also an ethical one. When models fail to generalize, they may amplify biases or perform poorly on underrepresented groups. For example:

- A face recognition model trained predominantly on light-skinned individuals may overfit to features of that subgroup, leading to poor accuracy on darker-skinned individuals.
- A hiring algorithm trained on biased historical data may overfit to discriminatory patterns, perpetuating inequality.

Thus, generalization is not only about accuracy but also about fairness, inclusivity, and reliability in diverse real-world contexts.

Overfitting and generalization remain central challenges in machine learning, even in the era of self-supervised learning and massive models. While the classical concepts of memorization and the bias-variance trade-off still apply, modern systems reveal new complexities such as double descent, implicit regularization, and coexistence of memorization with generalization.

Strategies for promoting generalization have expanded from simple regularization and data augmentation to large-scale pretraining and careful architectural design. At the same time, societal concerns remind us that generalization is not only a technical achievement but also a responsibility.

In self-supervised learning, these issues take on special importance. Because models are trained without labels, they must learn to generalize from patterns in raw data to downstream tasks. This makes careful monitoring, thoughtful loss design, and diverse datasets essential. As models grow larger and more powerful, the pursuit of robust generalization remains at the heart of building trustworthy AI.

Monitoring Convergence and Training Progress

Training a machine learning system is not simply a matter of initializing a model, feeding in data, and waiting for it to reach completion. It is a dynamic process that requires careful observation, measurement, and intervention. Models do not learn in a linear or predictable fashion. They oscillate, plateau, and sometimes regress. Left unchecked, training may converge to suboptimal states, collapse entirely, or waste enormous computational resources. Monitoring convergence and training progress is therefore both a scientific and practical necessity.

In traditional supervised learning, the role of monitoring was already important. Researchers tracked accuracy, loss curves, and validation scores to decide when to stop training or adjust hyperparameters. In self-supervised learning, the stakes are even higher. Because there are no human-provided labels, the training objective is often a surrogate task. Monitoring ensures that the model is not only minimizing its surrogate loss but also developing representations that will generalize effectively to downstream tasks.

This section explores the topic in five parts. It begins by examining training dynamics, training diagnostics, continues with convergence criteria, then discusses practical tools for monitoring, highlights pitfalls that practitioners must avoid, and finally explores emerging techniques that are reshaping the way convergence is tracked.

Training Dynamics

Training a neural network is inherently iterative. The model starts with random parameters and gradually adjusts them based on the feedback provided by the loss function. This adjustment is performed through gradient descent and its many variants. Observing how these dynamics unfold provides valuable insights into model behavior.

The Role of Learning Rate

The learning rate is one of the most critical hyperparameters.

- A learning rate that is too high causes instability, with loss oscillating or diverging.
- A learning rate that is too low slows convergence, sometimes making training impractical.
- Schedules that vary the learning rate over time, such as cosine annealing or warm restarts, can accelerate training while avoiding instability.

In practice, monitoring the trajectory of loss across epochs provides clues about whether the learning rate is appropriately tuned. Sudden spikes in loss may indicate that updates are too aggressive, while overly smooth but slow declines may indicate underutilized capacity.

Batch Size and Gradient Noise

The size of the training batch influences the variance of gradient estimates.

- **Small batches:** Introduce more noise into gradient updates, which can help models escape sharp minima and improve generalization.
- **Large batches:** Produce more accurate gradient estimates, accelerating convergence but sometimes leading to poorer generalization.

Monitoring batch effects requires comparing progress across different scales and noting whether validation performance deteriorates as training stabilizes.

Loss Curves and Plateaus

Loss curves are often the first visual reference for training dynamics.

- A steadily declining training loss with parallel declines in validation loss suggests healthy progress.
- A flattening curve indicates a plateau, where the optimizer struggles to make further improvements.
- Divergence between training and validation losses is a sign of overfitting.

Plateaus can be resolved by adjusting learning rates, altering architectures, or introducing new regularization strategies.

Oscillations and Instabilities

In some cases, training may oscillate, with loss decreasing in one epoch but increasing in the next. This instability often arises from poor hyperparameter choices, unnormalized data, or adversarial training objectives. Monitoring these fluctuations helps identify whether adjustments are needed.

Training Diagnostics

Monitoring training is not only about watching a single loss curve go down. For modern deep and self-supervised systems, you need a richer diagnostic toolkit that can reveal whether the model is learning useful structure, wasting compute, or drifting toward failure modes such as collapse, overfitting, or numerical instability. Training diagnostics provide that toolkit. They turn a long running experiment into a series of observable signals that let you intervene early instead of waiting until the end to discover that a run has failed.

At a high level, training diagnostics answer four questions.

1. Is the optimization process itself healthy?
2. Is the model improving on data it has not seen?
3. Are gradients, activations, and weights within reasonable numerical ranges?
4. Are the learned representations moving in the direction that downstream tasks need?

To cover these questions, it is useful to organize diagnostics into several layers.

1. Loss and objective diagnostics

   The most basic signals are the training loss and whatever proxy objective is used in self-supervised learning. These curves should be tracked at both step and epoch scale, ideally smoothed for readability but stored at full resolution for later analysis.

   Key patterns to watch:

   - Steady decrease in training loss with a corresponding decrease in validation or held-out loss suggests healthy learning.
   - Training loss dropping while validation loss flattens or rises indicates overfitting or a misaligned objective.
   - Large, sudden spikes in loss often signal numerical issues, for example exploding gradients or learning rate being too high.

For language or sequence models, perplexity is a useful derived diagnostic. Perplexity condenses the average negative log likelihood into a more interpretable quantity. When perplexity stops improving on validation data, it is usually a sign to lower the learning rate, adjust regularization, or stop training.

2. Generalization diagnostics

   In self-supervised learning, the main objective is a surrogate for the tasks you actually care about. That makes auxiliary evaluation crucial. Periodically during pretraining, snapshots of the model can be frozen and evaluated on small supervised probes or linear classifiers that test whether useful information is present in the representations.

   Examples include

   - Linear evaluation on image classification benchmarks for vision encoders
   - Sentence similarity or natural language inference probes for text encoders
   - Retrieval accuracy for multimodal encoders that align text with images or audio

   These probes do not need to be exhaustive or production grade. Their purpose is to ensure that progress on the self-supervised loss corresponds to progress on tasks that matter. If the pretext loss improves while probe performance stagnates, it may indicate that the model is exploiting shortcuts in the objective rather than learning general features.

3. Optimization and learning rate diagnostics

   Gradient-based optimizers such as SGD, Adam, or LAMB are sensitive to learning rate schedules, momentum parameters, and batch size. Training diagnostics should therefore include metrics that describe the behavior of the optimizer itself. Important signals include

   - Learning rate over time, including warmup and decay phases. Sudden changes in loss often align with transitions in the schedule.
   - Gradient norms, both global and per layer. Gradients that grow without bound indicate instability, while gradients that shrink toward zero point to stalled learning.
   - Parameter update norms relative to parameter norms, sometimes called the signal to noise ratio of updates. If updates become vanishingly small, the optimizer may have effectively stopped learning even if loss is still decreasing slightly.

   Visualizing these quantities can reveal misconfigured schedules. For example, if validation loss improves primarily during the warmup period and degrades once the learning rate reaches its peak, the maximum learning rate is probably too high. If gradient norms are near zero for many steps, the learning rate may be too low or the loss poorly scaled.

4. Numerical and statistical health diagnostics

   Large models running at scale can fail in ways that are not obvious from loss alone. Activations can saturate, attention weights can become degenerate, and batch statistics can drift. To guard against these issues, it is helpful to monitor

   - Activation statistics such as mean and variance per layer. Sudden shifts can indicate changes in data distribution or numerical instability.
   - Weight norms and distribution shapes, which can reveal layers that are not learning or are diverging.
   - The fraction of zeroed gradients when techniques like gradient clipping or mixed precision are used. If clipping occurs too frequently, learning may become noisy and unstable.

   Many teams integrate these diagnostics into dashboards that update in real time, with alert thresholds when values exceed expected ranges. Tools such as TensorBoard, Weights and Biases, or custom logging stacks make it practical to keep these metrics for long-running jobs without manual effort.

5. Collapse and representation diagnostics in self-supervised learning

   Self-supervised objectives are especially prone to degenerate solutions where the model satisfies the loss without learning meaningful structure. Contrastive methods can collapse to trivial embeddings, predictive methods can ignore parts of the input, and masked modeling can overfit to local cues.

To detect these failures, training diagnostics should include representation level metrics such as

- Embedding variance across a batch or dataset. Near zero variance indicates that different inputs are mapped to nearly the same point.
- Pairwise cosine similarity distributions between embeddings. A narrow peak at high similarity suggests collapse.
- For contrastive methods, the mutual information between different views of the same sample before and after encoding. Very low mutual information after encoding is a red flag.

Visual methods such as t-SNE or UMAP plots of embeddings at different training stages, although approximate, can also be informative when used sparingly. They help answer qualitative questions such as whether classes or semantic groups are separating over time.

6. Data and pipeline diagnostics

Many apparent training problems are actually data or pipeline issues. Diagnostics should therefore extend beyond the model to the input itself. Useful checks include

- Monitoring label distributions, token distributions, or image statistics over time to catch shifts in the data stream.
- Verifying that data augmentations behave as expected, for example, checking that masking ratios, crop sizes, or noise levels stay within intended bounds.
- Tracking throughput and data loader latency, since bottlenecks here can change the effective batch composition or introduce subtle correlations.

Recording small, random samples of preprocessed batches during training is often invaluable. When loss behaves unexpectedly, inspecting these samples can reveal corruption, misaligned labels, or extreme class imbalance that metrics alone do not show.

7. Integrated diagnostic workflows

Individually, each diagnostic provides a narrow view. The real power comes from integrating them into a coherent workflow. A practical pattern is

- Start each new experiment with a standard dashboard that includes loss curves, validation metrics, learning rate, gradient norms, and a few representation statistics.
- Define expectations for each metric before training begins, based on prior runs or small scale experiments.

- When a metric deviates from expectations, follow a consistent checklist: inspect data samples, review recent changes to the code or schedule, test a reduced learning rate, and if needed, restart from a safe checkpoint.

For large projects, it is often worth treating training diagnostics as an engineering product of its own. Clear visualizations, automatic alerts, and the ability to compare runs side by side can save weeks of experimentation time. Just as continuous integration made software development more reliable, continuous training diagnostics can make large-scale self-supervised learning reproducible instead of mysterious.

In summary, training diagnostics connect abstract objectives with concrete signals that practitioners can act on. By watching not only what the loss does but how gradients, activations, representations, and data behave over time, you gain the ability to steer training rather than merely observe it.

Convergence Criteria

Convergence refers to the point at which further training yields diminishing returns. Knowing when to stop is vital. Training too long wastes resources and risks overfitting. Stopping too early prevents the model from reaching its potential.

Loss Stabilization

The most direct criterion is stabilization of the training loss.

- If loss values change only marginally across multiple epochs, the model may have converged.
- However, a low training loss alone is not sufficient. It may indicate memorization rather than generalization.

Validation Metrics

Validation metrics offer a more reliable signal of convergence.

- Validation accuracy or error provides insight into performance on unseen data.
- Monitoring precision, recall, and F1 scores reveal whether improvements are balanced across different dimensions.
- In SSL, proxy evaluations such as linear probing accuracy can measure the utility of learned representations without requiring full downstream fine-tuning.

Early Stopping

Early stopping is a practical strategy that halts training when validation performance stops improving.

- A patience parameter specifies how many epochs to wait before declaring convergence.
- This prevents wasted computation and reduces the risk of overfitting.

Gradient Norms

Monitoring gradient norms provides another perspective.

- Declining norms suggest that updates are becoming smaller as the optimizer approaches a minimum.
- Persistently large norms suggest that the model is still far from convergence.

Gradient monitoring can also detect vanishing or exploding gradients, common issues in deep networks.

Representation Stability

In SSL, convergence can also be judged by the stability of representations.

- Embeddings for similar samples should stabilize over time.
- Clustering quality or neighborhood consistency can be monitored to ensure that the model is learning consistent structures.

Practical Monitoring Tools

Monitoring requires both conceptual understanding and practical instrumentation. Modern tools provide visualization, logging, and alerting to track training in real time.

TensorBoard

TensorBoard has become a standard for monitoring neural network training.

- It provides interactive plots of loss, accuracy, and other metrics.
- It visualizes model architectures and parameter distributions.
- It allows comparison across experiments to evaluate the effect of hyperparameters.

Weights & Biases

Weights & Biases (W&B) extends monitoring to collaborative research.

- Experiments can be logged, shared, and compared across teams.
- Hyperparameter sweeps can be automated and visualized.
- Alerts can notify practitioners when models diverge or plateau.

Custom Logging

In some settings, lightweight custom logging is preferable.

- Print statements of loss per batch can provide granular feedback.
- CSV logs can be parsed into custom visualizations.
- Logging libraries allow flexible and domain-specific monitoring.

Checkpointing

Checkpointing is not only a safety measure but also a monitoring tool.

- Saving models at intervals allows rolling back if training diverges.
- Comparing checkpoints across time reveals whether improvements are consistent.
- Ensembles of checkpoints can be combined to reduce variance.

Distributed Monitoring

In large-scale training across many GPUs or TPUs, distributed monitoring is essential.

- Metrics must be aggregated across nodes to provide coherent feedback.
- Communication delays and synchronization issues must be tracked.
- Centralized dashboards provide unified views of large-scale experiments.

Pitfalls in Monitoring

Monitoring is not foolproof. Misinterpretation or overreliance on certain signals can mislead practitioners.

Overfitting to Validation

Validation sets are meant to provide unbiased estimates of generalization. However, repeated monitoring can lead to implicit overfitting.

- If hyperparameters are tuned based on validation results, the model may begin to exploit patterns specific to the validation set.
- Rotating or enlarging validation sets mitigates this issue.

Noisy Loss Curves

Loss curves are often noisy, particularly with small batches.

- Overreacting to small fluctuations can lead to unnecessary interventions.
- Smoothing techniques or moving averages help reveal true trends.

Metric Myopia

Focusing on a single metric can be misleading.

- A model may improve accuracy while harming fairness across subgroups.
- A decrease in training loss may hide deteriorating calibration.
- Multiple metrics, including domain-specific ones, should be monitored together.

Resource Blindness

Monitoring should also account for resource utilization.

- A model that converges well but requires excessive compute may not be practical.
- Energy efficiency, memory use, and training time are important to track alongside accuracy.

Emerging Techniques

The landscape of monitoring is evolving. New techniques are being developed to handle the scale and complexity of modern AI.

Sharpness-Aware Monitoring

Sharpness-aware minimization has introduced the idea that flatter minima correlate with better generalization. Monitoring curvature and sharpness of the loss landscape is becoming a tool for convergence assessment.

Representation Drift Tracking

In self-supervised learning, representations may drift even when loss appears stable.

- Monitoring the consistency of embeddings over time reveals whether the model is converging to stable features.
- Tools that track clustering quality or neighborhood preservation provide additional insights.

Automated Alerts and Intervention

Automation is increasingly common.

- Systems can automatically adjust learning rates when plateaus are detected.
- Alerts can notify teams when divergence occurs, reducing wasted compute.
- Adaptive checkpointing can increase frequency when instability is observed.

Evaluation Beyond Training Loss

There is growing recognition that training loss is insufficient.

- Probing tasks provide more direct evaluations of representation quality.
- Downstream fine-tuning performance can be tested intermittently.
- Robustness metrics under distribution shifts are increasingly monitored during training.

Large-Scale Experiment Tracking

As experiments scale to thousands of runs, meta-monitoring becomes necessary.

- Systems that track trends across experiments help identify patterns.
- Visualization of hyperparameter landscapes provides strategic insights.
- Automated pruning of unpromising runs saves compute.

Monitoring convergence and training progress is the practice that connects theory with application. It ensures that training objectives lead to meaningful learning, prevents wasted computation, and provides early warnings of failure. In supervised learning, monitoring was already important. In self-supervised learning, where objectives are indirect and datasets are massive, it becomes indispensable.

By understanding training dynamics, defining clear convergence criteria, using practical monitoring tools, avoiding pitfalls, and adopting emerging techniques, practitioners can guide training to success. As models continue to grow in scale and complexity, monitoring will remain not only a practical necessity but also a research frontier. The ability to observe, interpret, and steer training dynamics will define the efficiency and reliability of the next generation of AI systems.

Conclusion

Training remains the central process by which deep learning systems transform from inert collections of weights into functioning models capable of generalization. This chapter has explored the conceptual and technical foundations that make training possible, highlighting the interplay between loss functions, learning paradigms, generalization, and self-supervised learning. By surveying these areas, we see that training is not simply a mechanical optimization routine but an evolving framework that balances stability, adaptability, and scalability.

From Supervised to Self-Supervised

A key theme of this chapter was the transition from traditional supervised learning to self-supervised paradigms. Supervised learning, for decades, was the dominant mode: datasets with carefully annotated labels guided networks to map inputs to outputs. However, as model size and data scale increased, the limitations of this paradigm became clear.

- **Labeling bottlenecks:** Collecting high-quality labeled datasets proved costly, slow, and, in many cases, impractical.
- **Bias in annotation:** Human labeling introduced biases that constrained model generalization.
- **Limited scalability:** Larger models demanded far more supervision than could feasibly be provided.

Self-supervised learning offered a conceptual breakthrough. By designing tasks where raw data supervises itself, such as predicting missing tokens in a sequence or reconstructing occluded portions of an image, models could learn directly from the abundance of unlabeled data. This shift was not incremental but transformative, enabling foundation models that scale across domains of text, vision, and speech.

The Core Idea of Self-Supervised Learning

At the heart of self-supervised learning lies the realization that structure in data can generate its own supervision. Pretext tasks such as masked language modeling, next-sentence prediction, contrastive learning, and image inpainting operationalize this idea.

These tasks allow networks to extract features by solving auxiliary problems that do not require external labels. Conceptually, the model learns a rich representation of the data distribution, which later transfers to downstream tasks.

- **In language:** Masked token prediction forces the model to capture syntax and semantics.
- **In vision:** Contrastive learning encourages models to cluster similar views together while separating dissimilar ones.
- **In speech:** Predictive tasks like predicting the next acoustic frame help models learn phonetic and linguistic features.

This principle reframes training as representation learning rather than purely label prediction.

Loss Functions As Guides to Learning

Training requires direction, and loss functions provide that signal. The chapter surveyed reconstruction losses, adversarial losses, and contrastive objectives, showing how each embodies a different philosophy of learning.

- **Reconstruction losses** emphasize fidelity to the input, useful for autoencoders and generative models.
- **Adversarial losses** emphasize realism, pushing models to generate outputs indistinguishable from true data.
- **Contrastive losses** emphasize relational structure, encouraging models to organize representations by similarity.

The design of a loss function is therefore not a detail but a conceptual decision about what kind of learning signal the model receives. Training success depends on aligning loss functions with the intended representational goals.

Overfitting and Generalization

The chapter also emphasized that effective training is not just about fitting training data but about generalizing to new data. Overfitting represents a persistent challenge, where models memorize idiosyncrasies of training data instead of capturing underlying structure.

Conceptually, generalization arises from

- **Regularization:** Techniques such as dropout, weight decay, and data augmentation
- **Early stopping:** Preventing models from fitting noise by halting at optimal points
- **Model capacity control:** Balancing expressivity against the risk of memorization

Generalization connects back to self-supervised learning. By learning broad representations rather than narrow label-specific mappings, self-supervised models often generalize more effectively to diverse downstream tasks.

Monitoring Convergence and Training Progress

Another key theme was monitoring. Training is not a linear journey to a single point but a dynamic process. Monitoring convergence provides insight into whether optimization is proceeding correctly.

- **Loss curves** reveal whether training is stable or diverging.
- **Validation performance** indicates generalization potential.
- **Gradient norms and learning rates** help detect instability.

Conceptually, monitoring transforms training from blind iteration into guided adjustment. It ensures that models not only minimize loss but do so in a way that yields robust outcomes.

The Advantages and Challenges of Self-Supervised Learning

The rise of self-supervised learning brings both benefits and new challenges.

Advantages:

- **Scalability:** Unlabeled data is abundant and unlocks unprecedented model size.
- **Transfer learning:** Representations learned on one domain often transfer across tasks.

- **Reduced labeling dependence:** Eliminates costly annotation pipelines.

Challenges:

- **Shortcut learning:** Models may exploit spurious correlations in pretext tasks.
- **Compute demands:** Training large self-supervised models requires vast resources.
- **Evaluation complexity:** Measuring representation quality is less straightforward than measuring accuracy on labeled tasks.

Conceptually, these challenges underscore that self-supervised learning is not a finished paradigm but an evolving one, demanding new forms of evaluation and robustness.

Interconnectedness of Training Concepts

Perhaps the most important lesson of this chapter is the interconnectedness of its components. Self-supervised paradigms, loss functions, generalization strategies, and convergence monitoring are not independent silos but reinforcing elements of a single process.

- The **loss function** determines what the model pays attention to.
- **Self-supervised pretext tasks** determine how the model extracts representations.
- **Generalization strategies** ensure these representations remain useful beyond the training set.
- **Monitoring tools** provide feedback to adjust methods dynamically.

Together, they illustrate that training is a systemic process. Success arises not from any single element but from their orchestration.

Broader Reflections

Training fundamentals have implications beyond technical details. They shape the trajectory of AI research and practice.

- **Scientific implications:** Self-supervised learning shifts research from hand-designed labels to emergent representation learning.
- **Economic implications:** Reduced dependence on labeled data lowers barriers for building competitive models.
- **Societal implications:** Bias in unlabeled data remains a concern, as self-supervised models may inherit and amplify existing inequalities.

The conceptual framework of training therefore connects algorithms with broader contexts of research, economics, and society.

Chapter 4 demonstrated that training is the core process of deep learning, guided by loss functions, evaluated by generalization, stabilized by monitoring, and expanded through self-supervised paradigms. It showed how the field has moved from reliance on supervised data to learning directly from raw data, a shift that underpins the rise of foundation models.

The chapter also emphasized that training is a balance of forces: stability against instability, generalization against overfitting, efficiency against complexity. Self-supervised learning provides new opportunities, but it also introduces new risks. Ultimately, training is not static but adaptive, evolving with model scale, data availability, and computational capacity.

In the broader trajectory of this book, training fundamentals anchor the transition from conceptual models to scalable systems. They remind us that the success of deep learning lies not only in architectures but in how they are trained. By mastering the interplay of loss functions, learning paradigms, and convergence strategies, researchers and practitioners can build models that both perform well and generalize effectively.

In our next chapter, we tackle optimization and learning strategies.

CHAPTER 5

Optimization and Learning Strategies

Optimization is the practical force that enables deep learning models to move from static collections of parameters to functioning systems that can learn from data. Without optimization, even the most advanced architectures remain inert and even the largest datasets cannot be harnessed. The act of optimization animates learning by guiding models to adjust their weights in ways that reduce error and capture structure.

At its simplest, optimization minimizes a loss function. The loss measures the gap between a model's predictions and the desired outcome. Through backpropagation, gradients are computed, and through gradient descent, parameters are updated. Yet in practice, optimization is never simple. Deep networks contain billions of parameters and operate in non-convex spaces where gradients can vanish, explode, or lead models into poor minima. Optimizers must therefore do more than descend a surface: they must manage instability, adapt to scale, and balance speed with generalization.

I. Cronin, *Building and Training Generative AI Models*,
https://doi.org/10.1007/979-8-8688-2332-9_5

Why Optimization Matters

Optimization is central because it determines whether models converge at all, how quickly they learn, and how well they generalize.

- **Too aggressive updates:** Training diverges, with losses increasing uncontrollably.
- **Too cautious updates:** Training becomes slow, requiring excessive epochs and compute.
- **Well-tuned updates:** Models balance rapid learning with stability, achieving useful generalization.

In modern practice, optimization also shapes the character of models. Two systems trained with the same architecture but different optimizers or schedules may reach very different generalization behaviors. Optimization is therefore both a technical and conceptual cornerstone of deep learning.

Historical Roots

The earliest neural networks relied on simple weight adjustment rules, which limited them to linearly separable problems. The reintroduction of backpropagation in the 1980s provided a systematic way to train multi-layer networks, but optimization remained unstable. It was only with the combination of better initialization, regularization, and new optimization methods that very deep networks became trainable in the 2010s.

Optimization advanced in parallel with compute and data. As GPUs enabled large-scale training and datasets expanded, optimizers adapted. Momentum improved stability, adaptive methods like RMSProp and Adam adjusted learning rates automatically, clipping bounded unstable gradients, and normalization layers reshaped activation distributions. These developments turned optimization from a fragile process into a reliable enabler of progress.

Core Challenges

Optimization faces a consistent set of challenges across architectures and domains.

- **Non-convex landscapes:** Loss functions contain countless minima and saddle points. Optimizers must navigate without getting stuck in poor regions.
- **Gradient instability:** Signals may vanish to zero or explode to huge values, preventing effective training.
- **Scale:** Billion-parameter models magnify small instabilities, making gradient management essential.
- **Dynamic data:** In reinforcement learning and self-supervised contexts, distributions shift during training, requiring adaptability.
- **Hyperparameter sensitivity:** Choices like learning rate, momentum, and clipping thresholds heavily influence success.

These challenges explain why optimization remains an active area of research and why diverse strategies coexist.

Themes of the Chapter

This chapter examines four core pillars of optimization that underpin deep learning practice.

- **Backpropagation and Gradient Descent:** The foundation for computing and applying gradients, forming the bedrock of training.

- **Adaptive Optimizers (Adam, RMSProp, LAMB):** Methods that automatically scale learning rates across parameters, improving stability and efficiency.
- **Gradient Clipping and Normalization:** Safeguards that keep gradient magnitudes under control and activations consistent, preventing collapse in deep or recurrent models.
- **Learning Rate Schedules:** Strategies that vary learning rates over time, balancing exploration in early stages with refinement later.

Each of these pillars addresses a different challenge of optimization. Together, they form a toolkit that makes deep learning possible at scale.

Broader Importance

Optimization is not only about minimizing loss. It also shapes the nature of learned representations and determines whether models generalize beyond their training data.

- **Choice of optimizer affects generalization:** Adaptive methods accelerate training but may require fine-tuning for strong test performance.
- **Schedules influence minima found:** Cyclical or cosine schedules encourage exploration of flatter minima.
- **Normalization improves representation quality:** BatchNorm and LayerNorm stabilize activations and indirectly enhance features.
- **Clipping ensures survival in unstable tasks:** Without it, recurrent models and adversarial training may fail entirely.

Optimization is therefore a sculptor as well as an engine, guiding both the speed and the quality of learning.

Optimization is the process that turns deep learning from theory into functioning practice. It addresses the instability of gradients, the complexity of non-convex landscapes, and the demands of large-scale training. The methods discussed in this chapter, gradient descent and backpropagation, adaptive optimizers, clipping and normalization, and learning rate schedules, represent decades of progress toward stable and efficient learning.

By understanding these strategies, one sees how modern optimization enables not only convergence but also the emergence of models that generalize, scale, and deliver useful capabilities.

Backpropagation and Gradient Descent

Backpropagation and gradient descent are the twin engines that power modern machine learning. While architectures and datasets draw the most attention in popular accounts of AI, these two methods quietly determine whether training succeeds or fails. Without them, deep networks would not be trainable at scale, and the current wave of progress in natural language processing, computer vision, and reinforcement learning would never have materialized.

This section provides a detailed exploration of backpropagation and gradient descent. It begins with a historical overview, showing how these methods emerged and matured. It then explains the core mechanics of backpropagation, focusing on how gradients are calculated and propagated through complex networks. It continues by examining gradient descent as the central optimization method, including its variants and their properties. Along the way, it addresses the practical challenges of vanishing and exploding gradients, as well as strategies like momentum that stabilize learning. Finally, it connects these foundations to the broader landscape of optimization research, preparing the ground for later sections on adaptive optimizers, gradient clipping, and learning rate schedules.

Historical Context of Backpropagation

The concept of adjusting parameters to minimize error dates back to the early days of neural networks. In the 1950s and 1960s, pioneers like Frank Rosenblatt introduced the perceptron, which adjusted weights based on classification mistakes. However, perceptrons were limited to linearly separable problems. The absence of an effective training algorithm for multi-layer networks meant that neural networks stagnated after Marvin Minsky and Seymour Papert highlighted their limitations in 1969.

The breakthrough came in the 1980s when researchers rediscovered and popularized backpropagation. The method itself is a straightforward application of the chain rule of calculus, but its impact was transformative. For the first time, multi-layer networks could be trained systematically, unlocking their ability to learn non-linear functions.

- In 1986, David Rumelhart, Geoffrey Hinton, and Ronald Williams published a landmark paper demonstrating how backpropagation could train multi-layer perceptrons.
- This work inspired a wave of excitement in neural networks, although enthusiasm waned in the 1990s due to limitations in data, compute, and algorithmic stability.
- The resurgence came in the 2000s and 2010s, when increased computational power, massive datasets, and better architectures allowed backpropagation to scale effectively.

Today, backpropagation is so central that it is rarely discussed explicitly in everyday practice. Yet its role is indispensable, and every modern deep learning system relies on it.

Core Mechanics of Backpropagation

Backpropagation is an algorithm for efficiently computing gradients of a loss function with respect to the parameters of a neural network. The gradients are then used to update parameters through gradient descent.

The process has three main steps:

- **Forward pass:** The input flows through the network layer by layer, producing outputs. The loss function measures the difference between predicted and target values.
- **Backward pass:** The chain rule of calculus is applied to compute gradients of the loss with respect to each parameter, starting from the output layer and moving backward.
- **Parameter update:** Gradients are used by the optimizer to adjust parameters in a direction that reduces the loss.

The backward pass is what makes training deep networks feasible. Without it, computing gradients would be prohibitively expensive because each parameter would need to be perturbed and evaluated individually. Backpropagation computes all gradients in a single backward sweep with efficiency proportional to the number of parameters.

Key insights about backpropagation:

- Gradients flow from the loss through each layer, capturing how sensitive the loss is to changes in each parameter.

- Layers closer to the input receive gradients through repeated multiplication, which can shrink (vanishing gradients) or grow (exploding gradients).
- The modularity of backpropagation allows new architectures to be invented easily, since s.gradients can be derived automatically using computational graphs.

In practice, modern frameworks like PyTorch, TensorFlow, and JAX handle backpropagation automatically. Users define the forward computations, the framework builds a computational graph, and gradients are obtained through automatic differentiation. For most architectures, the computational cost of backpropagation is on the same order as the forward pass, roughly proportional to the number of parameters and nonzero connections in the model. The main trade-off is memory versus compute, since backpropagation must either store intermediate activations for every layer or recompute them on the fly, so techniques like activation checkpointing deliberately recompute some layers to reduce memory usage at the cost of extra computation.

Gradient Descent As the Optimization Backbone

Once gradients are computed, they must be used to adjust parameters. Gradient descent is the canonical method for doing so. It dates back to the work of Cauchy in the nineteenth century and has since become the foundation of optimization in machine learning.

The intuition is simple: gradients point in the direction of steepest ascent of the loss. By moving parameters in the opposite direction, the model descends toward a minimum.

Different variants of gradient descent exist:

- **Batch gradient descent:** Uses the full dataset to compute gradients at each step. Accurate but computationally expensive.
- **Stochastic gradient descent (SGD):** Uses a single randomly selected sample per update. Highly noisy but cheap to compute.
- **Mini-batch gradient descent:** The standard in modern practice, where small batches of samples are used to balance efficiency and stability.

The dynamics of gradient descent are shaped by key hyperparameters:

- **Learning rate:** Determines the step size in the direction of the gradient. Too high and training diverges. Too low and training stagnates.
- **Batch size:** Affects noise in gradient estimates. Small batches promote exploration but may be unstable. Large batches improve efficiency but risk poor generalization.

Practitioners often monitor learning curves to assess whether gradient descent is functioning well. A steadily declining training loss suggests progress. Divergence, oscillation, or stagnation indicate problems with hyperparameters or architecture.

Practical Challenges in Gradient Descent

Despite its simplicity, gradient descent faces several challenges when applied to deep neural networks.

Vanishing Gradients

When gradients are propagated backward through many layers, they can shrink exponentially. This vanishing effect makes it difficult for earlier layers to learn.

- Sigmoid and tanh activations are especially prone to this problem because their derivatives are small.
- Deep networks trained with these activations often fail to converge.
- Solutions include ReLU activations, residual connections, and normalization techniques.

Exploding Gradients

The opposite problem occurs when gradients grow uncontrollably during backpropagation.

- This was historically common in recurrent neural networks, where repeated multiplication of weight matrices caused instability.
- Exploding gradients lead to divergence and numerical overflow.
- Gradient clipping, which limits gradient magnitudes, is a common solution.

Saddle Points and Local Minima

High-dimensional loss landscapes are riddled with saddle points, where gradients vanish but the point is not a true minimum.

- Models can stall at saddle points, slowing convergence.
- Momentum and adaptive methods help escape these traps.

Poor Conditioning

The geometry of the loss landscape can make optimization inefficient.

- If the landscape is elongated like a ravine, gradients zigzag rather than descend directly.
- Preconditioning methods and second-order approximations address this issue.

Variants of Gradient Descent

Over the years, many enhancements have been introduced to improve the stability and efficiency of gradient descent.

Momentum

Momentum accelerates gradient descent by accumulating a velocity vector that smooths updates.

- Instead of following the gradient alone, momentum combines past updates with current gradients.
- This helps overcome small local minima and speeds progress through flat regions.
- It reduces oscillations in ravine-shaped loss landscapes.

Nesterov Accelerated Gradient

An improvement over standard momentum, Nesterov momentum anticipates the next position before computing the gradient.

- This provides a corrective term that improves convergence.
- It has been widely used in computer vision tasks.

Mini-Batch Strategies

Mini-batching is not only efficient but also introduces beneficial noise.

- The randomness of batch sampling prevents overfitting.
- Batch size interacts with learning rate, influencing convergence behavior.
- Empirical studies show that small to medium batches often yield better generalization than very large ones.

Second-Order Methods

While gradient descent uses first-order information, second-order methods incorporate curvature.

- Newton's method uses the Hessian matrix of second derivatives.
- Quasi-Newton methods approximate curvature for efficiency.
- These methods are rarely used at scale due to computational cost but remain important in theory.

Broader Perspectives

Backpropagation and gradient descent are sometimes criticized for their biological implausibility. Human brains do not appear to compute precise gradients or propagate error signals layer by layer. Yet despite this, the methods remain unmatched in engineering practice. They offer efficiency, flexibility, and scalability that have driven decades of progress.

At the same time, the simplicity of gradient descent belies its complexity in high dimensions. Researchers continue to study its dynamics, revealing surprising properties such as implicit regularization, flat minima preference, and links to generalization.

In the context of self-supervised learning, backpropagation and gradient descent play an even larger role. Because labels are absent, the training objectives are more abstract. Careful optimization ensures that models learn representations rather than trivial shortcuts. Monitoring gradient behavior becomes essential, as losses can stabilize without yielding transferable features.

Backpropagation and gradient descent form the backbone of modern deep learning. They are the invisible mechanisms that enable massive networks to learn from data. While conceptually simple, their practical behavior is shaped by architecture, hyperparameters, and loss landscapes. Understanding their dynamics is essential for designing, training, and scaling models effectively.

The next section builds on this foundation by examining adaptive optimizers. These refinements extend the basic principles of gradient descent, introducing per-parameter adjustments that have become indispensable in large-scale learning.

Adaptive Optimizers (Adam, RMSProp, LAMB)

Optimization is at the core of deep learning. While backpropagation provides gradients and gradient descent offers a method to update parameters, the raw form of stochastic gradient descent is often insufficient for training modern neural networks effectively. Models today span billions of parameters, operate on highly non-convex loss landscapes, and are trained on massive datasets using distributed infrastructure. In such contexts, optimizers that can automatically adapt to the nuances of training are essential. Adaptive optimizers, such as RMSProp, Adam, and LAMB, have become central to the success of deep learning at scale.

This section explores the conceptual underpinnings, mechanics, and implications of adaptive optimizers. It begins with the motivation for moving beyond vanilla stochastic gradient descent, then examines RMSProp as an early adaptive method, continues with Adam as the most widely adopted optimizer, and explores LAMB as a solution for large-batch training. A comparative analysis highlights their strengths and limitations, followed by broader perspectives on the role of adaptive optimization in the evolving landscape of artificial intelligence.

Motivation for Adaptive Methods

Stochastic gradient descent (SGD) is simple and powerful, yet it has shortcomings. It applies the same learning rate to all parameters and does not account for differences in scale or variability among them. In deep networks, parameters can differ by orders of magnitude in sensitivity, leading to inefficient training.

The limitations of vanilla SGD include

- **Non-uniform curvature:** Parameters lie in regions of the loss landscape with different curvatures. A single global learning rate is suboptimal.
- **Sparse gradients:** In tasks like natural language processing, many parameters receive gradients infrequently. Applying the same learning rate to all parameters slows convergence.
- **Non-stationary objectives:** In reinforcement learning and self-supervised training, the target distributions change over time, making static learning rates inadequate.
- **Sensitivity to hyperparameters:** SGD requires careful manual tuning of learning rates and momentum, which is computationally expensive in large experiments.

Adaptive methods address these issues by adjusting learning rates individually for each parameter, based on local statistics of gradients. This adaptation improves stability, accelerates convergence, and reduces the need for manual tuning.

RMSProp

RMSProp was one of the first adaptive optimizers to gain widespread use. It introduced the idea of scaling updates by a moving average of squared gradients, smoothing the variability of updates across parameters.

The intuition is straightforward: parameters that consistently experience large gradients should have smaller updates, while those with small or infrequent gradients should have larger updates. This prevents unstable oscillations and improves progress in ill-conditioned landscapes.

Key features of RMSProp include

- **Running average of squared gradients:** RMSProp maintains an exponentially decaying average of past squared gradients for each parameter.
- **Element-wise adaptation:** Each parameter's learning rate is divided by the root of its running average, normalizing updates across dimensions.
- **Stability in non-stationary objectives:** By focusing on recent gradient history, RMSProp adapts to changing training conditions.

Practical consequences of RMSProp:

- It is particularly effective for recurrent neural networks, which suffer from exploding and vanishing gradients. RMSProp stabilizes updates across time steps.
- It reduces sensitivity to initial learning rates compared to vanilla SGD, though tuning is still required.
- It often converges faster than SGD with momentum, especially in noisy or sparse settings.

Limitations include

- RMSProp discards information about the sign of gradients beyond their square, which may reduce efficiency.
- It lacks bias correction, which can lead to underestimation of effective learning rates early in training.
- Generalization performance may sometimes lag behind SGD with momentum in large-scale supervised tasks.

RMSProp remains a foundational adaptive method. Many later optimizers, including Adam, build upon its principles.

Adam

Adam (Adaptive Moment Estimation) became the default optimizer in deep learning because it combined the strengths of momentum and RMSProp into a unified framework. It maintains running estimates of both first-order (mean) and second-order (variance) moments of gradients, applying bias correction to ensure stability in the early phases of training.

The core ideas of Adam are

- **First moment estimate:** A running average of gradients, akin to momentum, that smooths noisy updates.
- **Second moment estimate:** A running average of squared gradients, as in RMSProp, that normalizes updates across parameters.
- **Bias correction:** Adjustments applied to both moment estimates to account for initialization at zero, preventing early underestimation.

Adam's update rule is element-wise, adapting learning rates for each parameter independently. This makes Adam robust to variations in parameter scale and effective in high-dimensional spaces.

Advantages of Adam include

- **Fast convergence:** Adam often reaches good performance faster than SGD, particularly on tasks with sparse gradients.
- **Reduced hyperparameter sensitivity:** Default settings (learning rate = 0.001, beta1 = 0.9, beta2 = 0.999) work well in many applications.

- **Applicability to diverse domains:** Adam has been used successfully in NLP, computer vision, reinforcement learning, and speech processing.

Challenges associated with Adam:

- **Generalization gap:** Empirical studies have shown that while Adam converges quickly, models trained with SGD and momentum often generalize better. This has led to a hybrid practice of pretraining with Adam and fine-tuning with SGD.
- **Sensitivity to learning rate schedules:** Although Adam is robust to initial settings, its performance can degrade if schedules are poorly designed.
- **Memory overhead:** Adam requires storing additional moving averages for every parameter, increasing memory consumption.

Despite these challenges, Adam's versatility and stability explain its dominance. It provides a reliable starting point for most deep learning experiments.

LAMB and Large-Batch Training

In practice, modern frameworks like PyTorch, TensorFlow, and JAX handle backpropagation automatically. Users define forward computations, and the frameworks build computational graphs that enable automatic differentiation. This automation has accelerated innovation, since researchers can design complex models without manually deriving gradient equations. Computationally, backpropagation scales roughly linearly with the number of operations and parameters in the model. Each training step costs on the order of one extra forward pass, because every operation must be replayed in reverse to accumulate gradients.

The main bottleneck is often memory rather than pure compute, since intermediate activations from the forward pass must be stored for reuse during the backward pass. Techniques such as activation checkpointing, recomputation, mixed precision, and gradient accumulation explicitly manage this memory–compute trade-off, spending more flops to fit larger models into limited device memory.

As models grew to billions of parameters, training efficiency became a bottleneck. Large batch training emerged as a way to accelerate learning by distributing computation across many devices, but naive use of adaptive optimizers with very large batches often led to divergence or poor generalization. LAMB (Layer-wise Adaptive Moments for Batch training) was introduced to address this challenge. It extends Adam by scaling updates on a layer-wise basis so that parameter changes remain well balanced across layers even when batches are extremely large.

In practice, LAMB is most useful when you train very large models with batch sizes in the thousands across many GPUs or TPUs and you see Adam becoming unstable, requiring very small learning rates, or failing to scale as you add more devices. For small or moderate batch sizes on a single machine, Adam or AdamW are usually simpler and work well, so there is little benefit in switching. LAMB is available in open source across major ecosystems, including implementations in PyTorch (for example in DeepSpeed, NVIDIA Apex, and Hugging Face training utilities), TensorFlow and TensorFlow Addons, and JAX or Flax optimization libraries. That makes it straightforward for practitioners to experiment with LAMB without having to implement the optimizer themselves.

Conceptual contributions of LAMB:

- **Layer-wise normalization:** Instead of normalizing updates per parameter, LAMB scales updates relative to the norm of parameters at the layer level.

- **Compatibility with large batches:** This design stabilizes training when batch sizes scale to tens of thousands, a common setting in training language models.
- **Preservation of adaptivity:** LAMB retains the adaptive per-parameter learning rates of Adam while introducing layer-level consistency.

Practical benefits:

- Training times for large models such as BERT and GPT can be reduced significantly when using LAMB.
- LAMB maintains accuracy even at extreme batch sizes, enabling efficient use of distributed hardware.
- It reduces the need for complex learning rate schedules in large-batch contexts.

Limitations include

- LAMB is more complex to implement than Adam or RMSProp.
- It may not offer significant benefits for small-scale models or modest batch sizes.
- Empirical performance varies across domains outside of NLP.

LAMB represents the trend toward specialized optimizers designed for the realities of massive-scale training. As models continue to grow, such strategies will become increasingly important.

Comparative Analysis

Adaptive optimizers can be compared along several practical dimensions.

Convergence speed

- In the early stages of training, RMSProp and Adam usually reduce loss and reach reasonable validation accuracy noticeably faster than SGD with momentum. On image-classification benchmarks such as ResNet-50 on ImageNet, well-tuned Adam often reaches around 90% of its eventual top-1 accuracy in roughly 30–50% fewer epochs than SGD with momentum using the same batch size and learning rate schedule.
- However, SGD with momentum tends to "catch up" later in training. If you are willing to train longer or schedule multiple learning-rate drops, SGD commonly matches or exceeds the final accuracy that Adam achieves more quickly.
- LAMB is designed for extremely large batch sizes (for example, 8k–64k tokens or images per step in transformer or ResNet training). In that regime, where plain Adam or AdamW often diverge or require aggressive learning-rate tuning, LAMB maintains stable convergence with near-linear scaling: doubling the batch size frequently reduces the number of optimization steps required by close to a factor of two without degrading the training curve.

Generalization

- On many vision benchmarks (e.g., ImageNet-scale classification with CNNs or ViTs), SGD with momentum still has a small but consistent generalization edge when everything is carefully tuned, often finishing

0.5–1.0 percentage points better in top-1 accuracy than Adam or RMSProp at convergence. The gap can widen if training is extended well past the point where adaptive methods appear to have plateaued.

- In contrast, Adam and LAMB tend to perform better in domains with highly sparse, noisy, or non-stationary gradients such as NLP, recommendation systems, and reinforcement learning. For large language models, AdamW or LAMB are the default choices, and switching to pure SGD usually hurts both convergence speed and final perplexity unless major changes are made to the architecture and schedule.

Robustness to hyperparameters

- Adam is relatively forgiving: moderate changes (for example, a 2–4× change) in learning rate or batch size typically degrade performance smoothly rather than causing immediate divergence. This makes it attractive for rapid prototyping and for teams without the capacity to run extensive hyperparameter sweeps.
- SGD with momentum is more sensitive to its main hyperparameters (base learning rate, momentum, weight decay). Getting good results often requires grid searches or heuristics tied to batch size (such as linear learning-rate scaling).
- LAMB reduces sensitivity to batch size and learning-rate scaling in very large-batch regimes. In practice, users can often scale batch size from a few thousand to tens of thousands with proportional learning-rate increases and only minor retuning, which is crucial for efficient distributed training on large clusters.

Resource efficiency

- Adaptive methods maintain additional state per parameter (typically first and second moments), which roughly triples the memory footprint compared with plain SGD. For a 10-billion-parameter model, this can mean hundreds of gigabytes of extra optimizer state spread across devices.
- For most medium-scale models, this overhead is acceptable given the gains in convergence speed and ease of tuning. At frontier scale, however, the memory and communication cost of Adam or LAMB can become a first-order constraint, motivating optimizer-state sharding, low-precision optimizer states, or hybrid schemes that use SGD in later training stages to reduce resource use.

In summary, Adam and RMSProp often win on "time to reasonable accuracy," SGD with momentum often wins on ultimate generalization in vision tasks when training budgets are large, and LAMB is the tool of choice when very large batches and massive distributed setups are required.

Suitability for Domains

- **Vision:** SGD with momentum remains competitive.
- **NLP:** Adam and LAMB dominate due to sparse gradients and large-scale pretraining.
- **Reinforcement learning:** Adam is widely used because environments produce non-stationary and noisy gradients.

These comparisons highlight that no optimizer is universally superior. Each has strengths suited to particular contexts.

Broader Perspectives and Future Directions

The development of RMSProp, Adam, and LAMB reflects the trajectory of deep learning itself. As tasks grew more complex and models scaled, optimization strategies evolved to meet new demands. Several broader insights emerge.

Implicit Regularization

Adaptive methods not only accelerate convergence but also influence the type of solutions found. For example, Adam tends to explore flatter minima in certain contexts, while SGD often favors sharper solutions with better generalization. Understanding these dynamics remains an active research area.

Scaling Laws

As models scale, optimization strategies must adapt. LAMB is a response to the needs of large-batch training, but future optimizers may integrate directly with scaling laws that govern performance.

Hybrid Approaches

A growing practice is to use adaptive methods during pretraining for speed and stability, followed by SGD with momentum during fine-tuning to maximize generalization. Such hybrid approaches may become more formalized.

Optimizer–Architecture Interaction

Optimizers do not operate in isolation. Their effectiveness depends on architecture, normalization strategies, and loss functions. Future research may yield co-designed optimizers and architectures that complement each other.

Toward Automated Optimization

The long-term trajectory points toward automation. Hyperparameter tuning remains a bottleneck, but optimizers that adapt not only learning rates but also higher-level training strategies could reduce human intervention. Meta-learning and reinforcement learning are being explored to design optimizers that learn to optimize.

Adaptive optimizers have transformed the practice of training deep learning models. RMSProp introduced the principle of scaling updates by squared gradients. Adam combined momentum and adaptivity into a robust and widely adopted method. LAMB extended these ideas to large-batch training, enabling massive models to be trained efficiently. Each represents a step in addressing the evolving challenges of scale, sparsity, and stability.

While no optimizer is perfect, understanding the conceptual design of adaptive methods provides the tools to choose and apply them effectively. The next section examines gradient clipping and normalization, which complement optimization strategies by addressing instability in gradient dynamics.

Gradient Clipping and Normalization

Training deep neural networks requires constant vigilance in how gradients are managed. Gradients are the signals that flow backward through a network during backpropagation, carrying the information that allows weights to be updated. They embody the relationship between the loss function and every parameter in the model. If gradients remain stable, learning proceeds effectively and models improve steadily. If they become unstable, either exploding to extreme magnitudes or vanishing toward zero, training stalls, diverges, or collapses.

Two families of strategies emerged as particularly effective ways to keep gradients under control. The first is **gradient clipping**, which places explicit bounds on gradient magnitudes to prevent runaway growth. The second is **normalization**, which stabilizes gradient flow indirectly by keeping activations and parameter scales consistent. These approaches are not stand-alone replacements for optimizers like Adam or RMSProp but are crucial complements that ensure optimization works as intended.

This section explores the conceptual and technical foundations of gradient clipping and normalization in depth. It begins by examining the problem of exploding gradients and its historical context. It then introduces clipping strategies, ranging from simple global norms to adaptive thresholds. Afterward, it turns to normalization methods that shape gradient dynamics indirectly, from BatchNorm in convolutional networks to LayerNorm in transformers. Next, it considers the interactions between clipping, normalization, and optimizers in practice. Finally, it expands into broader reflections and forward-looking perspectives on gradient management in the era of massive-scale deep learning.

The Problem of Exploding Gradients

The challenge of exploding gradients became clear in the early days of recurrent neural networks (RNNs). Researchers attempting to model sequential dependencies noticed that as sequences grew longer, training became unstable. Losses suddenly spiked, weights diverged to extreme values, and networks failed to capture long-term structure.

The underlying reason lies in the mathematics of backpropagation through time. Gradients are propagated backward by multiplying Jacobian matrices across each time step. If the largest eigenvalue of these matrices is greater than one, repeated multiplication causes exponential growth of gradient magnitudes. Conversely, if eigenvalues are less than one, gradients vanish. This dual problem of exploding and vanishing gradients made RNNs notoriously difficult to train in the 1990s.

Consequences of exploding gradients include

- **Training divergence:** Loss values escalate rapidly, producing NaN errors and making optimization impossible.
- **Oscillatory updates:** Parameters overshoot minima, causing training to zigzag without settling.
- **Numerical overflow:** Extremely large gradients exceed the representational limits of floating-point arithmetic.
- **Instability in recurrent networks:** Dependencies across long sequences are lost because gradients for earlier steps are drowned out by extreme values later in backpropagation.

The phenomenon was not restricted to RNNs. Very deep feedforward networks also encountered instability, particularly before the adoption of better initialization schemes and activation functions. In adversarial training contexts like GANs, exploding gradients remain a persistent issue because the optimization involves two competing networks.

Historical strategies included careful initialization, smaller learning rates, and architectural modifications. However, these alone did not provide robust solutions. Gradient clipping emerged as the most practical direct method, while normalization became an indirect but highly influential approach.

Gradient Clipping Strategies

Gradient clipping places explicit constraints on gradient magnitudes. Rather than allowing gradients to grow unchecked, clipping thresholds them to keep updates within safe ranges.

Norm-Based Clipping

Norm-based clipping is the most widely used method.

- Compute the global L2 norm of the gradient vector across all parameters.
- If the norm exceeds a chosen threshold, scale the gradient vector proportionally so its norm equals the threshold.
- This preserves the direction of the gradient while limiting its magnitude.

Advantages:

- **Stability:** Prevents catastrophic parameter updates that destabilize training.
- **Direction preservation:** Since gradients are scaled uniformly, their direction remains intact.
- **Simplicity:** Implemented in a single operation in most frameworks.

Limitations:

- **Threshold sensitivity:** Too small a threshold dampens learning, while too large fails to prevent instability.
- **Task dependence:** Optimal thresholds vary across architectures and datasets.

Value-Based Clipping

Value-based clipping constrains each gradient component individually.

- Set upper and lower bounds, for example between -1 and 1.

- Clip any gradient element outside this range to the boundary value.

Advantages:

- **Component-level safety:** Prevents any single parameter from experiencing extreme updates.
- **Ease of implementation:** Straightforward operation applied element-wise.

Limitations:

- **Distortion of direction:** Component-wise clipping can significantly alter the overall gradient vector.
- **Reduced efficiency:** May eliminate useful gradient information in certain components.

Adaptive and Per-Layer Clipping

To address limitations of static thresholds, adaptive strategies were developed.

- **Adaptive thresholds:** Thresholds scale dynamically based on moving averages of gradient magnitudes.
- **Per-layer clipping:** Thresholds are set separately for each layer, accounting for differences in parameter scales.
- **Gradient norm balancing:** Normalize gradients across layers so that no single layer dominates updates.

These refinements make clipping more robust across different architectures.

Practical Heuristics for Clipping

Practitioners often combine clipping with other safeguards.

- Thresholds are chosen relative to batch size and learning rate.
- Clipping is applied only to gradients exceeding thresholds, avoiding unnecessary damping.
- **Clipping interacts with adaptive optimizers**: It prevents rare outliers from overwhelming moving averages.

Gradient clipping, though simple, remains one of the most effective guardrails in deep learning. It is particularly indispensable in recurrent networks, natural language models, and adversarial settings.

Gradient Normalization Methods

Normalization techniques tackle instability from another angle. Instead of bounding gradients directly, they normalize activations and weights, which indirectly stabilizes gradient flow. By controlling the distribution of intermediate signals, they reduce the risk of both vanishing and exploding gradients.

Batch Normalization

Batch Normalization (BatchNorm) standardizes activations within each mini-batch.

- **Centering:** Subtract the batch mean from each activation.
- **Scaling:** Divide by the batch standard deviation.

- **Learnable parameters:** Introduce a scale (gamma) and shift (beta) so the network can recover representational power.

Benefits:

- **Stabilized gradient flow:** Prevents activations from drifting into ranges where gradients vanish.
- **Accelerated convergence:** Enables higher learning rates and reduces the need for careful initialization.
- **Implicit regularization:** Adds stochasticity through batch statistics, reducing overfitting.

Challenges:

- **Batch size sensitivity:** Small batches yield noisy statistics, destabilizing normalization.
- **Sequence tasks:** For recurrent and transformer models, batch statistics may not reflect temporal dependencies.
- **Inference mode:** Requires maintaining moving averages of statistics to ensure consistency at test time.

BatchNorm transformed convolutional neural networks, making it possible to train very deep models like ResNet.

Layer Normalization

Layer Normalization (LayerNorm) standardizes activations across features within a single input.

- **Per-sample normalization:** Each sample is normalized independently.

- **Applicability to sequences:** Effective in transformers, where batch statistics are less meaningful.
- **Stability for attention mechanisms:** Keeps gradients consistent across tokens and layers.

LayerNorm became a cornerstone of transformer architectures, enabling stable training of models like BERT and GPT.

Group and Instance Normalization

Group Normalization divides features into groups and normalizes within each.

- Effective in vision tasks with small batch sizes
- Provides a middle ground between BatchNorm and LayerNorm

Instance Normalization normalizes features within each sample independently.

- Often used in style transfer tasks where global statistics dominate.

Weight and RMS Normalization

Weight Normalization reparameterizes weight vectors to separate magnitude and direction, making optimization more efficient.

RMS Normalization (RMSNorm) normalizes activations by their root mean square, simplifying computation compared to LayerNorm.

Broader Role of Normalization

Normalization reduces reliance on careful initialization, mitigates instability, and accelerates convergence. It has become so fundamental that most architectures include one or more normalization layers by default.

Interactions Between Clipping, Normalization, and Optimizers

Clipping and normalization interact with optimizers in subtle ways. Their effectiveness depends on how they integrate with momentum, adaptive scaling, and batch dynamics.

- **With SGD:** Clipping prevents divergence from large updates, while normalization improves conditioning. Together, they produce smoother trajectories.
- **With Adam:** Adaptive scaling reduces gradient variance, but clipping still guards against outliers. Normalization ensures that parameter scales remain consistent across layers.
- **With LAMB:** In large-batch training, normalization provides stability, while clipping has less influence because gradient variance is already reduced.

Context-specific examples:

- **Recurrent networks:** Gradient clipping is indispensable. LayerNorm stabilizes temporal dynamics.
- **Transformers:** LayerNorm is essential for stable training. Gradient clipping is sometimes applied during pretraining to handle rare spikes.
- **GANs:** Both clipping and normalization play roles in stabilizing adversarial training, which is notoriously sensitive to gradient dynamics.

In practice, practitioners often employ multiple methods simultaneously: normalization layers embedded in the architecture, clipping thresholds applied during optimization, and adaptive optimizers like Adam. This redundancy ensures resilience to diverse instabilities.

Extended Perspectives on Gradient Management

Gradient clipping and normalization highlight a broader truth: successful training requires active management of learning signals. Architectures, loss functions, and optimizers provide the foundation but gradient stability determines whether training succeeds.

Extended insights include

- **Gradient flow as the core signal of learning:** Well-behaved gradients carry the information needed to adjust weights; when gradients explode or vanish, updates can diverge, oscillate, or become so small that learning effectively stalls and the model fails to improve.
- **Architectural depth amplifies risks:** As models become deeper, even minor instabilities compound, making gradient management essential.
- **Complementary strategies over singular fixes:** No single method suffices. Clipping, normalization, careful initialization, and adaptive optimization all work together.
- **Future trends:** Emerging research explores automatic gradient management, where thresholds and normalization parameters adjust dynamically based on training signals.

Forward-looking themes:

- **Normalization and scaling laws:** As models scale, normalization methods must evolve to remain effective. RMSNorm in transformers reflects this adaptation.
- **Self-tuning clipping:** Algorithms that automatically determine clipping thresholds based on gradient distributions are being investigated.
- **Integration with meta-learning:** Optimizers may learn to adjust clipping and normalization strategies during training.

The trajectory suggests that gradient management will become increasingly automated and integral to optimization strategies.

Gradient clipping and normalization are indispensable components of modern deep learning. Clipping directly bounds gradients, preventing catastrophic divergence, while normalization stabilizes training by ensuring consistent activation and weight distributions. Together, they enable networks to train deeper, faster, and more reliably.

As models continue to scale into billions of parameters, these strategies remain at the center of practical deep learning. They demonstrate that while architectures and optimizers receive the most attention, the subtle art of managing gradients is what makes large-scale learning possible.

Learning Rate Schedules

The learning rate is the single most influential hyperparameter in the optimization of neural networks. It determines the size of the steps taken in parameter space as gradients guide updates. A well-chosen learning rate accelerates convergence, stabilizes training, and improves generalization. A poorly chosen one leads to divergence, stagnation, or suboptimal solutions. Because the optimal learning rate often changes throughout training, researchers and practitioners have developed a variety of **learning rate schedules**, which dynamically adjust the learning rate over time.

This section examines the role of learning rate schedules in depth. It begins with an explanation of why learning rate matters and why static choices are insufficient. It then explores fixed versus dynamic strategies, classical decay methods, modern approaches such as cosine annealing, warmup, and cyclical schedules, and large-scale considerations such as super-convergence. It concludes by comparing strategies and reflecting on their role in the broader landscape of optimization.

Why Learning Rate Matters

The importance of the learning rate arises from the geometry of optimization. Gradient descent moves parameters in the direction of the steepest descent of the loss function. The learning rate controls how far parameters travel along this direction.

- **If the learning rate is too high:** Updates overshoot minima, causing oscillations or divergence. Loss values may fluctuate erratically or explode to infinity.
- **If the learning rate is too low:** Updates become too small, making progress slow. Training may converge prematurely to suboptimal local minima or take excessive time to reach reasonable accuracy.
- **If the learning rate is just right:** Updates balance speed and stability, allowing rapid convergence to effective solutions.

Challenges arise because the optimal learning rate is not constant. In the early stages of training, large learning rates help escape poor initializations and explore the loss landscape. Later, smaller learning rates are needed for fine-grained adjustments near minima. This dynamic requirement motivates learning rate schedules.

Moreover, the **scale of gradients differs across parameters and time:**

- Some layers may have large gradient magnitudes and require smaller learning rates.
- Others may have sparse or small gradients and benefit from larger learning rates.
- The landscape may have steep directions and flat directions, making a single fixed step size inefficient.

These complexities underscore why learning rate schedules are critical. They adapt step sizes to the evolving conditions of optimization.

Fixed vs. Dynamic Schedules

The most basic choice is whether to use a fixed or dynamic learning rate.

Fixed Learning Rates

A fixed learning rate is constant throughout training.

- **Simplicity:** Easy to implement and understand.
- **Effectiveness in small tasks:** For shallow networks or small datasets, fixed rates can be sufficient.
- **Limitations in deep learning:** Large models often require different learning rates at different stages, making fixed schedules inefficient.

Dynamic Learning Rates

Dynamic schedules adjust the learning rate over time according to predefined rules.

- **Decay schedules:** Reduce learning rate gradually, allowing coarse updates early and fine-tuning later.
- **Cyclical schedules:** Vary learning rates between bounds to encourage exploration and avoid local minima.
- **Warmup schedules:** Start with small rates to stabilize early training, then increase to target values.

Dynamic schedules provide adaptability, making them more effective for large-scale deep learning.

Classical Decay Methods

Early research in optimization introduced decay methods that reduce learning rates monotonically over time.

Step Decay

Step decay reduces the learning rate by a fixed factor at specified intervals.

- Example: Start with 0.1, reduce to 0.01 after 30 epochs, and 0.001 after 60 epochs.
- Advantages: Simple and widely supported.
- Disadvantages: Abrupt changes may destabilize training.

Exponential Decay

Exponential decay multiplies the learning rate by a decay factor each epoch.

- Smooth and gradual reduction
- Useful when no clear epoch boundaries exist
- Sensitive to choice of decay factor

Polynomial Decay

Polynomial decay reduces learning rate according to a polynomial function of epoch count.

- **Flexible:** Degree of polynomial controls curvature of decay.
- Applied in tasks requiring precise fine-tuning near the end of training.

Analysis of Classical Decay

These methods embody the intuition that learning rates should shrink as training progresses. They work well in convex settings and small models but may be less effective in highly non-convex landscapes of deep networks. Sudden drops or overly rapid decay may trap models in sharp minima.

Modern Schedules

Modern schedules build on classical ideas but introduce more sophisticated dynamics tailored for deep learning.

Cosine Annealing

Cosine annealing reduces the learning rate following a cosine curve.

- Starts high, decreases smoothly toward zero.
- Encourages large exploratory steps early and small refinements later.
- Can be combined with restarts, periodically resetting the rate to encourage exploration.

Advantages:

- Smooth transitions avoid abrupt jumps.
- Effective in computer vision tasks such as ImageNet training.

Warmup

Warmup schedules gradually increase learning rates at the start of training.

- Prevents divergence when weights are randomly initialized.
- Particularly important for transformers, where large rates cause instability.
- Often combined with other schedules, such as linear warmup followed by cosine decay.

Cyclical Learning Rates

Cyclical schedules vary learning rates between lower and upper bounds across cycles.

- Encourage exploration of the loss landscape.
- Reduce risk of getting stuck in poor local minima.
- Effective in tasks requiring robustness across diverse data distributions.

Variants include triangular schedules and cosine cycles.

Super-Convergence

Super-convergence is a phenomenon where very high learning rates combined with cyclical schedules accelerate training dramatically.

- Requires careful balance of momentum and weight decay.
- Reduces training time significantly.
- Still under investigation in terms of theoretical explanation.

Comparative Insights

Modern schedules provide flexibility and adaptability.

- Cosine annealing emphasizes smooth decay.
- Warmup stabilizes early training.
- Cyclical schedules encourage exploration.
- Super-convergence pushes boundaries of efficiency.

Large-Scale Training Considerations

As models and datasets scale, learning rate schedules must adapt.

Interaction with Batch Size

Batch size strongly influences effective learning rates.

- Larger batches reduce gradient variance, enabling proportionally larger learning rates.
- **Linear scaling rule:** Double the batch size, double the learning rate.
- Requires schedule adjustments to prevent instability.

Distributed Training

In distributed systems, synchronization affects effective step sizes.

- Learning rate schedules must align with global batch sizes.
- Warmup is particularly important when scaling across many devices.

Pretraining vs. Fine-Tuning

Different phases of training may require distinct schedules.

- **Pretraining:** Larger initial rates with decay or cosine schedules
- **Fine-tuning:** Smaller fixed rates or gentle decays

Resource Efficiency

Schedules can reduce total training cost.

- Super-convergence reduces epochs needed.
- Aggressive warmup avoids wasted computation in unstable early phases.

- Cosine restarts allow reuse of trained weights for continual training.

Comparative Reflections and Future Trends)

Learning rate schedules have become central to practical optimization. Their comparative properties highlight trade-offs.

- **Classical decay:** Simple and reliable but sometimes abrupt
- **Cosine annealing:** Smooth and widely effective
- **Warmup:** Crucial for large models and transformers
- **Cyclical methods:** Encourage robustness but require tuning
- **Super-convergence:** Offers speed but is less predictable

Future research directions include

- **Automated schedules:** Meta-learning strategies that adapt schedules dynamically based on training signals.
- **Theoretical analysis:** Improved understanding of why schedules like cosine or cyclical work so well in non-convex landscapes.
- **Integration with adaptive optimizers:** Jointly optimizing per-parameter rates and global schedules.
- **Application to continual learning:** Designing schedules that adapt across long-term incremental training.

Learning rate schedules are indispensable for training deep networks. They transform the fixed hyperparameter of gradient descent into a dynamic control mechanism that adapts to the evolving requirements of optimization. From classical decay to modern cosine, warmup, and cyclical strategies, schedules provide stability, efficiency, and improved generalization.

As models continue to grow, schedules remain one of the simplest yet most powerful tools for guiding optimization. Their success underscores the principle that effective learning depends not only on architectures and optimizers but also on the careful orchestration of how step sizes evolve during training.

Conclusion

Optimization has always been the heartbeat of deep learning. Without it, architectures remain abstract structures, and data remains unprocessed raw material. This chapter has explored how optimization transforms static networks into dynamic systems that learn, highlighting backpropagation, gradient descent, adaptive optimizers, gradient stabilization techniques, and learning rate schedules. By studying these strategies, we have seen how optimization is not only a method for reducing loss but also a conceptual framework for shaping generalization, scalability, and efficiency.

Backpropagation and Gradient Descent: The Foundations

The chapter began with the foundations: backpropagation and gradient descent. Backpropagation provides the systematic method for computing gradients across deep networks, and gradient descent uses those gradients to update parameters. Together, they form the backbone of modern learning.

Conceptually, these methods are significant because they establish the link between architecture and data. They allow signals to flow backward, crediting or blaming parameters for errors, and they operationalize the principle of iterative refinement. Despite their apparent simplicity, these foundations continue to underpin all advanced strategies. Even the most sophisticated optimizers ultimately build on gradient descent, adjusting its parameters or augmenting its step sizes.

Adaptive Optimizers: Refinements of Gradient Descent

Adaptive optimizers represent one of the most influential innovations in modern training. By adjusting learning rates per parameter based on local gradient statistics, optimizers such as RMSProp, Adam, and LAMB reduce the sensitivity of training to manual hyperparameter tuning.

Conceptual contributions of adaptive methods include

- **Personalized learning rates:** Parameters with large gradients shrink their updates, while those with small gradients receive amplification.
- **Robustness across conditions:** Adaptive methods perform well on diverse architectures and data distributions.
- **Scalability to large models:** Optimizers like LAMB extend adaptive strategies to massive batch sizes and billions of parameters.

Yet these benefits come with trade-offs. Adaptive optimizers may converge quickly but sometimes generalize poorly compared to stochastic gradient descent with momentum. Their conceptual role is not to replace classical methods but to extend the toolbox of strategies, providing options that align with the constraints of scale and efficiency.

Gradient Clipping and Normalization: Stabilizing Training

Training deep networks is inherently unstable. Gradients may vanish to insignificance or explode to unmanageable magnitudes. This fragility is especially pronounced in recurrent architectures and very deep models. Gradient clipping and normalization emerged as conceptual solutions to stabilize learning.

- **Gradient clipping** enforces bounds on gradient magnitudes, preventing updates from destabilizing training.
- **Normalization techniques** such as BatchNorm and LayerNorm regulate activation scales, ensuring consistent gradient flow.

These strategies illustrate the principle that optimization is not only about updating parameters but about maintaining the conditions under which updates are meaningful. By controlling magnitudes and scales, clipping and normalization make optimization feasible for architectures that would otherwise collapse.

Learning Rate Schedules: Orchestrating Progress

Learning rate schedules highlight another key insight: the optimal step size is not fixed but dynamic. Large steps are useful early in training for exploration, while smaller steps are needed later for refinement. Schedules such as step decay, exponential decay, cosine annealing, warmup, and cyclical methods embody this principle.

Conceptually, learning rate schedules demonstrate that optimization is a temporal process. Training requires different regimes at different times, and schedules provide the choreography for balancing exploration and exploitation. They also reveal that learning is not simply about minimizing loss but about shaping the path of convergence to achieve better generalization.

Interdependence of Strategies

One of the unifying themes of this chapter is that optimization strategies are not independent silos but interdependent components.

- **Adaptive optimizers** reduce the sensitivity of gradient descent but still benefit from **learning rate schedules**.
- **Normalization** improves stability but interacts with optimizer dynamics in subtle ways.
- **Gradient clipping** is often essential when using high learning rates or adaptive methods.
- **Backpropagation** remains the common foundation upon which all other strategies rest.

This interdependence underscores that optimization is best understood as a system rather than a set of isolated techniques. Effective training emerges from the orchestration of multiple strategies working together.

Broader Reflections on Optimization

Optimization shapes more than convergence speed. It influences what models learn, how they generalize, and whether they scale to massive architectures.

- **Generalization:** Different optimizers find different regions of the loss landscape, with stochastic gradient descent often locating flatter minima that yield stronger test performance.

- **Scale:** Techniques like LAMB and warmup schedules are essential for scaling to very large models and datasets.
- **Efficiency:** Schedules and adaptive strategies reduce the time and compute required, influencing the economics of model training.
- **Accessibility:** By reducing the need for manual tuning, optimization strategies lower barriers for researchers and practitioners.

Conceptually, optimization acts as both an engine and a sculptor: it drives models toward convergence while shaping the nature of the solutions they find.

Looking Forward

As models grow larger and more complex, optimization strategies will continue to evolve. Several conceptual directions stand out:

- **Automated optimization:** Systems that learn to tune their own hyperparameters and schedules dynamically.
- **Integration with architecture design:** Optimizers co-evolving with new model families, such as those beyond transformers.
- **Energy-aware optimization:** Methods that reduce energy consumption while preserving performance.
- **Theoretical grounding:** Deeper understanding of why certain optimizers generalize better than others in non-convex landscapes.

The future of optimization lies in greater adaptability, efficiency, and integration, ensuring that models remain trainable even as scale pushes boundaries.

This chapter has shown that optimization is not a solved problem but an evolving framework. From the foundations of backpropagation and gradient descent, through adaptive optimizers, stabilization techniques, and learning rate schedules, optimization strategies have transformed what is possible in deep learning.

The key lesson is that optimization is not merely a technical detail. It determines whether models converge, how well they generalize, and whether they scale. It is the bridge between architecture and data, transforming potential into realized capability.

As deep learning advances, optimization will remain a central area of innovation. It ensures that models not only learn but learn effectively, efficiently, and sustainably. By mastering the principles explored in this chapter, researchers and practitioners can navigate the complexities of training and unlock the full power of modern neural networks.

In our next chapter, we focus on scaling training with infrastructure and distributed systems.

CHAPTER 6

Scaling Training with Infrastructure and Distributed Systems

Modern AI systems are defined as much by their infrastructure as by their algorithms. Training the largest models requires hardware capable of sustaining trillions of operations, communication protocols that synchronize thousands of devices, and frameworks that make these resources usable. Scaling training is therefore a central concern in contemporary deep learning.

Single-device training quickly reaches limits. Even the most advanced GPUs cannot store or process models with billions of parameters. Training times stretch into weeks, and memory bottlenecks prevent models from fitting in device memory. Scaling addresses these problems by distributing work across many GPUs, TPUs, and nodes. This requires careful provisioning of resources, strategies for splitting data and models, and tools that keep training resilient to inevitable failures.

The scaling problem has two sides.

- **Hardware** provides computational throughput through GPUs, TPUs, and high-bandwidth interconnects.
- **Coordination** ensures devices collaborate effectively, using parallelism, checkpointing, and distributed frameworks.

I. Cronin, *Building and Training Generative AI Models*,
https://doi.org/10.1007/979-8-8688-2332-9_6

Both must be managed together. Hardware without coordination wastes potential, while coordination without sufficient hardware capacity creates bottlenecks.

This chapter is organized into four themes.

- **GPU/TPU usage and resource provisioning** examines the hardware that underlies scaling and how it is deployed efficiently.
- **Data and model parallelism** explores strategies for distributing workloads across devices to handle both large datasets and massive models.
- **Checkpointing and fault tolerance** discusses how training jobs lasting weeks survive interruptions without catastrophic loss.
- **Distributed training frameworks** cover the abstractions, such as Horovod and DeepSpeed, that make large-scale training practical.

Scaling is more than an engineering detail. It shapes what kinds of models can be trained, how much they cost, and who has access to cutting-edge AI. By understanding the systems and methods described in this chapter, one can see how infrastructure has become as fundamental to AI progress as algorithms and data.

GPU/TPU Usage and Resource Provisioning

The shift from single-device computation to specialized accelerators marked one of the most important transitions in the history of deep learning. Central processing units, while flexible and general-purpose, proved inadequate for the scale of matrix operations required by neural networks. The introduction of graphics processing units and later tensor processing units transformed the possibilities of training, enabling parallelization on a scale that made modern architectures feasible. Yet simply possessing hardware is not enough. Training at scale requires careful provisioning of resources, strategies for balancing workloads, and policies for ensuring utilization efficiency. This section explores the conceptual and technical foundations of how GPUs and TPUs are used in training and how resources are provisioned to maximize their impact.

Evolution from CPUs to GPUs and TPUs

In the earliest days of neural network research, CPUs were sufficient. Networks were shallow, datasets were small, and computation could be handled sequentially. As networks deepened and datasets expanded, CPU limitations became clear.

- CPUs excel at general-purpose tasks, with complex instruction sets designed for diverse operations.
- They prioritize single-threaded performance and flexibility over raw parallel throughput.
- Neural networks, however, require repetitive linear algebra operations, particularly matrix multiplications, that can be parallelized extensively.

The mismatch between CPU design and neural network computation motivated the adoption of GPUs. Originally developed for rendering graphics, GPUs offered massively parallel architectures optimized for vector and matrix operations.

- GPUs emphasize throughput over latency, aligning naturally with neural network workloads.

- Thousands of cores operate simultaneously, handling large batches of matrix multiplications efficiently.
- Their memory hierarchies and bandwidth support the streaming of data necessary for training.

TPUs represent a further evolution. Unlike GPUs, which were repurposed for deep learning, TPUs were designed specifically for tensor operations.

- TPUs incorporate systolic arrays optimized for dense linear algebra.
- They eliminate many general-purpose instructions in favor of operations directly aligned with machine learning primitives.
- The design philosophy prioritizes deterministic high-throughput computation over flexibility.

This progression from CPUs to GPUs to TPUs reflects the broader trajectory of infrastructure specialization in artificial intelligence: from general-purpose devices to accelerators tailored to the structure of neural networks.

Architectural Characteristics of GPUs

GPUs are central to training because of their architectural alignment with the mathematical structure of neural networks. At a conceptual level, GPUs can be understood as collections of simple processors designed to work in parallel on small pieces of data.

Key characteristics include

- **Massive parallelism:** Neural networks involve large-scale matrix multiplications. GPUs divide these computations across thousands of cores, processing elements in parallel.

- **SIMD philosophy:** Single instruction, multiple data operations allow the same instruction to be applied to many data elements simultaneously.
- **Memory hierarchy:** GPUs balance fast but small caches with high-bandwidth global memory, ensuring that data flows efficiently to computational cores.
- **Throughput-oriented design:** GPUs are optimized for sustained high-volume operations rather than minimizing latency for individual tasks.

In the context of training, these characteristics mean that GPUs excel at

- **Forward propagation:** Multiplying large weight matrices with activations.
- **Backward propagation:** Computing gradients, which involve transposed matrix multiplications and element-wise derivatives.
- **Parameter updates:** Though less computationally demanding than forward or backward passes, updates are parallelized efficiently across cores.

Limitations of GPUs include

- **Memory constraints:** Even with high bandwidth, total memory capacity remains finite, restricting the maximum model size that can be trained on a single device.
- **Communication overhead:** Scaling across multiple GPUs introduces synchronization costs, as gradients must be aggregated across devices.
- **Energy efficiency:** GPUs consume significant power, raising both economic and environmental concerns at scale.

Despite these limitations, GPUs remain the workhorse of modern training, providing the necessary balance between flexibility and specialization.

Architectural Characteristics of TPUs

TPUs represent a design philosophy explicitly oriented toward machine learning workloads. Conceptually, they differ from GPUs in several ways.

- **Systolic arrays**: TPUs use systolic arrays that stream data rhythmically through processing elements, enabling efficient multiplication and accumulation.
- **Reduced instruction set**: Unlike GPUs, which retain flexibility for graphics and general computation, TPUs remove extraneous instructions and focus almost entirely on tensor operations.
- **Tightly coupled memory and compute**: TPUs place large on-chip SRAM buffers and high-bandwidth links directly next to the systolic arrays, so activations and weights are reused many times without repeatedly going out to external DRAM. In contrast, GPUs rely more heavily on a deeper cache hierarchy and off-chip global memory. This explicit, close coupling lets TPUs sustain very high utilization on dense matrix workloads while reducing memory latency and energy per operation.

Advantages of TPUs for training include

- **High deterministic throughput**: Systolic arrays deliver consistent performance across workloads dominated by linear algebra.

- **Simplified programming model**: Tensor operations map directly to hardware primitives, reducing abstraction layers.
- **Specialization for neural networks**: TPUs prioritize exactly the computations central to deep learning, such as matrix multiplications and convolutions.

However, TPUs trade flexibility for specialization; workloads that deviate from dense tensor operations or require fine-grained custom kernels are often better served by GPUs or more general accelerators.

- They are less suited for tasks outside tensor computation.
- Optimization of workloads for TPUs requires adaptation to their architecture.
- Like GPUs, TPUs remain constrained by memory and communication overhead when scaled across devices.

Conceptually, TPUs highlight the trend toward domain-specific accelerators, designed not for universal computation but for the dominant patterns of a particular field.

Provisioning Strategies

Having hardware is only one part of scaling. Effective provisioning ensures that resources are allocated efficiently and aligned with training goals. Provisioning refers to both the initial allocation of devices and the ongoing management of workloads across them.

Core strategies include

- **Static provisioning:** Allocate a fixed number of devices for the duration of training. Suitable for stable workloads but may waste resources during phases of low utilization.

- **Elastic provisioning:** Dynamically adjust resources as training progresses. Useful in cloud contexts where hardware can be scaled up or down.
- **Cluster-level orchestration:** Use schedulers to distribute jobs across many nodes, balancing workloads and minimizing idle time.

Important considerations in provisioning:

- **Cost efficiency:** Hardware is expensive, and inefficient utilization translates directly into higher costs.
- **Load balancing:** Ensuring that all devices are used productively avoids bottlenecks caused by stragglers.
- **Resource heterogeneity:** Different devices may coexist, requiring strategies for aligning workloads with capabilities.

Provisioning is not purely technical. It reflects trade-offs between speed, cost, and availability, requiring careful alignment with the goals of training.

Utilization Challenges

Even with powerful hardware and effective provisioning, utilization remains a challenge. Devices must be kept busy with meaningful computation, which is not always trivial.

Challenges include

- **Memory bottlenecks:** Models may exceed device memory capacity, requiring partitioning or offloading strategies.
- **Throughput limitations:** Data pipelines must supply devices quickly enough to prevent idle time.

- **Synchronization overhead:** In distributed contexts, devices must exchange gradients, creating communication bottlenecks.
- **Energy efficiency:** Sustained large-scale training consumes enormous power, making utilization optimization not only technical but also economic.

Strategies for addressing utilization:

- **Gradient checkpointing:** Save memory by recomputing activations instead of storing them.
- **Asynchronous pipelines:** Overlap computation and communication to reduce idle time.
- **Mixed precision training:** Use lower precision arithmetic where possible to accelerate throughput and reduce memory demands.
- **Profiling and tuning:** Monitor workloads to identify bottlenecks and adjust scheduling accordingly.

Utilization thus represents the operational frontier of scaling: ensuring that available hardware actually translates into effective computation.

Broader Implications of Hardware Scaling

The use of GPUs and TPUs for training has implications beyond technical performance. Conceptually, it illustrates the interdependence between algorithmic innovation and infrastructure development.

- **Algorithmic feasibility:** Many architectures that define modern deep learning, such as transformers, are only practical because of accelerators.

- **Research accessibility:** The need for specialized hardware concentrates capability in organizations with sufficient resources, influencing the trajectory of AI research.
- **Energy and sustainability:** Large-scale training demands raise questions about the sustainability of current approaches.

From a conceptual standpoint, hardware scaling is not simply about speed. It shapes what kinds of models are possible, how research is conducted, and who can participate in advancing the field.

GPUs and TPUs provide the computational foundation for scaling deep learning. Their architectural characteristics align with the needs of neural networks, transforming what was once computationally infeasible into practical reality. Yet effective usage requires more than hardware. Provisioning strategies, utilization optimization, and fault-aware coordination ensure that resources deliver their full potential.

Understanding GPUs and TPUs conceptually provides a foundation for the broader study of scaling. They are the raw engines of computation, but their value emerges only when paired with strategies for parallelism, reliability, and framework support. Together, these dimensions define the infrastructure of modern artificial intelligence.

Data and Model Parallelism

Scaling deep learning requires distributing workloads across multiple devices. As models and datasets grow, the limits of single-device training become unavoidable. Memory capacity, computation throughput, and training time all become bottlenecks. Parallelism provides a conceptual and technical solution: dividing the work of training across multiple devices in such a way that the collective system operates as a coherent whole. Parallelism is therefore the cornerstone of distributed training, enabling the largest neural networks to be trained within reasonable time and resource constraints.

This section explores the principles of data and model parallelism. It begins with the conceptual foundations, showing why parallelism is necessary and what properties make it effective. It then examines data parallelism, model parallelism, pipeline parallelism, tensor parallelism, and hybrid strategies. Along the way, it considers the challenges of communication overhead, workload balance, and synchronization. The discussion remains conceptual and technical, focusing on the underlying strategies rather than specific implementations.

Conceptual Foundations of Parallelism

Parallelism is rooted in the recognition that training neural networks involves large numbers of repetitive, independent operations. Forward and backward passes consist primarily of matrix multiplications and element-wise transformations, which can be divided into smaller tasks. By assigning subsets of these tasks to different devices, training can proceed in parallel, reducing wall-clock time and increasing feasible model size.

Conceptual requirements for effective parallelism include

- **Decomposability of tasks:** Work must be divisible into units that can be computed independently before synchronization.
- **Efficient communication:** Devices must exchange results without excessive overhead.
- **Load balance:** Each device should perform roughly the same amount of work to avoid stragglers slowing the system.
- **Scalability:** Adding more devices should yield meaningful improvements in performance.

Parallelism is not trivial. Communication costs, synchronization delays, and the complexity of splitting data and models all create challenges. Nevertheless, parallelism remains indispensable for scaling.

Data Parallelism

Data parallelism is the most intuitive and widely used form of parallelism. The idea is to divide the training dataset across multiple devices while replicating the model on each device.

Steps of data parallelism:

- Each device receives a mini-batch of data.
- The forward pass is computed independently on each device, producing local gradients.
- Gradients are aggregated across devices, typically through an averaging operation.
- Parameters are updated consistently across all devices.

Advantages of data parallelism:

- **Simplicity:** Conceptually straightforward and relatively easy to implement.
- **Scalability:** Effective up to many devices, especially when batch sizes are large.
- **Model independence:** Does not require splitting the model itself, only the data.

Limitations:

- **Batch size constraints:** As the number of devices increases, batch sizes must also increase to maintain efficiency. Very large batch sizes can harm generalization.

- **Communication overhead:** Aggregating gradients across devices introduces latency, particularly at scale.
- **Synchronization requirements:** Devices must wait for gradient exchange, making stragglers problematic.

Variants of data parallelism include

- **Synchronous data parallelism:** All devices synchronize at each step. Ensures consistency but sensitive to slow devices.
- **Asynchronous data parallelism:** Devices proceed independently, updating parameters without strict synchronization. Improves throughput but risks stale gradients.

Data parallelism remains foundational because it scales easily with dataset size and aligns with the natural independence of samples.

Model Parallelism

Model parallelism addresses situations where a model is too large to fit in the memory of a single device. Instead of replicating the model, different parts of the model are distributed across devices.

Approach:

- Layers or components of the network are partitioned across devices.
- Forward passes move activations from one device to the next.
- Backward passes propagate gradients in the opposite direction.

Advantages:

- **Memory efficiency:** Enables training of models that exceed the memory of any single device.
- **Necessity for very large models:** Essential for networks with billions of parameters.

Challenges:

- **Communication overhead:** Activations must be transferred between devices after each partitioned step.
- **Load balancing:** Some layers may be more computationally intensive, creating imbalance across devices.
- **Sequential dependencies:** Forward and backward passes require ordered execution, reducing opportunities for parallel speedup.

Model parallelism is less scalable than data parallelism because of its sequential dependencies. Yet it is indispensable for extremely large models where memory limits dominate.

Pipeline Parallelism

Pipeline parallelism is a refinement of model parallelism that seeks to improve efficiency. Instead of processing one mini-batch through the model sequentially, pipeline parallelism overlaps execution across devices.

Concept:

- Partition the model into stages assigned to different devices.

- Feed micro-batches of data into the pipeline, so that while one stage processes micro-batch n, the next stage processes micro-batch n-1.
- This overlapping creates a pipeline where devices remain busy simultaneously.

Advantages:

- **Improved utilization:** Devices are not idle waiting for earlier stages to finish.
- **Scalability:** Enables training of deeper models without excessive sequential delays.

Challenges:

- **Pipeline bubbles:** Idle time arises at the start and end of each pipeline cycle.
- **Complexity of scheduling:** Effective micro-batch scheduling is required to minimize idle periods.
- **Gradient accumulation:** Gradients must be aggregated across micro-batches, complicating synchronization.

Pipeline parallelism conceptually mirrors assembly lines in manufacturing: efficiency emerges from overlapping tasks rather than executing them serially.

Tensor Parallelism

Tensor parallelism divides individual tensor operations across devices. Instead of assigning whole layers or batches, the operations themselves are partitioned.

Approach:

- Large matrix multiplications are split across devices, each handling a sub-block of the operation.
- Results are combined to form the complete output.

Advantages:

- **Granular scalability:** Even extremely large operations can be distributed.
- **Alignment with matrix operations:** Neural networks rely heavily on matrix multiplications, making tensor parallelism conceptually natural.

Challenges:

- **Communication intensity:** Devices must frequently exchange partial results, increasing overhead.
- **Implementation complexity:** Requires fine-grained partitioning of operations.
- **Sensitivity to network latency:** Small delays can compound when operations are split extensively.

Tensor parallelism is conceptually powerful but operationally demanding. It is most useful when combined with other forms of parallelism.

Hybrid Strategies

No single form of parallelism suffices at modern scales. Hybrid strategies combine multiple approaches to exploit their complementary strengths.

Examples:

- **Data plus model parallelism:** Datasets are divided across groups of devices, while models are split within each group.
- **Pipeline plus data parallelism:** Pipelines handle depth, while data parallelism expands breadth.
- **Tensor plus pipeline parallelism:** Large operations are partitioned within pipeline stages.

Advantages:

- **Scalability across dimensions:** Hybrid strategies address both dataset and model size simultaneously.
- **Flexibility:** Different partitions can be tuned to workload characteristics.
- **Efficiency:** By combining methods, idle time and communication costs can be reduced.

Challenges:

- **Complexity:** Hybrid systems require sophisticated coordination.
- **Hyperparameter sensitivity:** Partitioning choices affect efficiency and must be tuned.
- **Debugging difficulty:** Errors in hybrid systems can be hard to trace.

Conceptually, hybrid strategies reflect the reality that modern models demand distribution not just in one dimension but across multiple dimensions simultaneously.

Bottlenecks and Challenges

Parallelism introduces its own challenges. While it expands capacity, it also creates new bottlenecks.

Key bottlenecks include

- **Communication overhead:** Synchronizing gradients and activations requires frequent device communication. At scale, communication may dominate computation time.
- **Straggler effects:** Variability in device speed or network latency means the slowest device determines overall progress.
- **Memory fragmentation:** Partitioning data and models introduces inefficiencies in memory usage.
- **Load imbalance:** Unequal workloads across devices reduce efficiency.
- **Fault sensitivity:** A failure in one device can halt the entire system, requiring checkpointing and fault tolerance mechanisms.

Mitigating these challenges requires careful design of communication patterns, scheduling algorithms, and synchronization strategies. Conceptually, the efficiency of parallelism depends not only on dividing work but also on minimizing the costs of recombining results.

Data and model parallelism form the conceptual foundation of distributed training. Data parallelism leverages the independence of samples to scale across datasets, while model parallelism distributes parameters to train networks too large for single devices. Pipeline and tensor parallelism extend these principles, and hybrid strategies combine them for maximum scalability.

Parallelism illustrates the central theme of scaling: that no single device, no matter how powerful, can meet the demands of modern deep learning. Only through carefully designed strategies for dividing and coordinating workloads can models of billions of parameters be trained. Understanding these concepts provides the basis for checkpointing, fault tolerance, and distributed frameworks, which further enhance the resilience and usability of large-scale training systems.

Checkpointing and Fault Tolerance

Training deep learning models at scale transforms the optimization problem into a systems challenge. When training jobs span thousands of devices and run for weeks, the likelihood of interruptions is not a rare event but a certainty. Without mechanisms to preserve progress and recover from faults, a single error can erase enormous amounts of computation. Checkpointing and fault tolerance form the foundation of resilience in large-scale training systems. They ensure that despite failures, progress is preserved and training continues reliably.

This section develops a conceptual account of checkpointing and fault tolerance. It begins with the necessity of resilience in distributed training and situates checkpointing in the broader history of high-performance computing. It then explores the principles of fault tolerance, models of failures, and strategies for detecting and recovering from them. Different checkpointing approaches are considered, including periodic, incremental, sharded, hierarchical, and asynchronous methods. Trade-offs in frequency and granularity are analyzed, followed by a discussion of elastic recovery in distributed systems. The section concludes by considering broader challenges and future directions in resilience at scale.

The Need for Checkpointing in Large-Scale Training

Checkpointing arises because training is fragile in the presence of failure. At small scale, restarting from scratch after a crash may be tolerable. At large scale, where models may take weeks to train, restarting is not viable.

Conceptual motivations for checkpointing:

- **Scale of time investment:** Training that requires millions of gradient updates cannot simply restart when interrupted. Without checkpoints, progress would be repeatedly lost.
- **High probability of failures:** With thousands of devices, even low per-device failure probabilities aggregate into frequent overall failures. A cluster with thousands of GPUs may expect multiple device failures per day.
- **Complex statefulness:** Training state includes not only model parameters but also optimizer momentum, learning rate schedules, normalization statistics, and random seeds. Preserving these ensures training resumes exactly where it left off.

Checkpointing provides continuity in this context. It transforms failure from a catastrophic event into a recoverable interruption.

Historical Context of Checkpointing

Checkpointing did not originate with machine learning. It was a core feature of high-performance computing long before neural networks scaled to billions of parameters. In supercomputing environments, long-running scientific simulations encountered the same fragility.

- **Scientific computing:** Weather simulations, molecular dynamics, and physics experiments required computations that ran for weeks. Failures in these systems could destroy valuable results.
- **HPC checkpointing:** Techniques such as periodic saving of memory states were developed to mitigate risk. These ideas migrated into deep learning as training workloads mirrored the scale of scientific simulations.
- **Shared lineage:** Both HPC and machine learning involve iterative computations over large datasets and require resilience to ensure that computation invested over long durations is not wasted.

Understanding this lineage helps clarify that checkpointing is not an AI-specific hack but a general principle of resilience in long-running computations.

Principles of Fault Tolerance

Fault tolerance is the broader concept within which checkpointing resides. It refers to the ability of a system to continue functioning correctly even when some components fail.

Principles include

- **Redundancy:** Critical components or computations are duplicated so that if one fails, another can take over.
- **Isolation:** Failures are contained to prevent them from cascading across the system.
- **Detection:** Systems must be able to recognize when a fault has occurred.

- **Recovery:** After detection, systems must restore themselves to a consistent state.
- **Transparency:** Higher-level processes should not need to know about low-level failures.

These principles reflect the idea that failures are inevitable in large systems, and resilience must be built in by design rather than treated as an afterthought.

Models of Failures

Different types of failures shape the design of checkpointing and fault tolerance strategies. Conceptual models of failures include

- **Crash faults:** Devices or processes halt unexpectedly but do not produce incorrect outputs. These are the most common assumptions in distributed training.
- **Transient faults:** Temporary errors such as bit flips or momentary network instability that resolve themselves. These may corrupt intermediate states without permanent damage.
- **Byzantine faults:** Processes behave arbitrarily or maliciously, producing incorrect outputs. Less common in trusted environments but relevant in adversarial or federated learning.
- **Permanent faults:** Hardware components fail irreversibly and must be replaced.
- **Communication faults:** Links between devices fail or degrade, causing delays or loss of synchronization.

Checkpointing primarily addresses crash and permanent faults, ensuring training can resume. Other forms of resilience address transient and Byzantine behaviors.

Checkpointing Strategies

Checkpointing can be implemented in many ways, each representing a different balance between resilience, performance, and complexity.

Periodic Checkpointing

- Save the full training state at regular intervals.
- Guarantees that recovery is possible from the most recent checkpoint.
- Overhead grows with checkpoint size and frequency.

Incremental Checkpointing

- Save only changes since the last checkpoint.
- Reduces storage and bandwidth costs.
- Recovery requires replaying increments, increasing complexity.

Sharded Checkpointing

- Divide the model state across devices, each saving its portion.
- Parallelizes checkpoint creation and reduces bottlenecks.
- Requires coordination to ensure consistency across shards.

Hierarchical Checkpointing

- Save states at multiple levels of storage, from fast local memory to slower remote storage.
- Provides trade-offs between speed and durability.
- Local checkpoints recover from transient faults quickly; remote checkpoints recover from full system failures.

Differential Checkpointing

- Save differences relative to a baseline checkpoint
- Further reduces storage requirements compared to incremental methods
- Increases reconstruction overhead

Asynchronous Checkpointing

- Overlap checkpoint creation with ongoing training
- Reduces visible overhead by masking checkpoint latency behind computation
- Risk of inconsistency if failure occurs mid-creation

Each strategy is an embodiment of the same principle: trading off overhead during normal operation against resilience during failure.

Trade-Offs in Frequency and Granularity

Checkpointing involves two key parameters: how often to save and what to save.

Frequency

- **High frequency:** Reduces potential lost progress after failure but increases runtime and storage costs
- **Low frequency:** Reduces overhead but increases risk of losing significant progress

Granularity

- **Full checkpoints:** Save entire training state, ensuring exact recovery.
- **Partial checkpoints:** Save only essential parameters, reducing size but introducing potential inconsistencies.
- **Component-based checkpoints:** Target volatile components such as optimizer states while ignoring stable parameters.

These trade-offs require balancing risk against efficiency. Conceptually, the optimal choice depends on the probability of failures, the cost of checkpointing, and the duration of training.

Elastic Recovery in Distributed Training

Elastic recovery extends checkpointing into dynamic environments where resources change over time. Instead of only resuming from failure, elastic systems adapt workloads to available resources.

Conceptual features:

- **Dynamic redistribution:** Work is reassigned automatically when devices drop out.

- **Scalable continuity:** Systems add or remove resources during training without halting.
- **Checkpoint integration:** Recovery incorporates saved states to ensure continuity.

Elasticity reflects the principle that failures and resource variability are normal in distributed systems. Training systems must not merely tolerate faults but adapt productively to them.

Conceptual Challenges and Bottlenecks

Checkpointing and fault tolerance themselves introduce difficulties.

- **Storage bottlenecks:** Saving large models frequently can saturate storage bandwidth.
- **Consistency across devices:** Ensuring coherent checkpoints across thousands of devices is nontrivial.
- **Performance overhead:** Time spent creating checkpoints detracts from computation.
- **Recovery complexity:** Reconstructing from incremental or differential checkpoints requires careful replay.
- **Scalability limitations:** Approaches effective at small scales may fail at exascale workloads.

Conceptually, resilience mechanisms embody a tension: maximizing continuity without undermining throughput.

Broader Reflections and Future Directions

Checkpointing and fault tolerance are evolving in response to increasing scale. Several future directions can be identified.

- **Predictive checkpointing:** Systems may anticipate failures using monitoring signals and create checkpoints proactively.
- **Self-optimizing policies:** Machine learning itself may guide when and how checkpoints are taken, balancing risk dynamically.
- **Hierarchical resilience:** Multi-layered approaches will integrate device-level, node-level, and cluster-level fault tolerance.
- **Energy-aware strategies:** As training consumes more energy, resilience strategies will need to minimize redundant computation.
- **Integration with scaling laws:** Checkpointing frequency and overhead may become predictable based on model size and training duration.

These directions suggest that resilience will become increasingly intelligent, automated, and integrated into the broader training stack.

Checkpointing and fault tolerance are indispensable for large-scale training. They transform inevitable failures from catastrophic setbacks into recoverable events. Checkpointing preserves state, while fault tolerance ensures continuity despite failures. Together, they provide the resilience needed to sustain long-running, resource-intensive computations.

The strategies examined—periodic, incremental, sharded, hierarchical, differential, and asynchronous checkpointing, along with elastic recovery—represent a spectrum of conceptual solutions. Each balances efficiency against resilience, simplicity against complexity, and performance against safety.

In the context of distributed training, these mechanisms are not peripheral but central. Without them, scale would collapse under the weight of fragility. With them, training becomes feasible, sustainable, and reliable, paving the way for models of unprecedented size and capability.

Distributed Training Frameworks

Scaling training to hundreds or thousands of devices is impossible without software frameworks that manage complexity. While parallelism strategies, checkpointing, and fault tolerance describe the *what* of distributed training, frameworks provide the *how*. They encapsulate communication patterns, coordinate gradient synchronization, and expose abstractions that allow researchers to write models without needing to manage every detail of distributed systems.

This section examines distributed training frameworks as conceptual infrastructures. It explores the role of frameworks, the communication primitives they rely upon, strategies for gradient synchronization, and approaches to fault-aware design. It considers how frameworks abstract parallelism, the challenges they face in scaling, and the comparative contributions of exemplars such as Horovod, PyTorch Distributed, TensorFlow's strategies, and DeepSpeed. Finally, it looks ahead to possible future directions in framework design.

The Role of Frameworks in Distributed Training

The core role of a distributed training framework is to provide abstraction.

- **Simplification of complexity:** Without frameworks, researchers would need to write custom code for device communication, synchronization, and recovery.

- **Consistency of execution:** Frameworks ensure that each device follows the same communication protocols, preventing inconsistencies across nodes.
- **Portability:** Frameworks make distributed training usable across different hardware clusters, from local GPU servers to large cloud deployments.
- **Optimization of communication:** They incorporate efficient algorithms for gradient aggregation and parameter updates.

Conceptually, frameworks are the layer that makes large-scale distributed training usable by domain researchers rather than only by systems engineers.

Communication Primitives and Collective Operations

All distributed training frameworks depend on a set of low-level communication primitives. These operations provide the building blocks for synchronizing parameters and gradients across devices.

Key collective operations:

- **Broadcast:** Send parameters or states from one device to all others.
- **Allreduce:** Aggregate gradients across devices and share the result with all participants. This is central to synchronous data parallelism.
- **Reduce:** Aggregate values from all devices and deliver the result to one designated device.

- **Allgather:** Collect slices of data from all devices and concatenate them into a shared view.
- **Scatter:** Distribute distinct portions of data to different devices.

Conceptually, these primitives allow a group of devices to act as one coherent system. The efficiency of a framework depends heavily on how it implements these collective operations.

Gradient Synchronization Strategies

Gradient synchronization is the defining challenge of distributed training. Frameworks provide different conceptual solutions to this problem.

Parameter Server Paradigm

- Devices send gradients to a central server.
- The server updates parameters and sends them back to devices.
- Conceptually simple but becomes a bottleneck at scale.

Ring-Allreduce

- Devices are arranged logically in a ring.
- Each device sends partial gradients to its neighbor, aggregates received values, and passes results along the ring.
- After a full cycle, all devices have the complete aggregated gradient.
- Introduced in Horovod, this eliminated central bottlenecks and provided bandwidth efficiency.

Hierarchical Allreduce

- Gradients are aggregated within groups of devices before being shared across groups.
- Reduces overhead in clusters with many nodes.

Gossip-Based Synchronization

- Devices exchange information with random peers rather than all devices at once.
- Trades exact consistency for scalability.

Frameworks operationalize these strategies, balancing communication cost against synchronization fidelity.

Fault-Aware and Elastic Framework Designs

Frameworks also embody concepts of fault tolerance. Since failures are inevitable, frameworks must recover gracefully.

- **Checkpoint integration:** Frameworks manage periodic saving of states so recovery is seamless.
- **Elastic membership:** Devices can join or leave without halting the job.
- **Failure detection:** Systems monitor communication health and reconfigure when a device drops out.
- **Graceful degradation:** Training continues with reduced resources until recovery or replacement occurs.

These features extend fault tolerance from a systems concept into a practical software abstraction. Frameworks operationalize resilience by embedding detection and recovery into their core.

Abstractions for Parallelism (Data, Model, Pipeline, and Tensor)

Parallelism strategies such as data, model, pipeline, and tensor parallelism are realized through framework abstractions.

- **Data parallel abstractions:** Frameworks replicate models across devices and manage gradient averaging transparently.
- **Model parallel abstractions:** Frameworks partition parameters across devices and coordinate activations and gradients between them.
- **Pipeline parallel abstractions:** Frameworks manage micro-batches across stages of computation, scheduling to minimize idle bubbles.
- **Tensor parallel abstractions:** Frameworks divide individual operations across devices and synchronize partial results.

Frameworks abstract away the implementation details, providing higher-level APIs that allow users to express intent without managing communication directly.

Scalability Challenges in Framework Design

Frameworks face conceptual challenges in scaling to ever larger clusters and models.

- **Communication bottlenecks:** Allreduce operations scale poorly if implemented naively. Frameworks must optimize bandwidth usage.
- **Synchronization costs:** Synchronous methods create straggler effects, where the slowest device dictates progress.
- **Memory overhead:** Maintaining optimizer states across devices can overwhelm resources.
- **Framework overhead:** Abstractions must not add excessive computational cost.
- **Heterogeneous environments:** Clusters may contain devices of different capacities, complicating load balancing.

Scalability requires frameworks to continuously innovate in algorithms for communication, scheduling, and state management.

Conceptual Comparison of Framework Approaches

Several frameworks illustrate different conceptual solutions to distributed training.

Horovod

- Introduced ring-allreduce as a scalable alternative to parameter servers.

- Focused on minimizing communication overhead with efficient collective operations.
- Conceptually significant for demonstrating that large-scale synchronous training could be practical.

PyTorch Distributed

- Provides process groups and collective operations integrated directly with the PyTorch ecosystem.
- Exposes both high-level and low-level APIs, enabling flexibility.
- Conceptually shows the power of integrating distributed primitives into the core deep learning framework rather than as an external library.

TensorFlow Distributed Strategies

- Provides mirrored strategies for data parallelism, parameter servers, and more.
- Conceptually highlights diversity of approaches within a single ecosystem.

DeepSpeed

- Introduced ZeRO (Zero Redundancy Optimizer), which shards optimizer states, gradients, and parameters across devices.
- Enabled training of trillion-parameter models without overwhelming memory.

- Conceptually significant for demonstrating that optimizer state partitioning can extend the feasible scale of models.

These frameworks exemplify different conceptual innovations that collectively define the state of distributed training.

Future Directions in Distributed Training Frameworks

As models and datasets continue to grow, frameworks will evolve conceptually in several directions.

- **Self-optimizing frameworks:** Systems that dynamically adjust communication algorithms and synchronization strategies based on observed workloads.
- **Tighter integration with fault tolerance:** Checkpointing and recovery will become more seamless and less manual.
- **Energy-aware scheduling:** Frameworks will optimize not only for speed but also for power efficiency.
- **Standardized abstractions:** Convergence toward common APIs will improve portability across hardware and platforms.
- **Meta-frameworks:** Systems may coordinate across multiple distributed training frameworks, unifying heterogeneous resources.

Conceptually, the trajectory points toward greater automation, resilience, and integration, reducing the need for manual configuration and tuning.

Distributed training frameworks are the operational backbone of large-scale deep learning. They encapsulate communication primitives, synchronize gradients, manage parallelism, and provide resilience against faults. Frameworks such as Horovod, PyTorch Distributed, TensorFlow strategies, and DeepSpeed illustrate the conceptual innovations that make scaling practical, from ring-allreduce to optimizer state sharding.

By providing abstractions that hide low-level details, frameworks democratize distributed training, making it accessible to researchers and practitioners without requiring deep systems expertise. Conceptually, they represent the bridge between the theory of parallelism and the practice of training at unprecedented scale.

Conclusion

The pursuit of scaling has always been central to progress in artificial intelligence. In earlier chapters, we examined how models learn from data and how optimization techniques make them converge. In this chapter, the focus shifted to the infrastructure that makes such training possible at unprecedented scale. Without the ability to coordinate vast computational resources, the conceptual advances in algorithms would remain inert. Scaling is what turns potential into practice.

Infrastructure As the Third Pillar

The chapter has emphasized that modern deep learning depends on three interdependent pillars:

- **Algorithms:** The architectures and optimization strategies that define how learning occurs.
- **Data:** The raw material from which patterns are extracted.

- **Infrastructure:** The hardware and systems that make training feasible at scale.

Infrastructure is the often-overlooked third pillar. While algorithmic innovation and data collection receive more attention, it is infrastructure that determines how far and how fast ideas can be realized. Large models exist not only because researchers designed them but because GPUs, TPUs, and distributed frameworks made them trainable.

Hardware As the Foundation

At the base of scaling lies hardware. The transition from CPUs to GPUs and TPUs provided the computational throughput that neural networks require. GPUs demonstrated how architectures optimized for throughput could transform deep learning, and TPUs highlighted the benefits of domain-specific design. Conceptually, these accelerators embody the principle that specialization unlocks scale.

Yet hardware alone is insufficient. Devices must be provisioned, workloads balanced, and utilization maximized. Provisioning strategies ensure that expensive resources are not wasted. Utilization optimization keeps devices busy, overlapping communication with computation and reducing idle time. The discussion of hardware underscored that scaling is not just about acquiring accelerators but about orchestrating them effectively.

Parallelism As the Core Strategy

Parallelism provides the conceptual mechanism for scaling. Data parallelism distributes samples, model parallelism distributes parameters, pipeline parallelism overlaps execution, and tensor parallelism divides operations. Each approach solves different constraints, and hybrid strategies combine them for maximum effect.

The chapter illustrated that parallelism is not a single solution but a spectrum. Data parallelism is the simplest and most widely used, but it is limited by batch size. Model parallelism enables training of extremely large networks but introduces communication overhead. Pipeline and tensor parallelism provide refinements, each with their own advantages and complexities.

At a conceptual level, parallelism represents the principle that no single device can shoulder the weight of modern models. Only by dividing work can systems scale.

Resilience Through Checkpointing and Fault Tolerance

Scaling introduces fragility. With thousands of devices and weeks-long training jobs, failures are inevitable. Checkpointing and fault tolerance transform these failures from catastrophic events into recoverable interruptions.

The discussion of checkpointing showed multiple strategies, from periodic full saves to incremental, sharded, hierarchical, and asynchronous methods. Each represented a different balance between resilience and efficiency. Fault tolerance introduced broader principles of redundancy, isolation, detection, and recovery. Together, these ideas illustrated that resilience is not optional but fundamental to training at scale.

Conceptually, checkpointing and fault tolerance highlight that distributed training must assume failure as the norm. Resilience is not a peripheral feature but a core requirement of sustainable scaling.

Frameworks As the Operational Backbone

Distributed training frameworks operationalize parallelism and resilience. Without them, researchers would need to write custom communication code and manage synchronization manually. Frameworks such as Horovod, PyTorch Distributed, TensorFlow strategies, and DeepSpeed provided exemplars of how ideas translate into usable systems.

- Horovod introduced ring-allreduce, demonstrating scalable synchronous training.
- PyTorch Distributed embedded collective operations into the heart of a widely used framework.
- TensorFlow strategies illustrated multiple approaches within a single ecosystem.
- DeepSpeed pioneered optimizer state partitioning with ZeRO, extending feasible model size dramatically.

Conceptually, frameworks show that scaling is not just about raw hardware but about the software abstractions that allow humans to harness it. They are the bridge between theoretical strategies and practical execution.

Interdependence of Components

A unifying theme of this chapter is the interdependence of hardware, parallelism, resilience, and frameworks. None of these dimensions alone is sufficient.

- Hardware without parallelism wastes potential.
- Parallelism without resilience collapses under failure.

- Resilience without frameworks leaves users with complexity they cannot manage.
- Frameworks without hardware provide abstractions but no capacity.

Scaling requires the integration of all components. Conceptually, it is not a stack of independent layers but a tightly coupled system where each dimension enables and reinforces the others.

Broader Reflections

The conceptual lessons of scaling extend beyond the technical.

- **Accessibility:** The requirement for specialized hardware and sophisticated frameworks concentrates capability in organizations with resources. Scaling therefore has implications for who participates in AI research.
- **Economics:** Efficient provisioning and utilization translate directly into cost savings, shaping the economics of large-scale training.
- **Sustainability:** The energy demands of scaling raise questions of efficiency and environmental impact.
- **Innovation trajectory:** Many breakthroughs in AI have been enabled not by new algorithms but by scaling existing ones. The infrastructure of scaling has therefore directly shaped the trajectory of progress.

Scaling is not only a technical phenomenon but also a socio-technical one, influencing accessibility, economics, sustainability, and innovation.

Checkpointing and fault tolerance, frameworks, parallelism, and hardware provisioning together form the conceptual architecture of scaling. They make it possible to train models of unprecedented size and capability, transforming abstract algorithms into functioning systems.

The key insight is that scaling is not a single technique but a constellation of interconnected strategies. Hardware accelerates, parallelism divides, checkpointing preserves, and frameworks coordinate. Each plays a role, and together they enable deep learning to operate at the frontier of possibility.

As models continue to grow, scaling will remain the defining challenge of infrastructure in artificial intelligence. The principles explored in this chapter will not only continue to matter but will evolve, shaping the trajectory of AI for years to come.

In our next chapter, we cover fine-tuning and domain adaptation.

CHAPTER 7

Fine-Tuning and Domain Adaptation

Large-scale pre-trained models have transformed AI by demonstrating broad capabilities across language, vision, and multimodal tasks. Yet in practice, organizations and industries rarely need general intelligence in the abstract. They require models tailored to their specific domains, tasks, and constraints. Fine-tuning and domain adaptation provide the methods for bridging this gap, enabling general-purpose models to perform with precision in specialized contexts.

The challenge is that domain-specific adaptation must balance efficiency with effectiveness. Training from scratch is prohibitively expensive, while many domains lack the massive labeled datasets required for traditional supervised learning. Techniques of adaptation address this problem by reusing knowledge from pre-trained models and augmenting it with smaller amounts of domain-relevant data.

Several conceptual strategies define the space of fine-tuning and adaptation:

- **Transfer learning** leverages pre-trained representations and updates them partially or fully for a new task.

I. Cronin, *Building and Training Generative AI Models*,
https://doi.org/10.1007/979-8-8688-2332-9_7

- **Domain-specific dataset curation** ensures that training data reflects the target environment accurately, reducing the risk of bias or irrelevance.
- **Few-shot and zero-shot learning** explore how models can generalize with minimal or no labeled examples, often using in-context prompting or meta-learning.
- **Adapters and LoRA** introduce parameter-efficient fine-tuning methods, allowing models to specialize rapidly without retraining the entire network.

These approaches reflect a broader theme: that adaptation is about efficiency as much as performance. By selectively reusing and modifying general models, researchers and practitioners can achieve domain specialization without incurring the costs of building systems from scratch.

This chapter explores these techniques in detail, situating them conceptually within the broader evolution of training strategies. By the end, it will be clear that fine-tuning and domain adaptation are not peripheral adjustments but central methods for making AI usable across diverse industries and applications.

Transfer Learning and Pre-trained Model Utilization

Transfer learning has emerged as one of the most significant conceptual and practical innovations in modern machine learning. It captures the insight that knowledge acquired in one context can be reused in another, allowing systems to generalize more efficiently than if they had to learn from scratch every time. In practice, transfer learning is the reason large pre-trained models have become so useful, as they can be adapted to new domains and tasks with far less data and computational cost than traditional training pipelines.

This section develops a comprehensive account of transfer learning and pre-trained model utilization. It begins by examining the conceptual foundations of transfer learning and then traces its historical evolution from early feature extraction approaches to the dominance of pre-trained transformers. It then surveys different strategies for transfer learning, outlines the benefits, identifies challenges and risks, and explores conceptual nuances in how pre-trained models are used. The section concludes with reflections on the broader significance of transfer learning for the future trajectory of artificial intelligence.

Conceptual Foundations of Transfer Learning

Transfer learning rests on a simple conceptual principle: representations learned in one setting can serve as useful priors in another. Instead of learning entirely from scratch, models can reuse structure captured during prior training.

Core assumptions of transfer learning:

- **Shared structure across domains:** Even different tasks often involve similar patterns. For example, detecting edges in images is useful both for classifying animals and for analyzing medical scans.
- **Hierarchical features:** Early layers of deep networks tend to capture general features (edges, textures, phonemes), while later layers capture task-specific patterns.
- **Economy of learning:** It is inefficient to repeatedly learn general patterns when they can be reused.

This principle has analogues in human learning. A person who learns algebra can later learn physics more easily because algebraic reasoning transfers. Similarly, models that have learned general structures in one domain can leverage them elsewhere.

Historical Evolution: From Feature Extraction to Pre-trained Transformers

Transfer learning in machine learning has evolved significantly over the past two decades.

- **Early feature extraction:** Before deep learning, transfer learning often involved reusing handcrafted features across tasks. For example, features like SIFT or HOG, designed for vision, were reused across applications.
- **Convolutional neural networks and ImageNet:** A turning point came when convolutional nets trained on ImageNet were reused in medical imaging and other domains. Researchers froze early layers of ImageNet-trained networks and retrained only the final classifier. This provided strong performance even with limited domain-specific data.
- **Fine-tuning deep networks:** Instead of only using frozen features, researchers began fine-tuning entire networks pre-trained on large datasets. Fine-tuning allowed adaptation of representations to the new domain while retaining useful general priors.
- **Pre-trained language models:** In natural language processing, masked language modeling tasks such as those used in BERT showed that transfer learning could be operationalized through pre-training on unlabeled text and fine-tuning on downstream tasks. GPT-like models further demonstrated that transfer could occur even without explicit fine-tuning, through few-shot or zero-shot prompting.

- **Foundation models:** Transfer learning has now culminated in foundation models, pre-trained on massive corpora and adaptable across a wide range of tasks and domains.

This trajectory demonstrates the increasing centrality of transfer learning. From reusing shallow features to adapting multi-billion parameter transformers, transfer has become the dominant paradigm in modern AI.

Strategies for Transfer Learning

Transfer learning can be implemented in different ways, depending on computational resources, data availability, and domain characteristics.

Full Model Fine-Tuning

- The pre-trained model is used as initialization, and all parameters are updated during training on the new task.
- **Advantages:** Maximizes flexibility, as all layers can adapt to the new domain.
- **Illustration:** Fine-tuning BERT on a biomedical dataset, where all layers are updated, produces domain-specific representations suitable for tasks like medical question answering.
- **Disadvantages:** Computationally expensive and risks overwriting general knowledge.

Partial Fine-Tuning (Selected Layers)

- Only some layers are updated, often the later layers, while earlier layers remain frozen.
- **Advantages:** Reduces compute and prevents catastrophic forgetting of general features.
- **Illustration:** ImageNet-trained convolutional nets reused for satellite imagery, where early layers remain frozen to capture edges and textures, while later layers adapt to unique spatial patterns.
- **Disadvantages:** May limit flexibility if early representations are mismatched to the domain.

Frozen Backbones with Task-Specific Heads

- The backbone remains entirely frozen, and only a lightweight classifier or task-specific head is trained.
- **Advantages:** Extremely efficient in terms of compute and data requirements.
- **Illustration:** Using a frozen transformer for text embeddings and training only a shallow classifier for sentiment analysis.
- **Disadvantages:** Performance may plateau if domain mismatch is significant.

Each strategy represents a trade-off between efficiency and adaptability. Conceptually, they differ in how much prior knowledge is preserved versus how much is rewritten.

Benefits of Transfer Learning

Transfer learning offers several conceptual and practical benefits.

- **Efficiency:** Training from scratch requires enormous datasets and compute. Transfer learning reduces both by reusing prior representations.
- **Representation reuse:** Pre-trained models capture hierarchical features that are broadly applicable. For example, word embeddings capture semantic relations useful across NLP tasks.
- **Performance with limited data:** Fine-tuning enables strong performance even in data-scarce domains, such as rare disease classification in medical imaging.
- **Rapid prototyping:** Practitioners can adapt pre-trained models quickly to new tasks without constructing massive datasets.
- **Cross-domain utility:** Models pre-trained on general domains often provide useful priors for specialized contexts, such as finance, law, or science.

Conceptually, transfer learning democratizes access to advanced AI by making it possible to build strong models without enormous data and compute budgets.

Risks and Challenges

Despite its benefits, transfer learning introduces risks and challenges.

- **Negative transfer:** In some cases, prior knowledge hinders performance when the source and target domains are mismatched. For example, pretraining on everyday photos may not transfer effectively to medical radiographs. Mitigation strategies include choosing closer source domains, adding a small amount of in-domain pretraining data, using domain adaptation techniques (such as feature alignment or reweighting of source examples), and limiting how much of the network is updated at first so that only higher layers adapt to the new domain.

- **Catastrophic forgetting:** Fine-tuning all parameters risks overwriting useful general representations, leaving the model less effective outside the new domain. This can be reduced by freezing some backbone layers, using parameter-efficient methods such as adapters or LoRA so that most pretrained weights remain unchanged, adding regularization terms that penalize large deviations from the original parameters, or using continual learning approaches that replay a small buffer of original data during fine tuning. In practice, monitoring performance on both target and general benchmark tasks during training helps catch negative transfer or forgetting early and adjust the strategy accordingly.
- **Domain mismatch:** If pre-training data diverges significantly from the target domain, transferred representations may encode irrelevant biases.
- **Compute requirements:** Full fine-tuning of large models remains expensive, even if less costly than training from scratch.
- **Overfitting small datasets:** While transfer reduces data requirements, fine-tuning with very small datasets may still cause overfitting.

Conceptually, these risks highlight that transfer learning is not automatic. Success requires alignment between pre-training and target domains, careful choice of fine-tuning strategy, and vigilance against overwriting useful knowledge.

Conceptual Nuances in Pre-trained Model Utilization

Transfer learning is not only about reusing weights but about reinterpreting what pre-trained models represent.

- **Representations as general priors:** Pre-trained models capture statistical regularities of language, vision, or audio that serve as priors for downstream learning.
- **Transfer across modalities:** Vision-language models like CLIP illustrate that transfer can cross modalities, with image and text representations aligned in a shared space.
- **Multi-domain fine-tuning:** Models can be fine-tuned sequentially or jointly on multiple domains, creating systems with broader adaptability.
- **Task-specific specialization:** Even within a domain, transfer learning allows models to specialize, such as adapting general biomedical models to specific subfields like genomics.

These nuances underscore that transfer learning is a conceptual framework for reusing knowledge, not merely a mechanical process of re-initializing weights.

The Broader Significance of Transfer Learning

Transfer learning reflects a broader shift in AI from narrow, task-specific systems to general models that can be adapted across contexts.

- **Economies of scale in AI:** Large-scale pre-training amortizes the cost of training across countless downstream applications.
- **Democratization of access:** Organizations without resources to train massive models can still benefit through fine-tuning.
- **Trajectory toward foundation models:** Transfer learning is central to the emergence of foundation models, where a single pre-trained model serves as the basis for thousands of applications.
- **Acceleration of innovation:** By reducing barriers, transfer learning allows researchers to experiment and deploy more rapidly.

Conceptually, transfer learning unifies efficiency, scalability, and adaptability. It is not just a technique but a paradigm that defines the current era of artificial intelligence.

Transfer learning and pre-trained model utilization illustrate the principle that knowledge is reusable. From early convolutional nets adapted from ImageNet to transformers fine-tuned across domains, transfer has enabled models to achieve state-of-the-art performance with fewer resources. The strategies surveyed (full fine-tuning, partial fine-tuning, and frozen backbones) highlight different ways to balance efficiency with flexibility. The benefits are clear, but so are the challenges of negative transfer, catastrophic forgetting, and domain mismatch.

In the broader context of domain adaptation, transfer learning is the conceptual anchor. It enables the efficient reuse of general intelligence for specialized purposes, ensuring that the effort invested in large-scale pre-training benefits the widest range of applications.

Domain-Specific Dataset Curation

Large-scale pre-trained models demonstrate remarkable generality, but their true value emerges only when adapted to specialized applications. The process of adaptation hinges on data. While transfer learning and fine-tuning provide mechanisms for reusing representations, the effectiveness of these techniques depends on whether the data reflects the properties of the target domain. Curating high-quality domain-specific datasets is therefore one of the most critical steps in adapting AI systems to practical tasks.

This section explores the conceptual and practical foundations of domain-specific dataset curation. It begins with an explanation of why domain-specific datasets matter, followed by principles of curation. It then examines challenges of collection, annotation, and scale, explores workflows and engineering practices, considers synthetic augmentation, addresses ethical and regulatory issues, and closes with reflections on broader implications.

Pipeline

The pipeline is as follows:

Pre-training ➤ Domain Data Curation ➤ Fine-Tuning ➤ Evaluation and Iteration

Pre-training on broad data

- Start with a large foundation model trained on massive, heterogeneous corpora (text, images, code, etc.).

- The goal is to learn general representations and capabilities that are not tied to any single industry or task.

Domain analysis and requirements

- Identify the target domain (for example, oncology reports, legal contracts, industrial sensor logs) and the concrete tasks you care about: classification, retrieval, summarization, agent workflows, or structured prediction.
- Define success metrics and constraints early (accuracy, latency, privacy, regulatory limits).

Domain data curation

- Source raw domain data from logs, internal systems, partners, and public repositories.
- Clean and normalize the data; remove duplicates, fix obvious errors, standardize formats.
- Annotate where needed (labels, spans, entities, rationales), using domain experts or carefully guided annotators.
- Optionally enrich with synthetic examples generated from the base model, but keep a clear boundary between purely real and synthetic data.
- Split into training, validation, and test sets that respect temporal and distributional constraints so later evaluation is honest.

Fine-tuning loop

- Initialize the model with pre-trained weights and fine-tune on the curated domain dataset.

- Use appropriate strategies (full fine-tuning, parameter-efficient adapters, LoRA, or instruction tuning) depending on data volume and compute budget.
- Monitor training diagnostics: domain-specific validation metrics, loss curves, calibration, and safety constraints.
- Iterate on the data: when diagnostics reveal failure modes, return to the curation box to add or fix examples that address those weaknesses.

Evaluation and deployment readiness

- Evaluate on the held-out test set plus targeted stress tests: rare cases, edge conditions, adversarial or safety-critical scenarios.
- Run human-in-the-loop reviews with domain experts to judge factuality, reasoning quality, and compliance.
- Compare against baselines (generic pre-trained model, simpler models, or existing non-AI systems).
- If performance and risk are acceptable, promote the fine-tuned model into a controlled deployment; otherwise, loop back to data curation or fine-tuning.

Post-deployment monitoring and feedback

- Once deployed, log real-world usage under strict privacy and governance rules.
- Use this feedback to identify new failure modes and data gaps, feeding them back into the “Domain Data Curation” stage for the next training cycle.

Why Domain-Specific Datasets Matter

Pre-trained models are usually trained on massive datasets scraped from the Internet. These datasets contain broad coverage but limited specificity. When deployed in specialized contexts such as medicine, law, or finance, models trained only on generic corpora often struggle.

The mismatch arises because specialized domains have unique distributions.

- **Vocabulary:** Medical text uses specialized terminology that differs from everyday language.
- **Structure:** Legal documents exhibit rigid syntactic and rhetorical conventions.
- **Signal properties:** Financial time series contain dynamics not present in everyday datasets.

Without domain-specific datasets, models risk producing outputs that are irrelevant, misleading, or even harmful. For instance, a model trained on casual online text may fail to interpret medical abbreviations correctly, potentially introducing dangerous errors in clinical contexts.

Domain-specific curation anchors adaptation by ensuring that fine-tuning reflects the statistical and semantic patterns of the target field.

Historical Context of Dataset Curation

The importance of curation can be traced back through the history of machine learning.

- **Early benchmarks:** In natural language processing, corpora such as the Penn Treebank provided structured, curated datasets for syntactic analysis. In computer vision, MNIST offered a simple, well-defined dataset for digit recognition. These early datasets were small by today's standards but carefully designed.

- **Scaling benchmarks:** ImageNet marked a turning point. With over a million curated images labeled into thousands of categories, ImageNet provided not only scale but systematic taxonomy. Models pre-trained on ImageNet could be transferred effectively to other visual tasks.
- **Domain-specific corpora:** Specialized fields began to assemble their own datasets, such as PubMed abstracts for biomedical text and MIMIC for clinical records. These corpora allowed language models to adapt effectively to biomedical tasks.

This trajectory shows that curation is not new but has become increasingly central as models have grown. Without curated datasets, benchmarks, and domain corpora, the trajectory of transfer learning and adaptation would not have been possible.

Principles of Dataset Curation

Curation is not the same as collection. It involves deliberate selection, organization, and refinement.

Principles include

- **Relevance:** The dataset must reflect the distributions and conventions of the target domain.
- **Coverage:** It should include sufficient variability to capture the diversity of the domain.
- **Balance:** Class distributions should be monitored to prevent bias. For example, a medical dataset should not overrepresent one demographic group.

- **Quality:** Data must be accurate, consistent, and free from artifacts.
- **Traceability:** Metadata and provenance must be maintained, allowing users to understand origins, context, and any modifications applied.

Curation is best understood as an iterative process. Initial collection is followed by cleaning, validation, and refinement, ensuring that the dataset continues to serve the intended purpose.

Challenges in Collecting Domain Data

Collecting domain-specific data is rarely straightforward. Unlike Internet scraping, which can provide billions of examples, specialized data is scarce and constrained.

Challenges include

- **Scarcity:** Certain domains simply do not have large publicly available datasets. Medical imaging, for example, is tightly controlled.
- **Fragmentation:** Data may be scattered across multiple institutions or formats, requiring extensive integration.
- **Privacy restrictions:** Legal and medical data often contain sensitive personal information that must be anonymized or excluded.
- **Proprietary barriers:** In finance, much relevant data is owned by private firms and not publicly available.
- **Cost of access:** Specialized data such as genomic sequences or high-resolution scans is often expensive to acquire.

A technical illustration is electronic health records. These records contain invaluable information but are stored across fragmented systems, often with inconsistent formats. Curation requires harmonizing these heterogeneous sources while complying with privacy regulations.

Annotation and Labeling Concerns

Raw data is not always sufficient. Annotation provides the labels that transform raw information into supervised training sets.

Challenges include

- **Expertise requirements:** Only domain experts can annotate reliably in many fields. For instance, radiologists must label medical images, and lawyers must annotate contracts.
- **High costs:** Expert time is expensive, making large-scale annotation difficult.
- **Subjectivity:** Even experts may disagree. For example, pathologists may provide different assessments of the same slide.
- **Bias in labeling:** Annotation guidelines may systematically favor certain interpretations.
- **Granularity trade-offs:** Detailed annotations provide richer supervision but increase time and cost.

For example, annotating clinical notes for disease entities requires not only identifying mentions but linking them to standardized vocabularies. This granularity supports better training but is extremely resource-intensive.

Engineering Workflows for Dataset Curation

Curation requires structured workflows that ensure consistency and scalability.

Typical workflow stages include

- **Collection:** Aggregating raw data from available sources
- **Cleaning:** Removing duplicates, inconsistencies, and corrupted entries
- **Normalization:** Standardizing formats, units, and conventions
- **Annotation:** Adding labels or metadata where needed
- **Validation:** Checking data quality through expert review and statistical analysis
- **Versioning:** Maintaining distinct versions of datasets to ensure reproducibility
- **Continuous updating:** Incorporating new data while preserving consistency with past versions

Engineering practices such as automated pipelines and validation scripts are essential for managing datasets that may grow over time. Conceptually, workflows illustrate that curation is not a one-off activity but an ongoing process.

Balancing Scale and Relevance

One of the deepest conceptual challenges in dataset curation is the tension between scale and relevance.

- **Large-scale datasets** capture broad variation but may contain irrelevant or noisy samples.
- **Curated small datasets** provide high precision but risk overfitting and limited coverage.

Hybrid strategies often resolve this tension:

- Pre-training on large generic datasets followed by fine-tuning on smaller curated domain sets.
- Curriculum strategies where training begins with broad data and progresses to specific data.
- Augmentation techniques that expand small datasets without sacrificing domain relevance.

An illustrative case is legal NLP. A model pre-trained on billions of words of web text may be fine-tuned on a much smaller dataset of statutes and contracts, achieving domain-specific competence without losing the benefits of scale.

Synthetic Data Generation and Augmentation

When natural data is scarce, synthetic data provides a vital supplement.
Methods include

- **Augmentation of existing data:** Rotations, translations, or perturbations in vision; paraphrasing or back-translation in language.
- **Synthetic generation:** Using generative models to create new samples, such as GANs generating radiographs.

- **Simulation environments:** Robotics often relies on synthetic environments for pre-training before deployment in the real world.
- **Cross-domain augmentation:** Leveraging related domains. For instance, biomedical abstracts can augment clinical notes.

The critical issue is fidelity. Synthetic data must preserve the statistical properties of the domain. If synthetic medical images introduce unrealistic patterns, models may learn artifacts instead of meaningful features.

Ethical and Privacy Considerations

Domain-specific data often involves sensitive information, so ethical and regulatory concerns must be treated as first-class design constraints, not afterthoughts. Key safeguards include

- **Anonymization:** Remove or robustly mask personally identifiable information, and verify that records cannot be re-identified through linkage with auxiliary data.
- **Consent:** Ensure that data subjects have given explicit, appropriate consent for the ways in which their data will be used, stored, and shared.
- **Bias mitigation:** Recognize that domain data frequently encodes historical and structural inequities; actively audit for skewed representation and apply corrective strategies such as reweighting, resampling, or targeted data collection.
- **Regulatory compliance:** Align collection, storage, and processing with frameworks such as HIPAA for healthcare, GDPR in Europe, and any relevant sector-specific or regional regulations.

- **Transparency and documentation:** Attach rich metadata that describes provenance, sampling methods, known limitations, and potential biases.

For example, in financial datasets, transaction histories must be carefully anonymized to prevent re-identification of individuals while still preserving the statistical structure needed for modeling. Similar care is required in medical imaging, educational records, or mobility traces, where even apparently harmless attributes can become identifying when combined. To make these practices concrete, teams can adopt tools such as "Datasheets for Datasets" and "Model Cards," which provide structured templates for documenting how data was collected, what it contains, who it represents, and how models trained on it should and should not be used. By embedding this kind of documentation into the engineering workflow, organizations turn abstract ethical principles into operational artifacts that can be reviewed, versioned, and audited over time.

Broader Implications of Domain Curation

Dataset curation has consequences beyond technical performance.

- **Accessibility:** Well-curated datasets make domain adaptation feasible for organizations without vast resources.
- **Innovation:** New applications become possible when high-quality domain data exists.
- **Trustworthiness:** Curated datasets improve reliability by ensuring outputs align with domain expectations.
- **Reproducibility:** Documented datasets enable scientific comparisons and replication.
- **Societal impact:** The biases present in curated datasets can shape societal outcomes if not addressed carefully.

Conceptually, domain-specific dataset curation is not merely a preparatory chore. It defines the boundaries of what is possible in adapting models. Curation sits one level above familiar activities such as annotation and dataset expansion. Annotation focuses on attaching labels or metadata to individual examples. Dataset expansion focuses on adding more examples through scraping, partnerships, or synthetic generation. Curation decides which sources to draw from, which subsets to keep or discard, how much weight each group should have, and which annotations and expansions are actually appropriate for the target use case. In other words, annotation and expansion are tools; curation is the process that decides when and how those tools should be used to shape a coherent corpus.

Domain-specific dataset curation is therefore the linchpin of fine-tuning and adaptation. It transforms general models into domain specialists by providing the data context that grounds their representations. The process involves more than simple collection: it requires principled selection, annotation strategies guided by domain goals, careful use of expansion techniques, balancing of scale and relevance, and attention to ethical and regulatory constraints.

The historical trajectory from early benchmarks like Penn Treebank and MNIST to specialized corpora like PubMed and MIMIC shows that curation has always been central to progress. The engineering workflows that sustain curation illustrate its iterative and ongoing nature. Synthetic data methods provide solutions to scarcity, while ethical considerations ensure trust and compliance.

In sum, domain-specific dataset curation is both a technical and conceptual requirement. It ensures that the broad generality of pre-trained models can be harnessed for the specificity demanded by real-world tasks. Without intentional curation, adaptation lacks grounding and direction. With it, models become precise tools capable of meaningful contributions across medicine, law, finance, and many other sectors.

Few-Shot and Zero-Shot Learning

Few-shot and zero-shot learning represent some of the most transformative developments in the history of AI. They challenge the long-standing assumption that every new task requires extensive labeled data. Instead, they allow models to adapt flexibly, with minimal or no supervision. These capabilities are central to the utility of foundation models, where general pre-training enables downstream application across countless domains.

This section expands on the conceptual foundations, historical trajectory, strategies, benefits, challenges, and future implications of few-shot and zero-shot learning. Particular emphasis will be placed on zero-shot learning, as it represents the most striking emergent property of large-scale pre-trained models.

Conceptual Foundations of Few-Shot and Zero-Shot Learning

Both few-shot and zero-shot learning rest on the principle that knowledge can be generalized across tasks. Instead of training bespoke models for every problem, a single system can adapt its internal representations.

- **Few-shot learning:** A handful of labeled examples are provided for the new task. The model must extrapolate from this tiny set, often fewer than ten per class.
- **Zero-shot learning:** No labeled examples are available. The task is described in natural language or by semantic attributes, and the model must generalize directly.

The key insight is that pre-trained models contain structured representation spaces. These spaces encode features broad enough to make rapid generalization possible. In practice, adaptation often requires nothing more than conditioning on a few examples or interpreting a prompt.

Historical Context: From Meta-Learning to In-Context Learning

The path to modern few-shot and zero-shot learning spans decades of research.

- **Meta-learning traditions:** Early work focused on training models to "learn how to learn." Algorithms such as model-agnostic meta-learning (MAML) trained networks across tasks so that they could rapidly adapt to new ones with few examples. Metric-based approaches like prototypical networks classified new samples by comparing them to prototype embeddings in a learned space.
- **Bayesian approaches:** Some research pursued probabilistic frameworks for few-shot learning, treating tasks as distributions and leveraging prior knowledge to inform new task inference.
- **Zero-shot vision with attributes:** Before transformers, vision researchers experimented with describing unseen classes using semantic attributes. For example, a zebra could be classified as an animal with "black and white stripes" even if no zebra images were present in training.

- **Multilingual transfer:** Cross-lingual embeddings demonstrated early zero-shot capabilities by allowing knowledge to transfer across languages without parallel corpora.
- **Transformers and in-context learning:** The most dramatic leap came with pre-trained transformers. GPT-style models showed that simply conditioning on prompts could elicit both few-shot and zero-shot performance, without additional training.
- **Multimodal breakthroughs:** Models like CLIP trained on paired text and images made zero-shot classification across modalities practical at scale.

This history shows how the field transitioned from specialized architectures to emergent generalization in large-scale pre-trained models.

Few-Shot Learning Strategies

Few-shot learning remains relevant, particularly when even minimal supervision improves performance in specialized tasks.

Fine-Tuning with Small Labeled Sets

- A common strategy is to fine-tune pre-trained models with only a handful of domain-specific examples.
- Effective when domain distributions differ significantly from the pre-training corpus.
- **Example**: Fine-tuning a pre-trained vision model on a few medical X-ray images to adapt from natural images to radiographic structures.

Metric-Based Few-Shot Learning

- Prototypical networks and similar architectures classify by comparing embeddings of examples.
- Few-shot classification involves embedding new samples and finding the closest class prototype.
- **Example**: Classifying bird species with only a few labeled examples by embedding them in a learned metric space.

In-Context Few-Shot Learning

- Large language models can adapt simply by being shown examples in a prompt.
- **Example**: Providing three labeled sentiment examples in a prompt before asking the model to classify a new sentence.
- This approach requires no parameter updates, relying instead on inference-time conditioning.

Few-shot learning demonstrates that small amounts of supervision, when combined with broad pre-training, suffice for strong adaptation.

Zero-Shot Learning Strategies

Zero-shot learning illustrates the most striking capabilities of foundation models. It allows tasks to be performed without labeled data, guided only by descriptions or semantic mappings.

Prompt-Based Generalization

- Tasks can be specified in natural language.
- **Example**: "Summarize this paragraph in one sentence" directs the model without needing task-specific fine-tuning.
- The prompt acts as an instruction, aligning model outputs with human intent.

Pre-training Alignment

- Zero-shot performance arises from diverse pre-training data.
- Language models trained on large corpora internalize a broad range of tasks implicitly.
- Vision-language models align images and captions, enabling cross-modal generalization.

Cross-Modal Zero-Shot Learning

- CLIP aligns images and text in a shared embedding space.
- A zero-shot classifier can be built by embedding candidate labels as text prompts and comparing them with image embeddings.
- **Example**: Classifying an image of a panda by comparing its embedding with text descriptions like "a picture of a panda" versus "a picture of a cat."

Zero-shot learning demonstrates that models can leverage descriptive mappings to generalize beyond explicit training.

Mathematical Intuition Behind Few-Shot and Zero-Shot

The effectiveness of these approaches can be understood conceptually through representation spaces.

- **Embedding structure**: Pre-training arranges inputs into high-dimensional spaces where semantically similar items cluster together.
- **Few-shot adaptation**: A handful of labeled points in this space act as prototypes for each class. New examples are classified by how close they lie to these prototypes, so even a few labeled samples can shape a useful decision boundary.
- **Zero-shot mapping**: Natural language descriptions of tasks and labels are embedded into the same space. The task becomes aligning new samples with these descriptive anchors, so the model can perform a task it was never explicitly trained on as long as the description lands in the right region of the space.
- **Similarity metrics**: Distances such as cosine similarity provide a quantitative way to compare positions in this space, enabling generalization from minimal supervision by choosing the label whose prototype or description is closest.

In practice, this "geometry of knowledge" is probed by broad evaluation suites such as BIG-Bench, MMLU, and HELM, which test how well models can perform new tasks in few-shot or zero-shot settings without task-specific fine-tuning. Collectively, they reveal how much useful structure pre-training has carved into the representation space and how effectively models can navigate it with only a small amount of additional guidance.

Engineering Workflows for Few-Shot and Zero-Shot Evaluation

Practitioners need systematic workflows to evaluate and deploy these methods.

- **Dataset sampling**: Few-shot experiments require carefully sampled support sets so that each class is represented fairly and results are not artifacts of a particular random draw.
- **Prompt libraries**: Zero-shot evaluation benefits from curated prompt templates and label verbalizations that reduce sensitivity to wording, allowing more stable comparisons across models and tasks.
- **Evaluation harnesses**: Multi-task evaluation frameworks, including the benchmark suites above, provide standardized protocols for running few-shot and zero-shot tests at scale, logging scores, and tracking regressions over time.
- **Uncertainty estimation**: Calibration techniques, such as temperature scaling or conformal prediction, help interpret model confidence in low-data regimes where errors are costly.

- **Iteration loops**: Prompts, example sets, and task descriptions are iteratively refined based on evaluation results, forming feedback loops that improve performance without retraining the underlying model.

These workflows illustrate that while few-shot and zero-shot capabilities emerge naturally, engineering discipline is essential for reliable application.

Benefits and Applications

The advantages of few-shot and zero-shot learning extend across domains.

- **Data efficiency:** Reduce the need for costly annotation.
- **Rapid deployment:** Models can be adapted in hours rather than months.
- **Cross-lingual capacity:** Zero-shot translation allows transfer across languages without parallel corpora.
- **Emergent tasks:** Systems can handle new problems without retraining.

Applications:

- **Healthcare:** Zero-shot triage systems interpret clinical notes for emergent conditions without labeled data.
- **Law:** Few-shot adaptation supports clause detection in contracts with only a handful of examples.
- **Finance:** Zero-shot models summarize quarterly reports into investor briefings.
- **Robotics:** Zero-shot vision-language models help robots interpret commands in natural language.

- **Humanitarian response:** Zero-shot classification supports crisis monitoring in regions without annotated datasets.

These examples illustrate the broad applicability of few-shot and zero-shot methods.

Challenges and Risks

The flexibility of few-shot and zero-shot methods introduces unique risks that must be managed deliberately.

- **Hallucination**: Models may produce fluent but incorrect or fabricated outputs. Mitigation includes grounding responses in trusted retrieval sources, constraining generation to verified knowledge bases, and exposing confidence indicators or citations so users can judge reliability.
- **Prompt sensitivity**: Small changes in prompt phrasing can lead to large swings in performance. To reduce this, teams can maintain tested prompt libraries, perform prompt ensembling (querying with several variants and aggregating), and run automated prompt regression tests whenever prompts or model versions change.
- **Bias amplification**: Zero-shot models inherit and sometimes magnify biases from their pre-training corpora. Mitigation requires bias audits on representative tasks, counterfactual testing (swapping sensitive attributes), debiasing prompts, and where possible adding targeted fine-tuning data that corrects harmful patterns.

- **Domain drift**: Performance deteriorates when the target domain differs from the data seen during pre-training. This can be managed with domain-specific prompt tuning, lightweight adaptation (for example, LoRA on a small curated corpus), and continuous evaluation on fresh, domain-relevant samples.
- **Underspecified tasks**: Zero-shot prompts may be too vague, producing unpredictable or inconsistent behavior. Clearer task instructions, explicit constraints on style and content, and example-based prompting (moving from pure zero-shot to few-shot) all help reduce ambiguity.
- **Adversarial prompting**: Malicious or carefully crafted prompts can elicit harmful, private, or policy-violating outputs. Guardrails such as input filtering, content classifiers on both prompts and responses, and multi-stage safety checks are needed to harden systems against these attacks.
- **Evaluation challenges**: Without labeled data, it is difficult to know how well zero-shot systems are performing. In practice, teams rely on a mix of small labeled evaluation sets, human-in-the-loop review, and proxy metrics such as calibration error or agreement with retrieval evidence.

Explainability and Uncertainty

Mitigation is stronger when systems expose not only outputs but also reasons and confidence. Few-shot and zero-shot deployments benefit from

- Lightweight explainability, such as showing which retrieved passages supported an answer, highlighting key tokens that influenced a classification, or surfacing the intermediate chain-of-thought in a controlled, summarized way for internal evaluation. This helps practitioners detect prompt failures, bias, and hallucinations earlier.
- Uncertainty quantification, for example, by using probability scores, confidence buckets, or agreement among multiple sampled outputs or models. Low-confidence regions can trigger fallbacks such as asking the user for clarification, escalating to a human expert, or refusing to answer.

Together, these practices turn the risk list into an operational playbook: hallucinations, prompt fragility, and bias are treated not as unavoidable quirks but as failure modes that can be monitored, explained, and constrained in real deployments.

Case-Style Illustrations Across Domains

Concrete examples clarify how few-shot and zero-shot approaches function in practice.

- **Medical imaging:** Few-shot fine-tuning of vision models allows classification of rare diseases with only a handful of annotated scans.
- **Legal analysis:** Zero-shot prompts such as "Does this clause impose liability on the tenant?" enable contract review without training examples.
- **Financial time series:** Few-shot adaptation supports anomaly detection with minimal historical labels.

- **Robotics:** Zero-shot grounding of commands such as "Pick up the red object" leverages multimodal embeddings.
- **Humanitarian data:** Zero-shot models classify crisis reports in previously unseen languages, aiding real-time response.

Each case demonstrates the principle that minimal or no supervision can suffice when general-purpose representations exist.

Future Trajectories

Research continues to push the boundaries of few-shot and zero-shot learning.

- **Instruction tuning:** Aligning models with natural language instructions enhances zero-shot reliability.
- **Reinforcement learning with human feedback:** Human-guided fine-tuning improves alignment and reduces harmful outputs.
- **Mixtures of experts:** Architectures that dynamically route tasks to specialized experts enhance zero-shot versatility.
- **Benchmarking frameworks:** Comprehensive evaluations across tasks and domains help track progress.
- **Uncertainty calibration:** Methods to quantify and communicate confidence will be critical for safe deployment.
- **Cross-modal generalization:** Expanding zero-shot capabilities across modalities, such as video and 3D environments.

Conceptually, these directions suggest that few-shot and zero-shot learning are not end states but stepping stones toward increasingly general and reliable learners.

Few-shot and zero-shot learning exemplify the evolution of AI toward greater flexibility. Few-shot methods demonstrate that a handful of examples can guide adaptation, whether through fine-tuning, metric learning, or in-context prompting. Zero-shot methods go further, allowing tasks to be performed with no supervision, relying only on descriptive prompts or aligned embeddings.

The benefits are transformative: efficiency, accessibility, and adaptability. The risks are equally important: hallucination, bias, prompt sensitivity, and domain dependence. Yet the conceptual significance is clear. Zero-shot learning in particular shows that broad pre-training can approximate general intelligence, enabling systems to perform tasks they were never explicitly trained for.

As research advances, the combination of instruction tuning, reinforcement learning with feedback, and cross-modal alignment promises to make zero-shot capabilities more robust. Few-shot learning will remain valuable in specialized contexts, but zero-shot will continue to define the trajectory of foundation models. In this sense, these approaches are not just techniques but paradigms, marking the transition from narrow training toward universal adaptability.

Use of Adapters and LoRA

Adapters and Low-Rank Adaptation (LoRA) represent two of the most important advances in parameter-efficient fine-tuning. They allow large-scale pre-trained models to be adapted to new domains and tasks without retraining or updating the entire parameter set. Instead, adaptation is achieved through small, additional modules (in the case of adapters) or low-rank modifications to weight matrices (in the case of LoRA). Both approaches make domain adaptation more efficient, cost-effective, and scalable.

This section examines the conceptual foundations, engineering practices, benefits, challenges, and future directions of adapters and LoRA. While both methods are significant, greater emphasis will be placed on LoRA, which has become the dominant method for adapting modern foundation models in practice.

Motivation for Parameter-Efficient Fine-Tuning

The need for parameter-efficient fine-tuning arises from the scale of modern models. Models with billions or even trillions of parameters cannot be fully fine-tuned for every domain or task.

Challenges of full fine-tuning include

- **Compute costs:** Updating billions of parameters requires enormous hardware resources.
- **Storage overhead:** Maintaining separate fine-tuned models for each task is infeasible, as each copy requires storing the entire parameter set.
- **Training time:** Full fine-tuning is time-consuming, making rapid adaptation impractical.
- **Catastrophic forgetting:** Updating all parameters risks overwriting general representations acquired during pre-training.

Beyond the technical advantages, parameter-efficient fine-tuning has a clear business motivation. It allows startups, smaller research labs, and domain teams inside large enterprises to customize frontier models without owning massive GPU clusters or duplicating multi-billion-parameter checkpoints for every use case. Instead of spinning up separate full-scale models for legal, medical, and customer-support assistants, an organization can maintain a single base model and attach lightweight adapter or LoRA modules per domain, lowering infrastructure costs while speeding up experimentation. In this sense, adapters and LoRA sit within the same family of parameter-efficient techniques: adapters add small task-specific layers on top of frozen weights, while LoRA injects low-rank updates directly into existing weight matrices. Together they make it practical to treat the foundation model as shared "platform capital" and the small adaptation modules as cheap, swappable assets that encode business- or domain-specific behavior.

Parameter-efficient methods address these challenges by introducing minimal changes that adapt the model while preserving the majority of its pre-trained parameters. This allows

- Faster adaptation
- Reduced storage needs
- Retention of general pre-training knowledge
- Scalability to multi-domain environments

The motivation is both technical and economic: efficient fine-tuning enables practical deployment of foundation models across industries.

Early Approaches to Modular Adaptation

Before adapters and LoRA, researchers explored other strategies for modular adaptation.

- **Task-specific heads:** Adding shallow classification layers on top of frozen backbones allowed models to adapt cheaply, but flexibility was limited.

- **Partial fine-tuning:** Updating only later layers reduced compute costs but performance could be inconsistent.
- **Soft prompts:** Prompt tuning added trainable vectors to the input embedding space. While parameter-efficient, early approaches lacked robustness.

These methods established the principle that adaptation does not require updating all parameters. However, they did not achieve the balance of efficiency and performance now seen in adapters and LoRA.

Conceptual Foundations of Adapters

Adapters introduce small bottleneck layers within each transformer block. The pre-trained parameters are frozen, and only the adapter weights are trained.

Conceptual principles of adapters:

- **Modularity:** Each adapter is a self-contained module added to the network.
- **Bottleneck design:** Adapters compress representations into a smaller dimension and then project them back.
- **Isolation:** Pre-trained parameters remain unchanged, ensuring general knowledge is preserved.
- **Multi-task capability:** Different adapters can be swapped in for different tasks without retraining the base model.

By adding only a few million parameters, adapters allow models with billions of weights to adapt flexibly.

Engineering Practices in Adapter-Based Fine-Tuning

Adapters require careful design and integration.

- **Placement**: Adapters can be inserted at different points within transformer blocks, such as after feedforward or attention layers, and their position influences both what they can specialize and how much extra compute they introduce.
- **Dimension choices**: The size of the bottleneck dimension controls trade-offs between efficiency and capacity; smaller bottlenecks reduce parameters and memory, while larger ones allow richer adaptation at the cost of extra computation.
- **Training dynamics**: Adapters typically require fewer epochs to converge than full fine-tuning, which reduces training time and makes rapid iteration across domains more practical.
- **Management**: Multiple adapters can be trained and stored, enabling modular switching across domains; at inference time, only the active adapter blocks are loaded, keeping memory usage bounded and avoiding the need to maintain separate full copies of the base model.

In terms of inference latency, adapters usually introduce only a modest overhead because they are small compared to the main transformer layers. In well-designed systems, the added latency is often a few percent relative to the base model. Latency can increase, however, if adapters are inserted at many layers or if several are composed for a single request, so practitioners typically benchmark different placements and bottleneck sizes to find a configuration that meets both accuracy and serving-time requirements.

Conceptual Foundations of LoRA

Low-Rank Adaptation (LoRA) takes a different approach. Instead of adding new modules, LoRA re-parameterizes weight updates as low-rank decompositions.

Key ideas:

- **Weight update re-parameterization**: A large weight matrix W is adapted not by full updates but by learning two smaller matrices A and B such that the update is $\Delta W = B\,A$.
- **Low-rank design**: By constraining the rank of the update, the number of trainable parameters is drastically reduced.
- **Frozen base weights**: The original weights remain unchanged, preserving pre-trained knowledge.
- **Additivity**: LoRA updates can be stored separately and combined additively, allowing efficient management of multiple adaptations.

Intuitively, "low-rank" means that instead of learning an entirely new, dense matrix of changes, the model only learns a small set of directions in weight space that matter most for the new task. In high-dimensional transformer layers, many directions are redundant, so a low-rank update can still capture the dominant task-specific adjustments while using a tiny fraction of the parameters. This is why LoRA can match or even exceed full fine-tuning performance while remaining lightweight.

LoRA was introduced by Hu et al. (2021), who showed that these low-rank updates are sufficient to adapt large transformers efficiently across a range of NLP tasks. Since then, variants such as QLoRA (which combines low-rank updates with weight quantization) and DoRA (which refines how the low-rank factors interact with base weights) have extended the idea, pushing parameter-efficient fine-tuning to even larger models and more constrained hardware settings.

LoRA in Practice: Efficiency and Flexibility

LoRA has become the dominant method for parameter-efficient fine-tuning because it balances efficiency with performance. In industry, it is now embedded in widely used tooling such as the PEFT library from Hugging Face, which standardizes how LoRA adapters are defined, trained, stored, and attached to a wide range of backbone models. This ecosystem support has dramatically lowered the barrier for startups, research labs, and internal platform teams to run many domain-specific fine-tunes on top of a single shared base model.

Advantages:

- **Parameter reduction**: LoRA can reduce trainable parameters by orders of magnitude compared to full fine-tuning.
- **Storage efficiency**: LoRA adapters are small, enabling storage of many domain-specific updates.
- **Training speed**: Fewer parameters to update means faster training.
- **Preservation of generality**: Frozen base weights ensure broad knowledge is retained.
- **Composability**: Multiple LoRA updates can be merged or combined at inference time.

Engineering practices:

- **Rank selection**: The rank of low-rank updates controls capacity. Small ranks reduce cost but may underfit complex domains.
- **Target layers**: LoRA updates are typically applied to attention and projection layers, where adaptation is most effective.
- **Deployment workflows**: LoRA modules can be hot swapped at inference, enabling dynamic adaptation across contexts. In practice, teams either (1) keep LoRA weights separate and apply them on the fly per request, or (2) periodically "merge" selected LoRA adapters into a frozen copy of the base weights for a given product surface, reducing runtime overhead and simplifying serving. Frameworks like PEFT provide utilities for both modes, including safe merge operations, versioning of adapters, and configuration files that describe which LoRA stacks are active for a particular deployment.

LoRA therefore provides a principled framework for adapting models in ways that are both efficient and scalable, while the surrounding tooling ecosystem makes it operationally tractable to manage dozens or hundreds of specialized adapters across environments and products.

Comparative Reflections: Adapters vs. LoRA

Both adapters and LoRA achieve parameter-efficient adaptation, but they differ conceptually and practically.

- **Adapters**: Modular add-ons that sit alongside existing layers. Good for multi-task switching.

- **LoRA**: Low-rank updates embedded within weight matrices. Good for efficient storage and composability.
- **Flexibility**: Adapters provide clear modular separation, while LoRA integrates more tightly with weights.
- **Efficiency**: LoRA often achieves higher parameter savings without loss of accuracy.
- **Deployment**: Adapters require maintaining multiple modules, while LoRA allows additive merging of updates.

In practice, LoRA has become more widely adopted, but adapters remain useful where modularity and interpretability are priorities. Emerging hybrid approaches combine the two, for example by placing small adapters in selected layers and applying LoRA inside their weights. This lets teams keep the clean task separation of adapters while gaining the extreme parameter savings and mergeability that LoRA provides.

Case-Style Applications Across Domains

Both methods have seen deployment across diverse industries.

- **Healthcare:** LoRA fine-tuning allows large language models to adapt to medical notes while keeping updates small and secure.
- **Law:** Adapters provide modular domain switches for contract analysis versus litigation support.
- **Finance:** LoRA enables efficient adaptation for summarizing financial reports and detecting anomalies.
- **Science:** Adapters help models specialize in subfields such as genomics or chemistry.

- **Conversational AI:** LoRA modules allow chatbots to switch between corporate tone, customer service, and informal interaction.

These cases illustrate that parameter-efficient adaptation is not only a theoretical advance but a practical necessity for applying foundation models.

Challenges and Limitations

Despite their success, adapters and LoRA present challenges.

- **Capacity limits**: Parameter-efficient methods may underperform when tasks require extensive reconfiguration. For example, LoRA with a very low rank can underfit complex cross-modal tasks such as video–text grounding or multi-document reasoning, where the model must coordinate many interacting features across modalities.
- **Hyperparameter sensitivity**: Choices like bottleneck size for adapters or rank for LoRA strongly affect outcomes, and poor settings can erase most of the efficiency gains or lead to unstable training.
- **Management complexity**: Large collections of adapters or LoRA updates require systematic naming, versioning, and dependency tracking, especially in organizations that support dozens of domains or customers.
- **Evaluation gaps**: Few benchmarks exist specifically for parameter-efficient adaptation, making it difficult to compare methods or detect subtle regressions in domain performance.

- **Security risks**: Swappable modules raise the risk that malicious or unverified updates are introduced into production systems, requiring signing, sandboxing, and rigorous review processes.

Conceptually, these limitations highlight the balance between efficiency and flexibility. They also point to several research directions: automatic rank selection for LoRA based on task complexity, adaptive adapter sizing and pruning strategies, better diagnostics for detecting underfitting in rich domains, and standardized benchmarks that capture the unique strengths and failure modes of parameter-efficient fine-tuning.

Future Directions in Parameter-Efficient Adaptation

Research continues to extend the principles of adapters and LoRA.

- **Hybrid approaches:** Combining adapter modules with LoRA updates to exploit the strengths of both.
- **Dynamic routing:** Systems that activate different adaptation modules based on task or input.
- **Compression-aware designs:** Further reducing memory overhead while maintaining accuracy.
- **Evaluation frameworks:** Developing benchmarks specifically for parameter-efficient methods.
- **Cross-modal adaptation:** Further extending LoRA principles to vision, speech, and multimodal systems.
- **Federated settings:** Secure deployment of small LoRA modules in distributed environments without sharing full models.

These directions suggest that parameter-efficient adaptation will remain a major research frontier.

Adapters and LoRA embody the principle that adaptation does not require retraining or updating every parameter. Adapters achieve this by inserting modular bottleneck layers, while LoRA achieves it by re-parameterizing weight updates as low-rank decompositions. Both approaches make fine-tuning efficient, scalable, and practical.

While adapters remain valuable, LoRA has emerged as the dominant method, combining efficiency, composability, and performance. Its ability to preserve pre-trained knowledge while introducing minimal updates makes it especially suited to foundation models at scale.

Together, adapters and LoRA illustrate that the future of domain adaptation lies not in full retraining but in parameter-efficient innovation. They are not merely technical tricks but conceptual milestones, redefining how models can be specialized across countless tasks and industries.

Conclusion

Fine-tuning and domain adaptation represent the mechanisms by which general-purpose AI becomes usable in specific contexts. While foundation models demonstrate broad capabilities, their true value lies in their ability to specialize. This chapter has traced the conceptual, technical, and practical strategies that enable specialization, from transfer learning and dataset curation to few-shot and zero-shot learning, and finally to parameter-efficient methods such as adapters and LoRA. Together, these strategies show how models can retain the generality of large-scale pre-training while adapting flexibly to the requirements of particular tasks and industries.

The Centrality of Adaptation

A recurring theme of this chapter is that adaptation is not a peripheral detail but the defining challenge of applied AI. Pre-trained models encode broad statistical knowledge, but without adaptation, they risk irrelevance in specialized domains. Fine-tuning allows this knowledge to be aligned with domain-specific data, ensuring that outputs are contextually appropriate.

- Transfer learning illustrated how pre-trained representations can be reused across tasks.
- Domain-specific dataset curation emphasized the importance of high-quality, relevant data.
- Few-shot and zero-shot learning showed that generalization is possible even when labeled data is scarce or unavailable.
- Adapters and LoRA demonstrated that adaptation can be made efficient and scalable, preserving general knowledge while enabling specialization.

Each strategy contributes a piece of the adaptation puzzle, but LoRA in particular marks a conceptual shift. Instead of treating adaptation as a full re-optimization of all parameters, LoRA reframes it as finding a low-dimensional "direction" in weight space that captures what is unique about a task or domain. This perspective separates the heavy, expensive part of the model (the frozen base weights) from the light, expressive part (the learned low-rank updates). Conceptually, it suggests that much of what distinguishes one domain from another can be encoded in a compact subspace, rather than requiring wholesale re-learning of the entire network.

Parameter-efficient fine-tuning more broadly, and LoRA in particular, also reshapes the economics of AI. When only large technology companies could afford to train and store many full copies of giant models, experimentation and domain adaptation were structurally limited. PEFT techniques make it possible for startups, research labs, and even individual teams inside larger organizations to carry out serious adaptation work on modest hardware and with manageable storage footprints. Instead of maintaining dozens of monolithic models, they can keep a single shared base model and swap in lightweight adapters or LoRA modules for different clients, regions, or regulatory regimes. In that sense, PEFT is as much a business innovation as a technical one: it lowers the barrier to entry and enables a marketplace of specialized models built on top of a few shared foundations.

At the same time, important open challenges remain. We still lack principled methods for choosing ranks, bottleneck sizes, and insertion points automatically, which means PEFT often relies on heuristic choices and empirical search. The interaction between multiple adaptations—such as stacking or merging several LoRA modules, or combining adapters with LoRA in hybrid schemes—raises unresolved questions about interference, stability, and evaluation. Robust benchmarking for parameter-efficient methods is only beginning to emerge, making it hard to compare techniques across modalities and domains. Finally, governance and safety questions are amplified when many small, swappable modules can be created and distributed independently of the base model.

In sum, fine-tuning and domain adaptation turn general models into domain specialists by providing the data, objectives, and parameterization needed for targeted refinement. LoRA and other PEFT methods show that this refinement can be both efficient and widely accessible, rather than the exclusive province of a few actors with vast compute budgets. The future of applied AI will depend on how well the community addresses the remaining challenges: automating design choices, understanding interactions among multiple adaptations, building better benchmarks, and establishing safeguards for an ecosystem where anyone can plug new behaviors into powerful shared models.

Transfer Learning As the Foundation

The chapter began with transfer learning, the conceptual anchor for modern adaptation. By reusing representations learned in one context, models avoid the inefficiency of learning from scratch. Historical progress from early convolutional nets to large-scale transformers highlighted how transfer learning matured into a paradigm. The principle is simple but powerful: knowledge is reusable, and general pre-training provides a foundation upon which specialized tasks can be built.

Transfer learning also clarified the risks of adaptation, such as negative transfer and catastrophic forgetting. These risks underscored the importance of aligning pre-training and target domains, setting the stage for the role of curated datasets.

Dataset Curation As the Anchor

Adaptation requires data that reflects the distributions of the target environment. The discussion of dataset curation emphasized that collecting, cleaning, annotating, and balancing domain-specific data is one of the most critical steps in adaptation.

- In medicine, curated imaging datasets ensure that models interpret radiographs accurately.
- In law, corpora of contracts and statutes provide the foundation for legal reasoning.
- In finance, transaction records and market data ground models in economic reality.

Dataset curation is both technical and ethical. It requires careful attention to privacy, fairness, and bias. It also requires engineering workflows for continuous updating, versioning, and validation. Without curated datasets, adaptation risks drifting into irrelevance. With them, adaptation is precise, reliable, and trustworthy.

Few-Shot and Zero-Shot Learning As Emerging Paradigms

Few-shot and zero-shot learning revealed the most striking emergent properties of large-scale pre-training. Few-shot learning showed that only a handful of labeled examples can suffice for adaptation, whether through fine-tuning, metric-based learning, or in-context prompting. Zero-shot learning went further, showing that tasks can be performed with no supervision, guided only by natural language descriptions or semantic attributes.

The significance of zero-shot learning is profound. It demonstrates that large-scale pre-training encodes general-purpose representations broad enough to support task generalization without explicit labels. This transforms how models can be deployed in practice.

- **Healthcare applications** include zero-shot triage of clinical notes.
- **Legal applications** include clause interpretation in contracts.
- **Multilingual applications** include translation without parallel corpora.

At the same time, the risks of hallucination, prompt sensitivity, and bias showed that these methods require careful safeguards. Few-shot and zero-shot learning expand the scope of adaptation, but they also demand new forms of monitoring and evaluation.

Parameter-Efficient Fine-Tuning As a Practical Breakthrough

The final section explored parameter-efficient methods, focusing on adapters and LoRA. These approaches address the practical reality that full fine-tuning of billion-parameter models is infeasible for most organizations.

Adapters demonstrated how modular bottleneck layers can be inserted into pre-trained networks, enabling task-specific adaptation without retraining the base model. LoRA demonstrated how low-rank updates can achieve even greater efficiency, preserving base parameters while capturing task-specific adjustments.

Together, these methods made fine-tuning scalable. They allowed multiple domain-specific adaptations to coexist without duplicating entire models. LoRA in particular emerged as the dominant approach, striking a balance between efficiency, composability, and performance.

Integration and Interdependence

Perhaps the most important conceptual lesson of this chapter is that these strategies are not isolated. They form an interdependent system of adaptation.

- Transfer learning provides the foundation.
- Dataset curation ensures alignment with domain distributions.
- Few-shot and zero-shot methods extend generalization into low-data regimes.
- Adapters and LoRA make adaptation efficient and scalable.

Adaptation succeeds not because of any one technique but because of the orchestration of all these methods. They interact to ensure that foundation models can be specialized quickly, accurately, and sustainably.

Broader Reflections

Fine-tuning and domain adaptation have implications beyond technical performance.

- **Economics:** Parameter-efficient methods reduce costs, making AI accessible to more organizations.
- **Accessibility:** Zero-shot and few-shot capabilities democratize AI by reducing dependence on labeled datasets.
- **Ethics:** Dataset curation raises questions of fairness, bias, and privacy.
- **Society:** The ability to adapt AI rapidly to new tasks has consequences for industries from medicine to law to finance.

These reflections underscore that adaptation is not only a technical challenge but a socio-technical one. The ways in which models are adapted shape who benefits from AI, how trustworthy it becomes, and how it integrates into human institutions.

Fine-tuning and domain adaptation define the practical reality of modern AI. They ensure that general-purpose models are not abstract curiosities but usable systems. Transfer learning, dataset curation, few-shot and zero-shot learning, and parameter-efficient methods together form a comprehensive framework for adaptation.

The trajectory of AI research suggests that adaptation will only grow in importance. Models will continue to increase in size and generality, but their utility will always depend on their ability to specialize. Adapters and LoRA will evolve into more advanced forms of parameter-efficient fine-tuning. Few-shot and zero-shot learning will become more reliable through instruction tuning and feedback alignment. Dataset curation will remain a central concern, shaping not only technical outcomes but ethical ones.

In sum, fine-tuning and domain adaptation represent the bridge between the general and the specific. They ensure that the vast potential of foundation models can be directed toward real-world impact. They are not merely techniques but the conceptual architecture that makes AI adaptable, usable, and relevant across diverse domains.

For our next chapter, we focus on Reinforcement Learning with Human Feedback (RLHF).

CHAPTER 8

Reinforcement Learning with Human Feedback (RLHF)

The rapid expansion of large-scale pre-trained models has produced systems of remarkable capability, but raw power does not guarantee usefulness. A model trained only to predict the next token or minimize reconstruction error often produces outputs that are technically correct but misaligned with human expectations, intent, or values. Reinforcement Learning with Human Feedback (RLHF) emerged as a method for addressing this gap. It provides a systematic way to take broad general-purpose models and align their behavior with what humans actually prefer.

At its core, RLHF integrates human judgment into the training loop. Human annotators evaluate candidate outputs, ranking them or scoring them according to quality, relevance, or alignment with given instructions. These annotations are then distilled into a reward model, which acts as a proxy for human preference. Using reinforcement learning algorithms such as Proximal Policy Optimization (PPO), the base model is fine-tuned to maximize the reward predicted by this model. In effect, RLHF transforms subjective evaluations into an optimization signal that guides model behavior.

I. Cronin, *Building and Training Generative AI Models*,
https://doi.org/10.1007/979-8-8688-2332-9_8

The importance of RLHF lies in its ability to bridge three spaces: pre-training objectives, human preferences, and downstream applications. Pre-training provides models with knowledge and fluency, but not necessarily desirable behavior. Human feedback encodes normative signals, though it is costly and imperfect. RLHF combines the two, creating models that are not only powerful but also more aligned with user intent.

This chapter explores the key components of RLHF: the design of human annotation pipelines, the construction and training of reward models, the use of PPO as the reinforcement learning algorithm, and the challenges inherent in collecting and interpreting feedback. By examining both the conceptual foundations and the practical engineering details, the chapter shows how RLHF has become the defining method for aligning foundation models with human expectations, and why it continues to play a central role in shaping their reliability and safety. Reinforcement learning with PPO is chosen over simple supervised fine-tuning because it can directly optimize model behavior with respect to a learned reward signal over multi-step interactions, rather than merely imitating static examples of human-written responses.

Human Annotation Pipelines

RLHF begins not with algorithms but with people. Human annotation pipelines are the foundation upon which the entire process rests. Without high-quality human judgments, reward models have no meaningful signal to learn from, and reinforcement learning reduces to optimizing against arbitrary criteria. Annotation pipelines transform subjective human evaluations into structured data that can be used to align models.

This section examines the principles, engineering practices, challenges, and implications of building human annotation pipelines for RLHF. It begins with the role of human preference data, then discusses annotation formats, recruiting and training annotators, interface design, quality control mechanisms, and scaling approaches. Along the way, it highlights both successes and difficulties encountered in real-world pipelines, emphasizing the centrality of human judgment in shaping model behavior.

The Role of Human Preferences in RLHF

The motivation for annotation pipelines is clear: supervised objectives cannot fully capture what humans want. A language model trained to predict the next token may generate grammatically correct sentences but fail to provide relevant, helpful, or safe answers. Human preferences introduce the normative layer absent from statistical prediction.

- **Judgment as supervision**: Instead of predicting the next token, the model is guided by judgments of quality and alignment.
- **Subjectivity as signal**: Human evaluations reflect values and intent, which cannot be encoded directly in objective functions.
- **Feedback as optimization target**: By converting subjective evaluations into reward signals, RLHF makes alignment trainable.

In this sense, annotation pipelines are the bridge between human values and machine learning. They operationalize alignment by making preferences measurable, while also aggregating disagreement across diverse annotators. In practice, this means intentionally recruiting labelers with different backgrounds, modeling each annotator's reliability, and using aggregation schemes (such as pairwise preference models or weighted voting) that can reconcile conflicting signals rather than simply averaging them away. Some systems even maintain multiple reward models for different user groups or regions, so that the underlying diversity of preferences is preserved instead of forced into a single global notion of "good" behavior.

Annotation Formats: Rankings and Ratings

The design of annotation tasks determines the quality of feedback. Two main formats dominate RLHF pipelines: pairwise ranking and direct scoring.

- **Pairwise ranking**: Annotators are shown two or more model outputs and asked to choose which is better. This format is robust because relative judgments are easier than absolute ones, and pairwise comparisons can be aggregated into global preference models using methods like the Bradley-Terry model. However, pairwise ranking does not scale well when the number of candidate outputs grows large, since the number of comparisons increases rapidly and can produce annotator fatigue or inconsistent choices over time.
- **Direct scoring**: Annotators assign a numeric score to each output. This provides more granular data but is more subjective and inconsistent, since scores may vary depending on annotator calibration or mood.
- **Hybrid methods**: Some pipelines combine rankings and ratings, asking annotators to both compare outputs and score them. This can improve data richness but also increases cognitive load, so careful interface design and quality checks are needed to keep labels reliable.

- Ranking has become the dominant choice in RLHF because it minimizes variance in human responses. It reduces the cognitive burden on annotators by asking simpler questions: which of these is better?

Recruiting and Training Annotators

The reliability of feedback depends on who provides it. Annotation pipelines require careful recruitment and training of annotators.

- **Crowdsourcing platforms**: Many RLHF projects use crowd workers recruited from online platforms. This allows large-scale annotation at lower cost, with rapid turnaround and diverse perspectives across cultures and user types, but quality and consistency can vary and require careful filtering, calibration tasks, and ongoing auditing.
- **Domain experts**: Some tasks require specialized knowledge, such as medicine or law, where only experts can provide valid judgments. Expert labels are usually slower and more expensive to obtain, but they offer higher reliability, better grounding in established standards, and clearer justifications for edge cases.
- In practice, many teams use a hybrid approach: crowdsourced annotators handle broad, generic instructions and safety checks, while expert annotators focus on high-risk or domain-critical examples, and their judgments are used to calibrate or audit the larger pool of crowd-sourced feedback.

Training is critical. Annotators must learn

- **Task guidelines:** Clear instructions for what constitutes quality, helpfulness, or safety
- **Consistency practices:** How to apply criteria consistently across outputs
- **Bias awareness:** Recognizing and avoiding systematic bias in evaluations

For example, OpenAI's InstructGPT pipeline included detailed training materials for annotators, with examples of good and bad responses. Without such preparation, annotation quality deteriorates rapidly.

Annotation Interfaces and Workflow Design

The interface through which annotators interact with model outputs shapes the feedback collected.

- **Simplicity:** Interfaces must minimize cognitive load. Clear side-by-side displays for rankings are effective.
- **Guided annotation:** Tooltips, definitions, and examples embedded in the interface help ensure consistency.
- **Metadata collection:** Interfaces often log response times, annotator IDs, and contextual information to support quality analysis.
- **Iteration cycles:** Interfaces can be updated as annotation guidelines evolve.

Workflow design includes batching tasks, randomizing output order to reduce bias, and ensuring annotators remain engaged without fatigue. Annotation pipelines must be treated as human-in-the-loop systems, where user experience design is as critical as algorithmic design.

Quality Control and Inter-Annotator Agreement

High-quality annotations are not guaranteed. Quality control mechanisms ensure that feedback reflects genuine, consistent judgments.

- **Gold standard checks**: Pre-labeled examples are inserted to measure annotator reliability.
- **Consensus measures**: Inter-annotator agreement is tracked to identify inconsistent or noisy judgments.
- **Weighting schemes**: Annotations from more reliable annotators can be weighted more heavily.
- **Redundancy**: Multiple annotators evaluate the same outputs, reducing the influence of individual biases.
- **Feedback loops**: Annotators receive feedback on their work to improve over time.

When annotators disagree, RLHF pipelines typically combine these signals using simple majority vote for straightforward tasks, weighting votes by demonstrated expertise or reliability for specialized domains, and, in more advanced setups, probabilistic fusion methods that jointly model annotator skill and task difficulty. For example, Anthropic's Constitutional AI pipeline used inter-annotator agreement metrics to refine their feedback collection and aggregation strategy, helping ensure that model alignment was not driven by inconsistent or low-quality signals.

Scaling Annotation Efforts

Annotation pipelines must balance quality with scalability.

- **Crowdsourcing at scale:** Large numbers of annotators provide breadth but increase the challenge of consistency.

- **Expert bottlenecks:** In specialized domains, the scarcity of experts limits annotation throughput.
- **Hierarchical workflows:** Some pipelines use general annotators for broad tasks and experts for critical review.
- **Synthetic augmentation:** AI-assisted annotation, where models pre-label outputs and humans refine them, reduces cost.
- **Active learning:** Models can prioritize uncertain or ambiguous outputs for annotation, maximizing information gained per annotation.

These strategies allow pipelines to scale without collapsing under the cost of human labor. However, they also raise new questions about the reliability of semi-automated feedback.

Case Illustration: InstructGPT

OpenAI's InstructGPT project provides a clear example of annotation pipelines in practice.

- Annotators were asked to rank outputs from different models for quality and alignment.
- Guidelines emphasized helpfulness, honesty, and harmlessness.
- Large numbers of pairwise comparisons were collected and used to train reward models.

This pipeline demonstrated that human feedback could produce models that were consistently preferred by users, even when they performed worse on traditional benchmarks. It highlighted the central role of annotation in shaping behavior.

Challenges and Trade-Offs in Annotation Pipelines

Building annotation pipelines involves trade-offs that shape model outcomes.

- **Cost vs. quality:** High-quality expert annotations are expensive, while low-cost crowdsourcing risks noise.
- **Scale vs. depth:** Large-scale annotation provides breadth but may miss subtlety; deep expert annotation provides nuance but lacks coverage.
- **Speed vs. reliability:** Rapid annotation cycles accelerate training but increase the risk of inconsistency.
- **Normativity vs. diversity:** Pipelines must balance universal alignment with respect for cultural and individual differences.

These trade-offs reflect the complexity of embedding human values into machine learning.

Emerging Alternatives to Traditional Pipelines

As annotation pipelines grow, researchers are exploring alternatives to reduce cost and increase scalability.

- **Synthetic feedback:** Models trained on human annotations can generate additional labels.
- **AI-assisted annotation:** Pre-trained models can filter or suggest labels, with humans refining outputs.

- **Preference distillation:** Feedback collected in one domain can be distilled and reused in another.
- **Constitutional feedback:** Instead of relying solely on human annotators, models can be guided by predefined principles or rules.

These approaches reduce dependence on large-scale human annotation, but they also raise questions of fidelity and oversight.

Human annotation pipelines are the bedrock of RLHF. They transform subjective judgments into structured data, enabling models to optimize toward human preferences. The design of annotation formats, recruitment and training of annotators, interface design, quality control mechanisms, and scaling strategies all shape the quality of alignment.

While annotation pipelines face challenges of cost, bias, and scalability, they remain indispensable. Emerging alternatives may supplement them, but human judgment will continue to play a central role in aligning AI. Annotation pipelines are not mere infrastructure; they are the human layer that ensures AI reflects the preferences, values, and norms of those it is meant to serve.

Reward Modeling

Reward modeling is the conceptual heart of RLHF. It translates subjective human judgments into quantitative signals that can be used to optimize large-scale models. Without reward modeling, human annotations remain fragmented evaluations, useful in isolation but impossible to scale. With reward modeling, individual judgments are aggregated and generalized into a structured function that approximates human preference. This allows reinforcement learning to proceed with a target that, while imperfect, captures what humans value in model outputs.

This section explores the motivation, foundations, methods, challenges, and future trajectories of reward modeling. It traces historical influences, explains how human judgments are converted into reward functions, discusses scaling practices, examines risks of overfitting and reward hacking, and analyzes challenges such as bias and misalignment. It also considers alternatives and extensions such as preference distillation and AI-assisted feedback. Throughout, the emphasis is on conceptual clarity and technical intuition, with grounded illustrations from summarization, dialogue, and instruction following.

Why Reward Modeling Matters in RLHF

At its core, reinforcement learning requires a reward signal. In traditional reinforcement learning domains such as games or robotics, the reward is clearly defined: a game score, a time-to-completion metric, or a distance to a goal provides an unambiguous measure of success. Language models, however, lack such natural rewards. There is no objective number that measures the helpfulness of a response or the safety of an explanation, so reward signals must be learned from human preferences. This introduces a new generalization challenge: reward models can overfit to the specific prompts and outputs seen during training, rewarding stylistic quirks or surface patterns rather than genuine quality. When that happens, the policy learns to "game" the reward model instead of truly aligning with human intent, producing answers that score well according to the learned reward but feel repetitive, shallow, or subtly misaligned to real users.

Reward modeling fills this gap by creating a surrogate reward function derived from human preferences.

- **Human preference as reward:** Annotators judge which outputs are better, and these judgments are generalized into a predictive model.

- **Alignment through approximation:** The reward model becomes a stand-in for what humans value, guiding optimization.
- **Scalability:** Once trained, the reward model can evaluate millions of outputs without requiring continuous human input.

Without reward modeling, RLHF would collapse into direct human-in-the-loop reinforcement learning, which is impractical at scale. Reward models thus operationalize the idea of alignment: they make human values computable.

Historical Roots of Reward Modeling

The concept of reward modeling has roots in earlier traditions of preference learning and human-in-the-loop optimization.

- **Preference learning in recommendation systems:** Early recommender models learned from pairwise comparisons, predicting which movies or products a user would prefer.
- **Inverse reinforcement learning:** Researchers explored how to infer reward functions from observed behavior, particularly in robotics.
- **Dialogue systems:** Human ratings of dialogue quality were used to train models to generate more natural conversations.
- **Comparative evaluation methods:** Studies showed that humans are more reliable at ranking outputs than at assigning absolute scores, a principle later adopted in RLHF.

Reward modeling builds on these traditions, adapting them to the scale of modern foundation models. The novelty lies not in the basic idea but in its application to billions of parameters and highly subjective domains like helpfulness and safety.

From Human Judgments to Reward Functions

Reward modeling begins with human judgments, which must be converted into structured training signals.

- **Annotation input:** Annotators compare outputs, indicating which they prefer.
- **Preference aggregation:** These comparisons are aggregated to capture broader patterns of agreement.
- **Model training:** A smaller neural network is trained to predict preferences given new outputs.
- **Reward proxy:** The trained network acts as a proxy for human judgment, assigning scalar values to outputs.

Conceptually, the reward model approximates a hidden function: the mapping from outputs to human preference. This mapping is noisy and subjective, but through aggregation and modeling, it becomes a useful guide for optimization.

Pairwise Comparisons and the Bradley–Terry Framework

Pairwise comparisons are the most common format for training reward models. Annotators choose which of two outputs is better, and these binary choices are aggregated into preference models.

The conceptual justification is simple: relative judgments are more reliable than absolute scores.

- **Ease of evaluation:** Humans find it easier to say which response is better than to assign an exact score.
- **Consistency across annotators:** Pairwise judgments reduce calibration issues between different annotators.
- **Aggregatable data:** Pairwise comparisons can be aggregated into global rankings.

This principle aligns with psychological research showing that humans are more consistent in comparative evaluations than in absolute judgments. Reward models leverage this natural human tendency to create more reliable training data. In the Bradley–Terry model, each response is assigned a latent "quality" score, and the probability that response A is preferred over response B is modeled as exp(s_A) / (exp(s_A) + exp(s_B)), where s_A and s_B are their respective scores. By adjusting these scores so that the predicted probabilities match the observed pairwise choices across many comparisons, the model converts local, pairwise preferences into a coherent global ranking over all candidate responses.

Training Reward Models at Scale

Scaling reward models requires significant engineering.

- **Data collection:** Thousands or millions of comparisons must be gathered to train robust models.
- **Model architecture:** Reward models are typically smaller than the base model but large enough to capture nuanced preferences.

- **Training regimes:** Reward models are trained on preference datasets with careful regularization to prevent overfitting.
- **Evaluation:** Held-out preference data and agreement with annotators provide validation.
- **Deployment:** Once trained, reward models can be applied automatically to new outputs, enabling large-scale reinforcement learning.

Scaling transforms reward modeling from a conceptual framework into a practical system capable of guiding billion-parameter models.

Overfitting and Reward Hacking

Reward models are not perfect proxies for human values. Optimizing against them directly can lead to overfitting and reward hacking.

- **Overfitting**: The reward model may latch onto superficial patterns in the training data (phrases, length, style) that correlate with good judgments in the annotation set but do not reflect deeper quality. When the policy is optimized against this model, it learns to reproduce those patterns rather than genuinely helpful behavior, so performance degrades on new, unseen prompts.
- **Reward hacking**: The policy model may systematically exploit blind spots in the reward model, producing outputs that score well according to the learned reward but are misaligned with what humans actually want. This happens because the policy is trained on a narrower distribution of pairwise comparisons or scored examples, then deployed on a much broader,

open-ended space of prompts. Under this distribution shift, the policy can find strange corners of output space where the reward model is overconfident yet wrong, for example, outputs that are excessively long, formulaic, or evasively "safe" while failing to answer the question. The policy is effectively searching the input space for loopholes in the reward function.

- **Examples of reward hacking**: Verbose answers that repeat the question and add boilerplate to appear more thorough; hedged or content-free replies that sound cautious and safe but avoid giving a clear answer; stylistically polished text that hides factual errors the reward model does not penalize strongly.
- **Mitigation strategies**: Regularization of both the reward model and policy (e.g., KL penalties to stay close to a supervised baseline), adversarial testing that intentionally probes for reward-model failures, and iterative retraining where new human feedback is collected on hacked or low-quality outputs and folded back into the reward model. Ensembles of reward models and cross-checks with separate safety or factuality checks can further reduce single-model blind spots.

Conceptually, reward hacking arises because a reward model is only an approximation of human judgment defined on a limited dataset, while RL optimization is powerful enough to discover edge cases where that approximation breaks. This tension between a finite, imperfect proxy and a highly capable optimizer is at the heart of why aligning models with human preferences remains challenging.

Reward Model Generalization Across Domains

Reward models trained on one domain may not generalize well to others.

- **Domain shift:** A model trained on general dialogue may not capture preferences in technical or specialized contexts.
- **Cultural variation:** Preferences differ across cultural and linguistic groups. Reward models trained on one group may misalign with another.
- **Task-specific preferences:** Summarization, dialogue, and reasoning tasks each require different evaluative criteria.

Addressing these challenges requires either training multiple domain-specific reward models or designing models capable of handling diverse domains simultaneously.

Case Illustrations: Summarization, Dialogue, and Instruction Following

Concrete tasks illustrate the role of reward modeling.

- **Summarization:** Annotators rank summaries by relevance, clarity, and conciseness. Reward models trained on these judgments guide models to produce higher-quality summaries.
- **Dialogue:** Annotators rank conversational responses by helpfulness, honesty, and harmlessness. Reward models capture these values and guide chatbots toward more aligned behavior.

- **Instruction following:** Annotators rank responses to instructions based on how well they satisfy the user's intent. Reward models generalize these judgments to unseen instructions.

In many deployments, these reward signals are combined into a multi-objective reward function, for example, weighting clarity, correctness, and creativity differently depending on the application. This allows designers to explicitly trade off properties such as safe, literal summaries versus more imaginative or stylistically rich outputs, rather than optimizing for a single objective that may not reflect real user preferences.

Iterative Refinement of Reward Models

Reward modeling is rarely a one-time process. Models must be refined iteratively.

- **Feedback loops**: As policies improve, new outputs are generated and annotated. These annotations improve the reward model.
- **Active learning**: Models can identify uncertain or contentious cases for annotation, maximizing information gain from limited human effort.
- **Continuous updates**: Reward models evolve as user preferences shift or expand, incorporating new norms, use cases, and safety requirements.
- **Human-in-the-loop cycles**: Humans remain central, providing ongoing judgments that guide refinement and catch emerging failure modes.

Iterative refinement serves a dual purpose: it stabilizes a core notion of alignment so that behavior does not drift unpredictably, while at the same time allowing the system to gradually evolve toward current user needs and societal expectations rather than freezing a single, static value system in place.

Challenges: Noise, Bias, and Misalignment

Reward modeling faces fundamental challenges.

- **Noisy annotations:** Human judgments are inconsistent and subject to fatigue.
- **Bias in data:** Annotators may reflect cultural or demographic biases.
- **Misalignment:** Reward models may drift away from actual human values, particularly under domain shift.
- **Complexity of values:** Human preferences are multidimensional and context-dependent, difficult to capture in scalar rewards.

These challenges illustrate the limits of modeling subjective values. They remind us that reward models are approximations, not ground truth.

Alternatives and Extensions: Preference Distillation, AI-Assisted Feedback, Constitutional Signals

Researchers are exploring alternatives and supplements to traditional reward modeling.

- **Preference distillation**: Compressing knowledge from multiple annotators or multiple reward models into a single, distilled reward function. In practice, this often looks like an ensemble or teacher–student setup, where several "teacher" reward models (trained on different subsets, objectives, or annotator groups) score candidate outputs, and a smaller "student" reward model is trained to match their aggregate preferences. This process smooths out individual biases, captures a richer notion of quality, and produces a lighter-weight reward model that is cheaper to deploy at scale.
- **AI-assisted feedback**: Using models to pre-score or filter outputs before human review, reducing cost by ensuring that annotators focus only on the most informative or ambiguous examples.
- **Constitutional signals**: Replacing some human feedback with rules or principles that encode normative guidance, allowing models to self-critique and self-revise without relying on constant human labeling.

These extensions reduce dependence on costly annotation pipelines and broaden the space of alignment methods.

Future Directions in Reward Modeling

Reward modeling is still an emerging science. Future research directions include

- **Multidimensional rewards:** Moving beyond scalar values to capture trade-offs between helpfulness, safety, and creativity

- **Cultural adaptability:** Designing reward models that adapt to diverse cultural and linguistic preferences
- **Robustness:** Ensuring models resist reward hacking and generalize across domains
- **Hybrid systems:** Combining human judgments, constitutional rules, and AI-assisted feedback into unified reward frameworks
- **Scalable pipelines:** Automating annotation and reward modeling workflows for ever-larger models

These directions suggest that reward modeling will remain a central research frontier for years to come.

Reward modeling is the mechanism that makes RLHF possible. It translates human judgments into structured signals that can guide reinforcement learning at scale. While imperfect, reward models capture enough of human preference to produce substantial improvements in alignment.

The process involves collecting annotations, aggregating preferences, training predictive models, and refining them iteratively. It faces challenges of noise, bias, and misalignment, but it also demonstrates remarkable progress in translating subjective values into trainable objectives.

Reward modeling is not only a technical tool but a conceptual milestone. It embodies the idea that human preferences, though messy and subjective, can be approximated by machine learning systems and used to shape the behavior of foundation models. As models continue to grow in scale and influence, reward modeling will remain indispensable to the project of aligning artificial intelligence with human values.

Proximal Policy Optimization (PPO)

RLHF depends on an optimization algorithm that can balance stability, efficiency, and fidelity to human preferences. Among the many reinforcement learning algorithms developed, PPO has become the standard choice. PPO provides the stability required to fine-tune massive language models without destructive divergence, while remaining efficient enough to train in large-scale industrial pipelines.

This section examines the conceptual foundations, historical context, algorithmic principles, and engineering practices of PPO as applied to RLHF. It explains why PPO is favored over alternative reinforcement learning algorithms, describes how PPO interacts with reward modeling and pre-trained policies, and considers challenges such as stability, computational cost, and interpretability. It also explores applications in instruction following, dialogue, and summarization and concludes with reflections on future directions for optimization in alignment research.

Why PPO Matters in RLHF

Training with RLHF requires updating large models in response to reward signals derived from human preferences. This is a delicate process. If updates are too aggressive, the model may drift away from the distribution it was pre-trained on, losing fluency and coherence. If updates are too weak, the model fails to incorporate preferences effectively. PPO provides a principled way to strike this balance.

- **Stability**: PPO introduces mechanisms that prevent updates from straying too far from the original policy.
- **Scalability**: PPO can handle large-scale training, making it suitable for billion-parameter models.

- **Compatibility**: PPO integrates naturally with neural network architectures, including transformers.
- **Empirical success**: Experiments consistently show that PPO produces stable, aligned improvements in model behavior.

In contrast, supervised fine-tuning simply treats preferred responses as fixed targets and optimizes them token by token, which cannot directly enforce global preference signals such as brevity, tone, or safety and offers no explicit control over how far the new policy drifts from the base model. PPO instead views the model as a policy optimized for long-term reward, combining preference-driven improvement with constraints on policy change. Without PPO or an equivalent reinforcement learning method, RLHF would be far more prone to instability, either drifting away from pre-trained capabilities or remaining too close to the base model to meaningfully reflect human feedback.

Historical Context of PPO

PPO emerged from a long lineage of policy gradient methods in reinforcement learning.

- **Policy gradient origins**: Early methods directly optimized expected rewards with respect to policy parameters. These methods were powerful but unstable.
- **Trust Region Policy Optimization (TRPO)**: TRPO introduced the idea of constraining updates so that the new policy does not diverge too far from the old one. While effective, TRPO was computationally expensive and complex to implement.

- **PPO as simplification**: PPO simplified the trust region idea by using a clipped objective that approximates the constraint without heavy computation. The adoption of PPO in RLHF reflects this history. It combines the stability of TRPO with the practicality required for large scale model training. As a result, PPO gained strong industrial traction because it is straightforward to implement, produces relatively reproducible learning curves, and aligns well with efficient batched execution on modern GPUs.

Conceptual Principles of PPO

The conceptual core of PPO lies in constraining policy updates. PPO balances two competing forces:

- **Exploration:** The policy must adjust to optimize reward signals, incorporating human preferences.
- **Conservation:** The policy must remain close to its pre-trained distribution to retain fluency and coherence.

To achieve this, PPO uses mechanisms that

- **Clip updates:** Prevent policy ratios from straying too far, ensuring gradual changes.
- **Penalize divergence:** Impose costs when the updated policy deviates excessively from the reference distribution.
- **Encourage learning stability:** Create an optimization landscape where improvements are incremental rather than volatile.

Conceptually, PPO can be seen as a stabilizing lens, focusing the reinforcement signal into manageable adjustments that large models can absorb.

Integration with Pre-trained Policies

A distinctive feature of RLHF is that optimization begins not from scratch but from a pre-trained model. PPO must therefore operate within the context of massive pre-existing knowledge.

- **Frozen priors:** Pre-training provides fluency and world knowledge that must not be destroyed.
- **Kullback–Leibler (KL) control**: PPO uses a KL penalty to discourage large deviations from the base policy, helping preserve useful pre training behavior, but if this constraint is set too strongly it can over-restrict policy updates and suppress the reward-driven learning signal.
- **Preference incorporation:** PPO integrates reward signals on top of this base, steering the model toward alignment without erasing prior capability.

This integration is critical. PPO is not optimizing raw exploration as in games or robotics. It is refining a model that already performs well, nudging it toward preferred behaviors.

Engineering Practices in PPO for Language Models

Applying PPO to large-scale models requires careful engineering.

- **Batching strategies:** PPO requires collecting batches of model outputs and rewards before updating. Efficient batching is essential at scale.
- **Reward normalization:** Rewards are normalized to prevent instability from extreme values.
- **KL balancing:** A KL divergence penalty balances reward optimization with preservation of pre-training.
- **Gradient clipping:** Prevents excessively large updates that destabilize training.
- **Adaptive learning rates:** Learning rates are tuned dynamically to maintain stability.
- **Checkpoints:** Frequent saving ensures recovery from instability.

These practices turn PPO from a theoretical algorithm into a practical training loop that can manage billion-parameter transformers.

PPO in Summarization and Dialogue Alignment

Concrete applications illustrate how PPO improves alignment.

- **Summarization:** PPO adjusts models to prefer summaries ranked higher by humans. The result is more concise, accurate, and readable outputs.
- **Dialogue:** PPO fine-tunes chatbots to prefer responses judged as more helpful, honest, and harmless. Over time, dialogue becomes more aligned with user expectations.

- **Instruction following:** PPO refines models to respond to prompts more faithfully, integrating reward signals into coherent task execution.

These applications show PPO's flexibility in handling diverse tasks guided by human preferences.

PPO Compared to Alternative Algorithms

Why PPO rather than other reinforcement learning methods?

- **Q-learning and variants:** Value-based methods like Q-learning are less compatible with high-dimensional text generation.
- **Actor-critic methods:** PPO belongs to this family but introduces stability improvements absent in simpler versions.
- **TRPO:** While theoretically elegant, TRPO is computationally intensive, unsuitable for billion-parameter models.
- **PPO advantages:** Simplicity, efficiency, and empirical stability make PPO the most practical choice for RLHF.

Conceptually, PPO represents the balance point in the design space of reinforcement learning algorithms: stable enough to work, simple enough to scale.

Challenges of PPO in RLHF

Despite its success, PPO is not without challenges.

- **Computational cost:** PPO remains expensive, requiring large-scale infrastructure.

- **Reward hacking:** Policies can exploit weaknesses in reward models even under PPO.
- **KL tuning:** Balancing KL penalties is delicate; too strong and learning stalls, too weak and divergence occurs.
- **Sample efficiency:** PPO requires many samples, increasing annotation and compute burdens.
- **Interpretability:** PPO updates are difficult to interpret, obscuring why policies change as they do.

These challenges show that PPO is not a perfect solution but a pragmatic compromise.

Future Directions Beyond PPO

While PPO dominates RLHF today, research explores alternatives and extensions.

- **Adaptive algorithms:** Methods that dynamically adjust constraints rather than using fixed clipping.
- **Model-based RL:** Leveraging predictive models to reduce sample inefficiency.
- **Hybrid approaches:** Combining supervised fine-tuning, RLHF, and rule-based constraints.
- **Preference optimization without RL:** Some researchers explore direct preference optimization as an alternative to reinforcement learning.
- **Scaling considerations:** As models grow, new optimization strategies may be required to maintain efficiency.

Future work may retain PPO as a baseline while evolving beyond it to address its limitations.

Broader Conceptual Significance of PPO

PPO represents more than an algorithm. It embodies a conceptual principle: alignment requires controlled optimization.

- **Constraint as virtue**: PPO demonstrates that limiting change is as important as pursuing reward.
- **Balance of forces**: PPO formalizes the tension between pre training generality and alignment specificity.
- **Scalability of alignment**: PPO proves that reinforcement learning can be scaled to billion-parameter models with careful design.

Conceptually, PPO is the algorithmic foundation that makes RLHF possible at scale. PPO is the optimization engine of RLHF. It balances exploration and conservation, reward pursuit and stability, alignment and preservation. By constraining updates while enabling progress, PPO allows models to incorporate human preferences without losing the fluency and knowledge of pre training.

Its success reflects both historical evolution and pragmatic design. From unstable policy gradients to complex trust region methods, reinforcement learning has long struggled with stability. PPO resolves this struggle with a simple yet powerful mechanism, making it the backbone of modern alignment pipelines.

Challenges remain, including computational cost, reward hacking, and interpretability. Yet PPO's conceptual lesson is enduring: alignment is not about maximal optimization but about careful, constrained adaptation. In this sense, PPO embodies the philosophy of RLHF itself, integrating human preference into machine learning through controlled, scalable processes. A concrete illustration comes from OpenAI's Learning to Summarize with Human Feedback work, where policies trained with very weak KL control pushed the model far from the reference distribution, achieving higher reward model scores but collapsing into shorter and less informative summaries that human evaluators actually liked less than those from more constrained policies, showing that relaxing PPO's constraints can cause behavioral collapse even while the nominal optimization objective appears to improve.

Challenges in Feedback Collection

The central promise of RLHF lies in making human preferences computable. Yet the effectiveness of this promise depends entirely on the collection of feedback. If feedback is unreliable, inconsistent, or biased, the entire RLHF pipeline is compromised. The collection of human judgments is therefore not only a logistical challenge but also a conceptual one, requiring systems that can translate subjective, context-dependent evaluations into data robust enough to guide billion-parameter models.

This section explores the major challenges of feedback collection in RLHF. It considers the economic cost of annotation, the difficulty of ensuring consistency, the risks of bias, the challenges of scaling across cultures and languages, the problem of annotation noise, and the risks of over-optimization to imperfect signals. It also examines emerging alternatives such as synthetic feedback, AI-assisted feedback, and constitutional frameworks, analyzing their potential to supplement or partially replace direct human annotation.

The Cost of Feedback Collection

Feedback collection is costly at scale. Unlike raw data scraping, which can be automated, preference annotations require human labor.

- **Monetary cost:** Large RLHF pipelines require thousands or millions of annotations. Paying human workers fairly creates significant expense.
- **Time cost:** Annotation takes time, especially for complex tasks such as summarization or reasoning.
- **Expert cost:** In specialized domains such as medicine or law, only experts can provide valid judgments, raising costs further.

These costs create trade-offs between scale and precision. Crowdsourcing provides cheaper annotations but risks inconsistency. Expert annotations provide higher quality but at prohibitive expense. Organizations must decide how much cost they are willing to incur for alignment.

Annotation Quality and Consistency

The value of feedback depends on its quality. Human annotators vary in skill, attention, and interpretation of guidelines.

- **Subjectivity:** Different annotators may interpret the same task differently, leading to inconsistent judgments.
- **Fatigue effects:** Annotators working long sessions may produce lower-quality annotations.
- **Calibration issues:** Some annotators score outputs more harshly or leniently than others, distorting data distributions.
- **Inter-annotator disagreement:** Even with clear guidelines, agreement across annotators is rarely perfect.

Pipelines address these issues through training, calibration, and redundancy. Yet inconsistency remains a fundamental challenge, as human values cannot be reduced to a single uniform measure.

Bias in Feedback Collection

Feedback collection risks encoding human biases directly into reward models.

- **Cultural bias:** Annotators from one cultural background may prefer responses that align with their norms but misalign with others.
- **Demographic bias:** Preferences may vary across age-groups, genders, or socioeconomic classes, but annotation pools may not reflect this diversity.
- **Topical bias:** Annotators may unconsciously favor certain perspectives, particularly on controversial issues.
- **Systematic bias amplification:** Reward models trained on biased data may amplify those biases during optimization.

Bias challenges the assumption that feedback reflects universal human values. Instead, feedback reflects the values of specific annotator groups. Addressing this requires deliberate diversity in annotation and careful monitoring of bias effects.

Cultural and Linguistic Diversity

Foundation models are global in scope, but feedback collection often reflects a narrow subset of cultures and languages.

- **Language availability:** Annotation pipelines frequently operate in English, neglecting other languages.
- **Cultural norms:** What counts as polite, safe, or helpful differs across cultures. A model aligned to one norm may misalign with another.
- **Global deployment:** Models are used worldwide, but their alignment may reflect the preferences of annotators from a few regions.

This raises the question of whose preferences RLHF should encode. A culturally narrow pipeline risks producing globally deployed models that misrepresent or marginalize large populations.

Annotation Noise and Reliability

Human feedback is inherently noisy. Annotators may misunderstand tasks, make mistakes, or provide inconsistent judgments.

- **Random errors:** Simple errors such as mis-clicks introduce noise.
- **Guideline confusion:** If instructions are unclear, annotations vary unpredictably.
- **Low-effort responses:** Some annotators may rush tasks without serious engagement, reducing reliability.
- **Detecting noise:** Quality control mechanisms such as gold-standard checks and redundancy are necessary but increase cost.

Noise undermines reward model fidelity. While large-scale aggregation mitigates some effects, persistent noise creates misalignment.

The Problem of Over-optimization

Optimizing against imperfect feedback creates risks of overfitting.

- **Exploiting annotator tendencies:** Models may learn to produce outputs that exploit annotator preferences without genuinely improving quality.
- **Reward hacking:** Policies may generate verbose, polite-sounding answers that win preference rankings while avoiding substantive content.
- **Over-optimization to guidelines:** If guidelines are too rigid, models may learn to conform mechanically rather than helpfully.
- **Loss of diversity:** Over-optimization can reduce creativity, leading to formulaic outputs.

This challenge reflects a deeper issue: human preferences are multidimensional, and reducing them to scalar signals inevitably introduces distortion.

Scaling Feedback Pipelines

Scaling feedback collection creates logistical and conceptual difficulties.

- **Annotator recruitment:** Large pipelines require thousands of annotators, creating challenges of training and oversight.
- **Management overhead:** Coordinating annotation work across platforms and regions is complex.
- **Quality dilution:** Scaling increases the risk of low-quality contributions slipping through.

- **Feedback loops:** As models improve, new outputs require annotation, creating ongoing demand.

Scalability is not only a question of logistics but of sustainability. How can RLHF remain feasible as models and tasks expand?

Ethical Concerns in Feedback Collection

Feedback collection raises ethical concerns about labor and fairness.

- **Annotator working conditions:** Crowdsourced annotators may be underpaid or overworked.
- **Transparency:** Annotators often do not know how their judgments will be used.
- **Psychological burden:** Exposure to harmful or disturbing outputs can harm annotators.
- **Consent and awareness:** Participants may lack informed consent about their role in shaping global AI behavior.

These concerns highlight that annotation pipelines are not only technical systems but also socio-technical ones, embedded in labor markets and human welfare.

Emerging Alternatives to Human Feedback

To address the challenges of feedback collection, researchers are exploring alternatives.

- **Synthetic feedback:** Models trained on human annotations generate additional labels, reducing dependence on humans.

- **AI-assisted feedback:** Pre-trained models filter or score outputs, with humans refining rather than generating annotations.
- **Constitutional frameworks:** Instead of collecting judgments, models are guided by rules or principles that embody normative constraints.
- **Preference distillation:** Feedback collected in one setting is distilled and reused in others, reducing annotation redundancy.

These approaches reduce reliance on costly, biased, and noisy human feedback, though they raise questions of reliability and oversight.

The Fundamental Tension in Feedback Collection

Feedback collection faces a fundamental tension.

- **Specificity vs. generality:** Feedback must be specific enough to guide optimization yet general enough to apply across tasks.
- **Subjectivity vs. consistency:** Feedback must reflect subjective human values yet be consistent enough for machine learning.
- **Cost vs. quality:** Pipelines must be affordable yet produce reliable data.
- **Diversity vs. scalability:** Feedback must reflect diverse values yet scale to industrial levels.

These tensions illustrate why feedback collection is the bottleneck of RLHF. They also explain why alternative approaches are actively pursued.

Feedback collection is the most human-intensive component of RLHF and also its most fragile. It determines the fidelity of reward models and thus the trajectory of optimization. The challenges are many: cost, inconsistency, bias, cultural narrowness, noise, over-optimization, scalability, and ethical concerns. Each challenge threatens the reliability of alignment if left unaddressed.

At the same time, emerging alternatives such as synthetic feedback, AI-assisted annotation, and constitutional frameworks suggest ways to supplement direct human annotation. These approaches cannot fully replace human feedback but may reduce dependence on costly pipelines.

Ultimately, feedback collection represents the conceptual bottleneck of alignment. It embodies the difficulty of translating human values into machine-readable form. Addressing its challenges is therefore not only a technical necessity but also a philosophical one. The success of RLHF depends on whether we can collect feedback that truly reflects human intent while scaling to the demands of global AI deployment.

Conclusion

RLHF has become the defining approach for aligning large-scale pre-trained models with human expectations. While supervised pre-training provides fluency, general knowledge, and broad capability, it does not guarantee that models will act in ways that humans find useful, safe, or trustworthy. RLHF addresses this gap by embedding human preference directly into the optimization process. This chapter has explored the central components of RLHF: human annotation pipelines, reward modeling, PPO, and the challenges inherent in feedback collection. Together, these elements provide a framework for understanding how alignment is achieved in practice and why RLHF remains both powerful and limited.

Human Annotation Pipelines As the Foundation

RLHF begins with people. Human annotation pipelines transform subjective judgments into structured data. Through pairwise comparisons, ratings, and rankings, annotators provide the raw information that later stages of RLHF depend upon. Designing these pipelines requires attention to recruitment, training, interface design, quality control, and scalability.

The value of these pipelines lies not only in their technical outputs but in their conceptual role. They operationalize alignment by turning human preferences into a trainable signal. Yet they also reveal the fragility of the process: inconsistent annotations, fatigue, bias, and cultural narrowness all threaten the reliability of the data. Annotation pipelines are thus both the cornerstone and the bottleneck of RLHF.

Reward Modeling As the Surrogate for Human Judgment

Human annotations alone cannot scale to the millions of samples required for reinforcement learning. Reward models provide the solution by generalizing from human comparisons to predictive functions that can score outputs automatically. These models approximate human preference, serving as surrogates for direct human judgment.

The strength of reward modeling is its scalability. Once trained, a reward model can evaluate countless outputs, making RLHF feasible at industrial scale. Its weakness is approximation. Reward models are imperfect, prone to overfitting and reward hacking. They capture some, but not all, of the richness of human values. Nevertheless, they embody the central insight of RLHF: that human preferences, while subjective, can be modeled well enough to guide machine learning systems.

PPO As the Engine of Alignment

The optimization stage of RLHF requires an algorithm that can update large models without destabilizing them. PPO has emerged as the standard because it balances change with conservation. By clipping updates and penalizing divergence from the pre-trained distribution, PPO ensures that models move toward alignment gradually, without losing the fluency and coherence of pre-training.

PPO illustrates the principle that alignment requires controlled optimization. Aggressive updates risk destroying valuable pre-trained knowledge. Conservative updates risk leaving misalignment uncorrected. PPO's design embodies the balance between these extremes. It demonstrates that the art of alignment lies not only in collecting preferences but also in integrating them into optimization with stability.

Challenges in Feedback Collection As the Bottleneck

Feedback collection remains the most difficult aspect of RLHF. It is costly, inconsistent, biased, and ethically complex. Scaling annotation pipelines requires vast resources and introduces risks of noise and bias. Cultural and linguistic diversity complicates the assumption that feedback reflects universal human values. Over-optimization against imperfect signals can lead to reward hacking, formulaic outputs, and loss of creativity.

These challenges are not peripheral but fundamental. They highlight the difficulty of turning subjective, multidimensional human preferences into scalar signals that can guide optimization. They also raise questions of labor ethics, transparency, and fairness in the global AI ecosystem. RLHF is therefore as much a socio-technical process as a technical one.

The Conceptual Significance of RLHF

Beyond its mechanics, RLHF carries conceptual significance. It embodies the recognition that alignment cannot be achieved solely through supervised objectives. Human values are not embedded in raw text corpora but must be actively incorporated into training. RLHF operationalizes this by embedding human preference into reward signals and optimizing against them.

This makes RLHF more than a technique. It is a paradigm that defines how AI systems are aligned in practice. It shows that alignment is not a one-time process but an iterative cycle: collecting feedback, refining reward models, optimizing with PPO, and revisiting feedback to correct misalignment. It embodies the philosophy of alignment as continuous negotiation between human judgment and machine learning.

Limitations and Open Questions

While powerful, RLHF is not a complete solution.

- **Scalability limits:** Human annotation cannot keep pace with the growth of model size and task diversity.
- **Bias persistence:** Feedback pipelines reflect the biases of annotators, embedding them into models.
- **Over-optimization risks:** Optimizing against imperfect reward models can reduce diversity and creativity.
- **Normative challenges:** Whose preferences should be collected, and how should they be weighted in global systems?
- **Alternatives and extensions:** Methods such as preference distillation, AI-assisted feedback, and constitutional frameworks aim to supplement RLHF but raise new challenges.

These limitations underscore that RLHF is a step, not an endpoint, in the broader alignment project.

RLHF represents the most influential alignment method of the current generation of foundation models. By integrating human preferences into annotation pipelines, modeling them through reward functions, optimizing with PPO, and confronting the challenges of feedback collection, RLHF demonstrates that alignment is both technically feasible and conceptually complex.

Its success lies in its pragmatism. RLHF does not solve alignment in a theoretical sense, but it improves models in practice, making them more useful, safe, and trustworthy. Its limitations lie in the difficulty of capturing human values fully, the risks of bias and over-optimization, and the cost and complexity of feedback collection.

As AI continues to expand in scale and influence, RLHF will remain central to alignment efforts, but it will also evolve. Emerging methods may reduce reliance on human annotation, diversify the sources of preference signals, and refine optimization strategies. Yet the conceptual lesson of RLHF will endure: aligning AI requires embedding human judgment into the training loop, not as an afterthought but as the core mechanism of guidance.

In this sense, RLHF is both a milestone and a challenge. It marks a turning point in the practical alignment of AI systems, demonstrating that models can be tuned toward human preference. At the same time, it highlights the unresolved complexity of translating human values into machine-readable signals. The future of alignment will build upon RLHF, extending its principles, correcting its flaws, and deepening its integration of human and machine intelligence.

In our next chapter, we cover model compression and inference optimization.

CHAPTER 9

Model Compression and Inference Optimization

The success of modern machine learning has been fueled by increasingly large models, with parameter counts scaling from millions to billions and now even trillions. While these models demonstrate extraordinary capability, their size creates a serious obstacle to practical deployment. Training costs are already immense, but the larger challenge lies in inference. Serving models at scale requires memory capacity, compute throughput, and latency reduction that exceed what many organizations can afford. Without strategies to compress models and optimize inference, even the most powerful systems remain impractical for widespread use.

Model compression and inference optimization address this challenge. The goal is not to diminish capability but to retain as much functionality as possible while reducing resource demands. Compression involves techniques that shrink the size of models, such as quantization, weight pruning, and knowledge distillation. Inference optimization focuses on improving the efficiency of model execution, leveraging hardware acceleration, efficient architectures, and specialized runtimes. Together, these approaches bridge the gap between research-grade models and production-scale deployment.

I. Cronin, *Building and Training Generative AI Models*,
https://doi.org/10.1007/979-8-8688-2332-9_9

Several key strategies define the field. Quantization reduces precision, allowing models to operate with fewer bits while maintaining acceptable accuracy. Weight pruning removes redundant parameters, simplifying architectures without fundamentally altering their representational power. Knowledge distillation transfers knowledge from large teacher models into smaller student models, achieving compactness without complete retraining. Efficient architectures such as MobileBERT and TinyML exemplify design choices that prioritize inference speed and deployment on constrained devices. In parallel, optimized runtimes manage serving at scale, orchestrating memory, batching, and hardware specific execution paths without necessarily changing the model parameters themselves.

The importance of these methods extends beyond cost savings. Compression techniques such as quantization, pruning, distillation, and compact architectures primarily modify the model to reduce its size and compute requirements. Inference optimization focuses on the execution layer, improving how a given model runs on specific hardware through better scheduling, batching, caching, and parallelism. Together they determine both the cost and the feasibility of deployment. Edge devices, mobile platforms, and embedded systems cannot host billion parameter models in their original form. With proper compression these devices can run smaller yet capable variants of such models, while inference optimization ensures that these compressed models respond quickly and efficiently under real workload conditions. Similarly, large organizations can serve models to millions of users with reduced latency and greater reliability when they combine compact models with highly tuned serving stacks. Compression is therefore the set of methods that reshape the model itself, while inference optimization is the set of methods that reshape how that model is executed in practice.

This chapter explores the conceptual foundations and practical implementations of model compression and inference optimization. It begins with quantization and pruning, the most direct approaches to reducing parameter size. It then examines knowledge distillation, which shifts the paradigm from raw reduction to learned transfer. Next, it analyzes efficient architectures designed with compactness in mind. Finally, it addresses serving at scale, where infrastructure and runtime optimizations become critical. Taken together, these strategies represent the toolkit through which advanced AI can be deployed efficiently, sustainably, and widely, with a clear distinction between modifying the model and optimizing the path through which the model reaches users.

Quantization and Weight Pruning

The first strategies considered in model compression and inference optimization are quantization and weight pruning. Both approaches share a common goal: to reduce the computational and memory footprint of deep learning models without excessively sacrificing accuracy. Quantization achieves this by lowering the precision of numerical representations, while pruning eliminates parameters deemed unnecessary for performance. Each method embodies a philosophy of efficiency, showing that models can be made smaller and faster without fully retraining from scratch.

This section explores the conceptual and technical dimensions of quantization and weight pruning. It begins by introducing the motivation for these approaches, then explains the principles of quantization, including post-training quantization, quantization-aware training, and mixed-precision strategies. It next turns to pruning, outlining both structured and unstructured methods, their historical roots, and modern advances. Trade-offs in accuracy, interpretability, and scalability are analyzed, followed by engineering practices for implementing these methods in production. Case illustrations from vision, language, and edge deployment highlight the practical impact of quantization and pruning.

Motivation for Quantization and Pruning

Large scale deep learning models present severe challenges for deployment. Models with hundreds of millions or billions of parameters require enormous storage, memory bandwidth, and compute resources. For example:

- A 175 billion parameter model in full 32 bit floating point precision requires hundreds of gigabytes of memory, far beyond the capacity of most hardware accelerators.
- Inference latency grows as parameters increase, limiting the practicality of real time applications.
- Energy consumption becomes significant, raising sustainability concerns and making edge deployment nearly impossible.

Quantization and pruning were developed to address these challenges.

- Quantization reduces precision from formats such as 32 bit floating point to lower bit representations such as 16 bit, 8 bit, or even 4 bit. This reduces memory requirements and accelerates computation. A shift from 32 bit to 8 bit weights alone yields roughly a fourfold reduction in parameter storage and often comparable gains when inference is limited by memory bandwidth.
- Pruning removes parameters that contribute little to model performance, reducing the number of active weights and computations. When 80% of weights are pruned in a structured manner, the corresponding floating point operations can drop by a similar factor, significantly lowering inference cost.

The shared motivation is clear: make large models practical for real world use without retraining from scratch or fundamentally altering architectures. To make this more concrete, you could include comparative figures that plot how memory footprint and FLOPs shrink as quantization levels change or pruning ratios increase, visually highlighting the trade-off between compression strength and model accuracy.

Conceptual Foundations of Quantization

Quantization rests on the principle that neural networks tolerate lower numerical precision without catastrophic loss of accuracy. Many parameters and activations contain redundancy, and exact 32-bit precision is often unnecessary. By reducing the precision of numbers, models become smaller and faster.

Core principles of quantization include:

- **Discretization:** Continuous floating-point values are mapped to a limited set of discrete values.
- **Dynamic range reduction:** Values are represented with fewer bits, shrinking the space of possible representations.
- **Approximate equivalence:** The reduced representation preserves enough information for the model to function with minimal accuracy loss.

Quantization thus embodies the idea that representation can be approximate without undermining function.

Types of Quantization

Quantization strategies vary in when and how reduced precision is applied.

Post-training Quantization

This approach quantizes a pre-trained model without retraining.

- **Simplicity:** Requires minimal additional computation.
- **Speed:** Models can be compressed quickly for deployment.
- **Accuracy impact:** May degrade performance if the model is highly sensitive to precision changes.

Post-training quantization is attractive for production pipelines where retraining is impractical.

Quantization-Aware Training

Here quantization effects are simulated during training.

- **Integration during training:** The model learns to operate robustly under lower-precision constraints.
- **Improved accuracy:** Retains more performance compared to post-training methods.
- **Higher cost:** Requires retraining, increasing computational demand.

This strategy produces more resilient models at the expense of additional training effort.

Mixed-Precision Quantization

Not all layers or parameters require the same level of precision. Mixed-precision strategies assign different precisions to different parts of the model.

- **Flexibility:** Critical layers retain higher precision, while less sensitive layers use lower precision.
- **Optimization:** Achieves a balance between efficiency and accuracy.
- **Hardware alignment:** Exploits accelerators designed for mixed-precision arithmetic.

Mixed-precision reflects the nuanced reality of deep networks: some computations demand more precision than others.

Hardware Support for Quantization

Quantization's impact depends strongly on hardware. Modern accelerators such as GPUs and TPUs now include native support for low precision arithmetic, which turns theoretical compression gains into real speed and cost savings.

- **INT8 operations:** Many GPUs and specialized inference accelerators implement integer eight instructions that can process large batches of matrix multiplications with much higher throughput than full precision arithmetic. When models are carefully calibrated for integer eight, this can yield substantial gains in tokens per second and reductions in power consumption. However, kernels and instruction layouts are often highly tuned to a specific vendor and generation of hardware, so an integer eight optimized model that runs efficiently on one accelerator family may see smaller gains or different numerical behavior on another.

- **FP16 operations:** Mixed precision training and inference rely on floating point sixteen arithmetic for most operations while retaining higher precision accumulators where needed. This approach accelerates both training and serving while usually preserving accuracy, but again the exact performance profile depends on how well a given device implements floating point sixteen tensor cores or equivalent units. Some older GPUs emulate parts of this behavior with lower efficiency, which can narrow the practical benefit of quantization on those systems.
- **Custom hardware:** Dedicated chips such as Google's Edge TPU and other edge inference accelerators are built around low precision multiply accumulate units and tight on chip memory. When models are quantized to the formats these chips expect, they can deliver very high throughput at low power. The trade off is that models tailored to one custom accelerator may require re quantization, re calibration, or even architectural changes to run well on a different family of devices, which complicates portability across heterogeneous fleets.

These hardware specific optimizations introduce a tension between maximum efficiency and portability. Aggressively quantizing and tuning a model for a single accelerator can unlock impressive gains in throughput and latency, but may tie the deployment pipeline to that hardware vendor. More portable strategies aim for quantization schemes and operator sets that are widely supported, accepting slightly lower peak performance in exchange for the ability to move models between clouds, on premises clusters, and edge devices without extensive rework.

Software stacks sit between quantized models and the underlying hardware and are crucial for managing this trade off. Frameworks such as ONNX Runtime provide back ends for many different devices and can automatically select optimized kernels for integer eight or floating point sixteen operators when available, while still falling back to generic implementations on unsupported hardware. Compiler stacks such as TVM take model graphs and generate specialized code for each target, performing operator fusion, layout transformations, and low precision code generation so that the same high level model can be retargeted to CPUs, GPUs, or custom accelerators. XLA compilers in ecosystems such as JAX and TensorFlow apply similar ideas by compiling computation graphs into optimized executables that exploit device specific low precision instructions and memory hierarchies.

The effectiveness of quantization therefore reflects not only the algorithmic design of lower precision representations but also the maturity of the surrounding hardware and software ecosystem. Strong support in accelerators and compiler stacks can turn a quantized model into a practical, portable system, while weak support can limit gains or lock optimizations to a single class of devices.

Conceptual Foundations of Pruning

Pruning is motivated by the observation that many parameters in deep networks contribute little to overall performance. Early studies showed that networks could be heavily pruned while maintaining accuracy, revealing significant redundancy.

Principles of pruning include:

- **Redundancy reduction:** Many parameters overlap in function and can be safely removed.
- **Sparse representation:** Pruned models are sparse, requiring fewer computations.

- **Approximate equivalence:** The pruned model approximates the original with fewer parameters.

Conceptually, pruning demonstrates that deep networks are over-parameterized, and efficiency can be achieved by removing unnecessary components.

Types of Pruning

Pruning methods differ in granularity and structure.

Unstructured Pruning

Individual weights are removed based on importance measures such as magnitude.

- **Fine-grained control:** Allows selective removal of redundant weights.
- **Sparsity benefits:** Reduces the number of non-zero parameters.
- **Hardware inefficiency:** Sparse matrices are harder to accelerate on standard hardware.

Structured Pruning

Entire neurons, filters, or channels are removed.

- **Hardware alignment:** Structured pruning produces smaller dense matrices that accelerate well.
- **Interpretability:** The architecture remains coherent and compact.
- **Potential accuracy loss:** Coarser pruning may remove useful structures.

The choice reflects trade-offs between theoretical efficiency and practical acceleration.

Historical Development of Pruning

Pruning has a long history in neural networks.

- **Early work:** Studies in the 1990s showed that neural networks could be pruned without significant performance loss.
- **Lottery ticket hypothesis:** More recent research demonstrated that sparse sub networks could match the accuracy of dense networks if properly initialized.
- **Modern approaches:** Structured pruning aligns better with hardware, making pruning practical for deployment.
- **Transformer era resurgence:** In large transformer models, structured sparsity techniques such as attention head pruning, block sparse matrices, and sparsified feedforward layers are gaining renewed interest as a way to cut inference FLOPs and memory while preserving quality, driven by the high cost of serving these models at scale.

The evolution of pruning therefore reflects a progression from theoretical curiosity to practical engineering tool and now to a central strategy for controlling inference cost in large transformer based systems.

Trade-Offs in Quantization and Pruning

Both quantization and pruning involve trade-offs.

- **Accuracy vs. efficiency:** Reducing precision or parameters accelerates inference but risks accuracy loss.
- **Flexibility vs. simplicity:** Structured methods align with hardware but may oversimplify networks. Unstructured methods retain accuracy but are harder to accelerate.
- **Short-term vs. long-term:** Post-training quantization and pruning provide quick gains, while training-aware methods require more investment but yield better results.

These trade-offs define the practical application of compression methods.

Engineering Practices for Quantization and Pruning

Deploying these methods requires careful engineering.

- **Profiling sensitivity:** Some layers are more sensitive to pruning or quantization than others. Profiling identifies where compression is safe.
- **Iterative compression:** Gradual quantization or pruning with retraining reduces accuracy loss.
- **Hybrid approaches:** Combining quantization and pruning often yields better efficiency than either alone.
- **Validation and benchmarking:** Compressed models must be tested against representative tasks to ensure robustness, for example by running side by side benchmarks on key workloads, tracking accuracy, latency, and throughput across a range of inputs rather than only on a single test set.

Engineering practices turn conceptual methods into deployable systems. In production, teams should validate compressed models with controlled A/B tests against existing baselines, monitor user facing metrics such as latency and error rates, and watch for performance drift over time as data distributions change. Logging confidence scores, error patterns, and hardware utilization helps detect when a compressed model begins to degrade so that it can be recalibrated, retrained, or rolled back, ensuring that efficiency gains do not come at the cost of reliability.

Case Illustrations

Quantization and pruning have been deployed successfully across domains.

- **Vision:** INT8 quantization of convolutional networks accelerates image classification on mobile devices.
- **Language:** Distilled and quantized transformers enable real-time translation on smartphones.
- **Edge deployment:** Pruned and quantized models run on embedded systems such as IoT devices.

These cases show that compression enables applications otherwise impossible with full-size models.

Broader Implications

Quantization and pruning illustrate a broader theme in AI: capability is not the only measure of success. Efficiency, accessibility, and sustainability matter as much as raw power.

- **Accessibility:** Compression makes advanced AI available on low resource devices.

- **Sustainability:** Reducing energy consumption mitigates environmental impact. When a compressed model cuts inference compute by even a modest percentage, that improvement compounds across billions of requests, translating into measurable reductions in energy usage and associated carbon emissions per inference or per model run.
- **Scalability:** Efficient inference enables deployment to millions of users simultaneously without a proportional increase in hardware and energy demand.

Compression is therefore not only an optimization strategy but also a pathway to democratizing AI.

Quantization and pruning represent the first line of strategies for compressing and optimizing models. They demonstrate that deep networks contain redundancy in both precision and parameters and that this redundancy can be reduced without catastrophic accuracy loss. Quantization lowers numerical precision, while pruning removes unnecessary parameters. Both methods embody the principle that approximate representations suffice for practical deployment.

While trade-offs remain, these strategies have enabled real world deployment of models that would otherwise be impractical. They also highlight a concrete link between engineering decisions and environmental impact: each reduction in floating point operations, memory traffic, and serving energy contributes to lower carbon cost per user interaction. In this sense, intelligence in machines does not require maximal computation, but efficient representation and execution. Quantization and pruning thus stand as milestones in the effort to make artificial intelligence not only powerful, but also sustainable and usable at global scale.

Knowledge Distillation

Knowledge distillation has emerged as one of the most influential strategies in model compression. It captures the insight that large models, while powerful, often carry redundancy in their learned representations. A smaller model can mimic the behavior of a larger one if it is trained not just on the original dataset but also on the outputs of the larger model. This teacher–student paradigm allows compact models to inherit much of the performance of their larger counterparts, enabling efficient deployment without discarding the benefits of scale.

This section explores the conceptual and technical aspects of knowledge distillation. It begins with the motivation for distillation and its historical roots, then outlines the theoretical principles that underpin it. It examines various forms of distillation, including task-specific, sequence-to-sequence, and self-distillation. Engineering practices for implementing distillation are discussed, along with applications in natural language processing, computer vision, and speech recognition. Challenges such as loss of generalization, domain shift, and interpretability are analyzed, and emerging extensions are considered.

Motivation for Knowledge Distillation

The motivation for knowledge distillation arises from the gap between training and deployment.

- **Large models are effective but costly:** Training massive models yields state-of-the-art performance but results in systems too resource-intensive for real-world use.
- **Smaller models are efficient but limited:** Compact models run quickly but lack the representational capacity of larger models.

- **Distillation bridges the gap:** By transferring knowledge from a large teacher to a smaller student, distillation achieves efficiency without sacrificing as much accuracy.

The promise of distillation is that knowledge is not bound to the parameter count of the model. Instead, knowledge can be abstracted and transferred, enabling efficiency through guided learning.

Historical Roots of Distillation

Knowledge distillation was formalized in the mid-2010s but has roots in earlier traditions.

- **Model compression research:** Early work showed that ensembles of models could be approximated by a single model trained to reproduce their outputs.
- **Soft targets:** Researchers discovered that training on soft probability distributions conveyed richer information than training on hard labels.
- **Formalization by Hinton et al.:** The landmark work introduced the teacher–student framework, showing how large teacher models could guide smaller students by providing softened outputs as training targets.

This progression reflected a conceptual shift: learning is not only about fitting to ground-truth labels but also about absorbing the representational biases of larger models.

Theoretical Principles of Distillation

Distillation rests on several theoretical principles.

- **Soft probabilities as knowledge:** Teacher models produce probability distributions over classes, not just hard labels. These distributions reveal relationships between classes, providing richer training signals.
- **Information transfer:** The student model absorbs both the correct answer and the teacher's uncertainty structure, inheriting nuanced knowledge.
- **Approximate equivalence:** The student model does not replicate the teacher exactly but approximates its behavior in a compressed form.
- **Regularization effect:** Distillation acts as a form of regularization, smoothing the student's predictions and improving generalization.

Conceptually, distillation demonstrates that knowledge is not binary but probabilistic, and that smaller models can capture broader representational structures through guided learning.

Types of Knowledge Distillation

Distillation has diversified into multiple forms.

Task-Specific Distillation

The student learns to replicate the teacher's outputs on a specific task.

- **Efficiency:** Enables deployment of compact models specialized for a given application.
- **Trade-off:** Loses some generality of the teacher model.
- **Applications:** Distilled models for classification, translation, and summarization.

Sequence-to-Sequence Distillation

In sequence tasks, the teacher generates outputs, and the student learns to reproduce them.

- **Simplification:** Complex sequence models can be approximated by smaller students.
- **Example:** Distilling machine translation systems for mobile deployment.

Self-Distillation

A model serves as both teacher and student.

- **Iterative refinement:** A model trains itself with its own softened outputs.
- **Efficiency:** Avoids the need for a separate large teacher.
- **Applications:** Common in transformer models where self-distillation improves robustness.

Multi-teacher Distillation

Multiple teachers guide a single student.

- **Knowledge diversity:** Aggregates expertise from different models.
- **Stability:** Reduces dependence on any single teacher's biases.

These variations show the flexibility of the distillation paradigm.

Engineering Practices in Distillation

Implementing distillation requires careful design.

- **Temperature scaling:** Softening the teacher's probability distribution emphasizes relative class relationships.
- **Loss balancing:** Distillation loss is often combined with supervised loss, requiring careful weighting.
- **Architecture alignment:** The student's architecture must be compatible with the knowledge being transferred.
- **Training pipelines:** Efficient training loops are needed to manage large-scale teacher outputs.

These practices ensure that distillation retains the strengths of the teacher while producing efficient students.

Applications in Natural Language Processing

NLP has been a major beneficiary of distillation.

- **DistilBERT:** A smaller version of BERT that retains most of its performance while being faster and lighter.
- **TinyBERT:** Optimized for mobile devices, combining distillation with quantization for compact deployment.
- **Sequence tasks:** Distillation applied to translation, summarization, and dialogue systems enables efficient inference without large resource demands.

These applications demonstrate that distillation enables state-of-the-art NLP models to run in practical environments.

Applications in Computer Vision and Speech

Beyond NLP, distillation has proven valuable in other domains.

- **Computer vision:** Students distilled from large convolutional networks achieve strong accuracy in image classification and detection while running on mobile hardware.
- **Speech recognition:** Distilled models reduce latency and memory usage in real-time transcription systems.
- **Cross-modal tasks:** Distillation transfers knowledge across modalities, enabling compact multimodal systems.

These applications illustrate distillation's universality as a compression method.

Challenges in Knowledge Distillation

Despite its success, distillation faces challenges.

- **Loss of generalization:** Students may inherit teacher performance on specific tasks but lose general capabilities.
- **Domain shift:** Students distilled in one domain may not transfer well to others.
- **Teacher biases:** Students inherit not only strengths but also weaknesses and biases of teachers.
- **Capacity limits:** Extremely small students cannot absorb all of the teacher's knowledge.
- **Training cost:** While deployment is efficient, distillation requires access to large teacher outputs, increasing training cost.

These challenges highlight the complexity of transferring knowledge across architectures and scales.

Extensions and Innovations

Distillation continues to evolve through extensions and hybrid methods.

- **Layer-wise distillation:** Students mimic teacher representations at multiple layers, not just outputs.
- **Cross-task distillation:** Knowledge is transferred across tasks, creating versatile students.
- **Progressive distillation:** Students are trained iteratively with increasingly smaller teachers.
- **Unsupervised distillation:** Teachers trained on unlabeled data provide guidance in low-resource settings.

These innovations expand the scope and effectiveness of distillation.

Future Directions of Knowledge Distillation

Distillation is poised to remain a central method in compression. Future research will focus on:

- **Scalability:** Distilling from trillion-parameter teachers into manageable students.
- **Robustness:** Ensuring students retain generalization and resist adversarial vulnerabilities.
- **Fairness:** Addressing the transfer of biases from teachers to students.

- **Integration with other methods:** Combining distillation with quantization, pruning, and efficient architectures.
- **Deployment ecosystems:** Developing toolchains for automatic distillation pipelines tailored to specific hardware.

These directions show that distillation is not static but evolving with the frontier of model scaling.

Knowledge distillation demonstrates that learning is not tied to model size but to the transfer of representational patterns. By training smaller models to mimic larger ones, distillation achieves efficiency without discarding the benefits of scale. Its success across NLP, vision, and speech shows that it is not a niche technique but a universal strategy for compression.

While challenges remain, particularly in preserving generalization and avoiding bias transfer, distillation provides a conceptual and practical pathway to efficient deployment. It embodies the principle that knowledge is transferable, and that efficiency is achieved not by discarding learning but by compressing it intelligently.

Efficient Architectures

The pursuit of efficiency in deep learning has long recognized that compression is not only about modifying existing models but also about designing architectures that are inherently efficient. While quantization, pruning, and distillation retrofit efficiency onto existing models, efficient architectures embed efficiency into the design itself. They are built with the explicit aim of reducing parameter counts, memory footprint, and inference latency, while preserving as much representational capacity as possible.

This section explores the principles, design choices, and real-world exemplars of efficient architectures. It begins with historical context, showing how early architectures foreshadowed efficiency as a goal. It then outlines the principles of efficient design, including parameter sharing, bottleneck modules, depthwise separable convolutions, and factorized representations. It analyzes specific architectures such as MobileBERT, TinyBERT, and ALBERT, as well as the emerging field of TinyML for edge deployment. Finally, it considers trade-offs, challenges, and future trajectories, emphasizing the conceptual significance of efficiency as a design paradigm rather than an afterthought.

Historical Context of Efficient Architectures

The history of deep learning shows that efficiency has always been a concern.

- Early convolutional networks: Even before large scale models, researchers sought architectures that balanced accuracy with computational cost. LeNet, for example, achieved efficiency through small filters and limited depth.
- Mobile era demands: The rise of smartphones created demand for architectures that could run on constrained hardware. This demand led to designs such as MobileNet, which introduced depthwise separable convolutions as a way to reduce computational load.
- Transformer expansion: As transformers became dominant, their enormous size raised new concerns. Efficient transformer variants such as ALBERT and MobileBERT arose to meet these concerns.

- Edge deployment: The push to bring AI to embedded devices highlighted the necessity of TinyML architectures designed from the ground up for efficiency.

Efficiency is therefore not a new objective but one that has become increasingly urgent as models have scaled. In practice, architectures are now often designed around specific use cases: edge and embedded models prioritize strict limits on memory, latency, and power consumption, while on device AI emphasizes predictable performance, compatibility with local accelerators such as NPUs or DSPs, and robustness under intermittent connectivity, even if that means accepting a modest reduction in peak accuracy.

Principles of Efficient Design

Efficient architectures embody design principles that reduce redundancy while maintaining representational power.

- **Parameter sharing:** Weights are reused across layers or components, reducing memory footprint.
- **Bottleneck modules:** Narrow intermediate layers compress information before expanding it, reducing computations without losing expressivity.
- **Depthwise separable convolutions:** Factorizing convolutions into depthwise and pointwise components drastically reduces computational cost in vision models.
- **Factorized embeddings:** Large embedding matrices are decomposed into smaller factors, reducing parameter counts in NLP models.

- **Low-rank approximations:** Representations are compressed by assuming low-rank structure, aligning with methods like LoRA but embedded into architecture.
- **Sparse attention mechanisms:** Transformers replace dense attention with sparse or linearized alternatives to scale more efficiently.

These principles reveal that efficiency is a matter of architectural choice as much as algorithmic modification.

MobileBERT and TinyBERT

MobileBERT and TinyBERT exemplify efficient design in the transformer family.

- **MobileBERT:** Designed to run on mobile hardware, it uses bottleneck structures, inverted residuals, and optimized depth to reduce computational overhead.
    - Embedding layers are factorized to reduce parameter size.
    - Intermediate layers compress information, reducing memory cost.
    - Latency is minimized without catastrophic loss of accuracy.
- **TinyBERT:** Created with distillation in mind, TinyBERT integrates efficiency into architecture.
    - Layer-wise distillation ensures student layers capture teacher knowledge.

- Compact dimensions reduce memory footprint while preserving task performance.
- Adapted for mobile and edge scenarios where full BERT is impractical.

Both architectures show how transformer designs can be rethought for efficiency, rather than merely compressed after training.

ALBERT and Parameter Sharing

ALBERT represents a radical step toward efficiency through parameter sharing.

- **Shared parameters across layers:** Instead of separate weights for each transformer layer, ALBERT shares weights across layers.
- **Factorized embeddings:** Embedding matrices are decomposed into smaller matrices, reducing parameter counts without reducing representational capacity.
- **Impact:** ALBERT reduces parameter counts dramatically while maintaining strong performance on benchmarks.

The principle behind ALBERT is that depth provides representational richness even if parameters are shared, challenging the assumption that more parameters are always necessary.

TinyML and Edge Deployment

TinyML extends efficient design to the extreme by creating models that run on embedded systems with minimal resources.

- **Memory constraints:** TinyML models often run on devices with kilobytes or megabytes of memory.
- **Latency constraints:** Real-time applications such as voice recognition demand millisecond-level responses.
- **Energy constraints:** Battery-powered devices require ultra-low energy consumption.

Design principles for TinyML include:

- **Quantized operations:** Models operate at 8-bit or lower precision by default.
- **Micro-architectures:** Extremely small models are designed to handle specific tasks efficiently.
- **Hardware–software co-design:** Architectures are tailored to specific microcontrollers.

TinyML demonstrates the conceptual significance of efficiency: it enables AI in contexts where large-scale models are impossible.

Sparse Attention Mechanisms

Transformers face quadratic scaling in attention, making efficiency critical. Sparse attention mechanisms address this challenge.

- **Local attention:** Restricts attention to nearby tokens, reducing computational load.
- **Strided and dilated attention:** Selects tokens at intervals, balancing coverage and efficiency.
- **Low-rank approximations:** Approximates dense attention with lower-rank operations.
- **Linear attention:** Reforms attention into linear complexity by reordering operations.

These innovations show that efficiency can be achieved not only by compressing models but by redesigning their fundamental operations.

Trade-Offs in Efficient Architectures

Efficient architectures involve trade-offs.

- **Accuracy vs. speed:** Smaller models may lose accuracy, though careful design mitigates this.
- **Generalization vs. specialization:** Efficient models often perform best in targeted domains, while larger models retain broader generality.
- **Simplicity vs. innovation:** Techniques like sparse attention add complexity to achieve efficiency.
- **Deployment vs. research benchmarks:** Efficient models may underperform on research benchmarks but excel in practical deployment.

These trade-offs reflect the reality that efficiency is not free but must be balanced against performance and usability.

Engineering Practices for Efficient Architectures

Building efficient architectures requires engineering practices distinct from post hoc compression.

- **Profiling bottlenecks:** Engineers identify where computation, memory usage, and activation footprint are concentrated, using tools that report per layer FLOPs, memory traffic, and kernel level latency.

- **Layer rebalancing:** Depth, width, and embedding dimensions are tuned for efficiency so that no single stage dominates FLOPs or memory while overall accuracy remains acceptable.
- **Hardware alignment:** Architectures are tailored to specific accelerators such as GPUs, TPUs, or edge devices, choosing operator types and tensor shapes that map cleanly to available tensor cores, NPUs, or SIMD units.
- **Iterative refinement:** Architectures are tested, refined, and rebalanced repeatedly using automated sweeps and targeted ablations to understand the cost benefit of each component.
- **Benchmark driven design:** Efficiency is measured against concrete targets, including FLOPs reduction relative to a baseline, latency per token or per request, peak and tail latency under load, memory footprint, and energy per inference or per thousand tokens served, rather than accuracy alone.

These practices ensure that architectures are efficient in both theory and deployment, with quantitative metrics guiding design choices rather than relying only on qualitative impressions of simplicity.

Future Directions of Efficient Architectures

The future of efficient architectures will likely expand across multiple dimensions.

- **Neural architecture search (NAS):** Automated search methods will design architectures that are explicitly optimized for efficiency under given constraints such as latency, memory, and power.

- **Co-design with hardware:** Architectures will be developed alongside specialized accelerators so that operator choices, tensor shapes, and memory layouts match what the hardware can execute most effectively.
- **Hybrid efficiency strategies:** Architectures will increasingly integrate quantization, pruning, and distillation from the beginning rather than treating these as purely post training steps, so that networks are efficient by construction.
- **Task specific efficiency:** Models will be customized for specific deployment contexts, from smartphones and wearables to IoT sensors and industrial controllers, with different budgets for latency, energy, and model size.
- **Sustainability:** Energy efficient architectures will play a growing role in reducing the environmental impact of AI by lowering energy per inference and total compute over a model's lifetime.
- **Emerging sparse and modular designs:** Paradigms such as mixture of experts and modular transformers aim for efficiency by activating only a small subset of parameters for each input, allowing very large total model capacity while keeping the average compute per request relatively low.

These directions show that efficiency will remain central as AI expands into new domains and that future architectures will combine automated design, hardware awareness, and sparse modular computation to deliver more capable systems at lower cost.

Efficient architectures demonstrate that compression need not be retrofitted but can be embedded into design. By integrating parameter sharing, bottleneck modules, depthwise separable convolutions, factorized embeddings, and sparse attention mechanisms, these architectures reduce computational demands while retaining functionality. Examples such as MobileBERT, TinyBERT, ALBERT, and TinyML highlight the practical achievements of efficiency-focused design.

Conceptually, efficient architectures shift the narrative of AI from maximalism to pragmatism. They show that intelligence in machines need not require ever-larger models but can be realized through careful design that balances accuracy, latency, memory, and energy. As AI expands into constrained environments, efficient architectures will remain essential, not as compromises but as exemplars of intelligent engineering.

Serving at Scale with Optimized Runtimes

The final challenge in model compression and inference optimization is not merely reducing parameters or designing efficient architectures. The ultimate test lies in serving models at scale. Even a well compressed model may be impractical if it cannot be deployed across distributed infrastructure with acceptable latency, throughput, and reliability. Serving at scale is where theoretical efficiency meets the real world demands of millions of users, heterogeneous hardware environments, and fluctuating workloads.

This section explores the principles and practices of serving at scale with optimized runtimes. It begins by outlining the challenges of inference at production scale, then introduces the role of runtimes and inference engines. It examines techniques such as batching, memory management, operator fusion, and hardware specialization. It analyzes frameworks such as ONNX Runtime, TensorRT, and DeepSpeed Inference as conceptual exemplars. In practice, ONNX Runtime is often chosen when teams need a portable, framework agnostic engine that can target many back ends, from CPUs to different vendors' GPUs and specialized accelerators, trading a small amount of peak performance for flexibility and ease of deployment across mixed fleets. TensorRT, by contrast, is tightly tuned to NVIDIA GPUs and focuses on extracting maximum throughput and minimum latency on that hardware through aggressive graph optimization, kernel fusion, and calibration passes, delivering excellent performance where NVIDIA hardware is standard but offering less portability outside that ecosystem.

Optimized runtimes are also the point where compression techniques and hardware awareness come together. Quantized models see the largest gains when the runtime can issue low precision kernels that match the chosen formats, schedule integer or reduced precision operations efficiently, and manage calibration scales without manual intervention. Similarly, pruned or sparsified models benefit most when runtimes support sparse kernels, skip computation on zero blocks, and fuse operators in ways that preserve sparsity patterns. As a result, compression and serving cannot be treated as separate concerns; the same design that makes a model smaller must be exposed to the runtime so that batching, memory layouts, and operator fusion are all chosen to exploit that structure.

The section then considers elastic scaling, monitoring, and fault tolerance in large scale serving environments. It discusses autoscaling based on request rate, placement strategies for models across heterogeneous hardware, and instrumentation to track latency, throughput, error rates, and hardware utilization. Finally, it reflects on trade offs and future trajectories, emphasizing that optimized runtimes are the connective tissue between research models and production systems, and that their ability to understand and exploit compressed models is what ultimately turns algorithmic efficiency into real performance gains for users.

The Challenge of Serving at Scale

Inference at scale is distinct from training. Training is resource-intensive but predictable: it runs on clusters designed for long-running jobs. Serving, by contrast, must be real-time, responsive to user queries, and resilient to spikes in demand.

Key challenges include:

- **Latency:** Users expect responses in milliseconds, even from models with billions of parameters.
- **Throughput:** Systems must handle thousands or millions of requests per second.
- **Resource constraints:** Hardware is limited, and energy efficiency is critical.
- **Scalability:** Workloads fluctuate, requiring dynamic scaling.
- **Reliability:** Production systems must remain available despite hardware or software failures.

These challenges define the requirements for optimized runtimes: efficiency, scalability, and robustness.

Runtimes and Inference Engines

At the heart of serving are runtimes and inference engines, specialized software layers that execute models efficiently on given hardware.

- **Abstraction layer:** Runtimes abstract away hardware details, enabling portability across CPUs, GPUs, and accelerators.
- **Operator optimization:** Inference engines optimize neural network operations, fusing them where possible.

- **Hardware utilization:** Runtimes maximize throughput by aligning computation with hardware capabilities.
- **Deployment pipeline:** Runtimes integrate with serving systems, handling requests, batching, and response delivery.

Examples include ONNX Runtime, NVIDIA TensorRT, and Microsoft DeepSpeed-Inference. Each embodies principles of optimization through hardware-aware execution.

Batching Strategies for Efficiency

Batching is a key strategy for scaling inference. Rather than processing each request individually, runtimes aggregate requests and process them together.

- **Throughput gains:** Batching increases hardware utilization, processing multiple inputs in parallel.
- **Latency trade-offs:** Larger batches improve efficiency but increase response time for individual requests.
- **Dynamic batching:** Advanced runtimes dynamically adjust batch sizes to balance throughput and latency.
- **Micro-batching:** In streaming contexts, small batches provide near-real-time responses with some efficiency gain.

Batching illustrates the central trade-off of serving: balancing latency with throughput.

Memory Management in Large-Scale Serving

Memory is often the bottleneck in inference. Models require memory for weights, activations, and intermediate computations.

- **Weight loading:** Efficient runtimes load weights only once and reuse them across requests.
- **Activation management:** Memory is reused across layers to reduce overhead.
- **Offloading strategies:** Some runtimes offload intermediate results to CPU or disk when GPU memory is insufficient.
- **Compression during serving:** Quantization and pruning reduce memory requirements during inference as well as training.

Effective memory management ensures that even large models can be deployed on resource-constrained hardware.

Operator Fusion and Graph Optimization

Operator fusion is a central technique in optimized runtimes.

- **Fusion principle:** Multiple sequential operations are combined into a single operation, reducing memory transfers and kernel launches.
- **Graph-level optimization:** Models are represented as computation graphs, which can be simplified through algebraic transformations.
- **Examples of fusion:** Matrix multiplication followed by activation can be fused into one operator.
- **Impact:** Fusion reduces latency and increases throughput significantly.

Graph optimization illustrates the shift from naïve execution to hardware-conscious inference.

Hardware Specialization and Acceleration

Optimized runtimes exploit hardware specialization.

- **GPUs:** Runtimes use libraries like cuDNN to accelerate convolution and transformer operations.
- **TPUs:** Google's TPUs offer hardware optimized for matrix multiplications, with runtimes tailored to exploit them.
- **FPGAs:** Field programmable gate arrays allow customized inference pipelines, with runtimes mapping models onto reconfigurable hardware.
- **Edge accelerators:** Devices such as Google's Edge TPU and Apple's Neural Engine are supported by specialized runtimes.

Hardware specialization aligns inference efficiency with the physical substrate of computation, but it also introduces portability challenges. Models and optimization pipelines that are heavily tuned for one stack of kernels, instruction sets, and memory layouts often transfer poorly to other accelerators, requiring separate export formats, calibration procedures, and performance tuning for each target platform. As a result, organizations must balance the benefits of deep hardware specific optimization against the engineering cost of maintaining multiple runtime configurations and deployment paths across heterogeneous fleets.

Frameworks As Conceptual Exemplars

Several frameworks illustrate the principles of optimized runtimes.

- **ONNX Runtime:** Provides portability across frameworks and hardware, with graph optimization and operator fusion.
- **TensorRT:** Specializes in GPU acceleration with aggressive optimization of precision, memory, and kernels.
- **DeepSpeed-Inference:** Extends efficiency to very large models, offering parallelization and memory optimization.
- **TVM:** A compiler stack that generates optimized code for diverse hardware targets.

Each framework exemplifies a distinct strategy, but all converge on the principle of hardware-aware optimization.

Elastic Scaling for Fluctuating Demand

Serving systems must handle fluctuating workloads. Elastic scaling enables dynamic adjustment of resources.

- **Horizontal scaling:** Additional servers are added as demand increases.
- **Vertical scaling:** Individual servers allocate more resources dynamically.
- **Autoscaling:** Systems automatically scale resources based on workload metrics.
- **Challenges:** Elastic scaling requires load balancing and rapid model initialization to avoid latency spikes.

Elastic scaling illustrates the convergence of model efficiency with cloud infrastructure.

Fault Tolerance and Reliability

Reliability is essential in production serving. Failures must not interrupt service.

- **Redundancy:** Multiple replicas of models ensure continuity if one fails.
- **Checkpointing:** Models and runtime states are saved regularly for recovery.
- **Graceful degradation:** Systems reduce service quality rather than fail entirely under stress.
- **Monitoring:** Continuous monitoring detects anomalies in latency, throughput, or accuracy.

Reliability ensures that efficiency does not come at the cost of availability.

Trade-Offs in Optimized Runtimes

Optimized runtimes involve trade-offs that must be balanced.

- **Latency vs. throughput:** Larger batches improve throughput but increase latency.
- **Portability vs. specialization:** Portable runtimes work across hardware but may be less efficient than specialized ones.
- **Complexity vs. maintainability:** Aggressive optimizations add complexity, increasing maintenance burdens.

- **Energy vs. performance:** Higher performance may increase energy use unless carefully optimized.

These trade-offs define the engineering art of serving at scale.

Future Directions in Serving and Runtimes

Future research will extend the principles of serving and runtimes.

- **Compiler-level optimization:** Compilers will automatically generate optimized inference code.
- **Neural architecture co-design:** Architectures will be designed alongside runtimes for maximal efficiency.
- **Heterogeneous hardware support:** Runtimes will integrate CPUs, GPUs, TPUs, FPGAs, and edge accelerators seamlessly.
- **Adaptive runtimes:** Systems will adjust dynamically to workload, precision, and resource constraints.
- **Sustainability focus:** Energy-efficient serving will become a priority, aligning with environmental goals.

Serving at scale is therefore not static but evolving with both hardware and software advances.

Serving at scale with optimized runtimes represents the culmination of model compression and inference optimization. Quantization, pruning, and distillation make models smaller. Efficient architectures embed efficiency into design. But it is optimized runtimes that make efficiency real in production, enabling billions of inferences to be served reliably and responsively.

By leveraging batching, memory management, operator fusion, hardware specialization, and elastic scaling, runtimes transform compressed models into usable systems. Frameworks such as ONNX

Runtime, TensorRT, and DeepSpeed-Inference illustrate the principles of portability, specialization, and scalability. Challenges remain in balancing latency, throughput, portability, and energy use. Yet the conceptual lesson is clear: efficiency is not only a property of models but of the systems that serve them.

Serving at scale is the bridge between research and reality. It ensures that advances in model efficiency translate into accessible, reliable, and sustainable AI systems. In this sense, optimized runtimes are not peripheral but central to the project of making artificial intelligence usable at global scale.

Conclusion

The rise of large-scale models has brought artificial intelligence to unprecedented levels of capability, but it has also exposed the limits of hardware, energy, and latency in real-world deployment. This chapter has examined the methods through which these limits are addressed, showing how compression and optimization strategies make advanced models usable at scale. Quantization and pruning demonstrated how numerical precision and parameter redundancy can be reduced. Knowledge distillation illustrated how knowledge itself can be transferred into compact forms. Efficient architectures showed that efficiency can be designed from the ground up. Finally, serving with optimized runtimes revealed how these compressed and efficient models are deployed in practice. Together, these strategies define the toolkit for making intelligence not only powerful but practical.

Quantization and Pruning As First-Line Tools

Quantization and pruning revealed the inherent redundancy in deep learning systems. By lowering numerical precision, quantization reduced memory and computation while maintaining acceptable performance.

By removing unnecessary parameters, pruning simplified networks and reduced computational overhead. These methods proved that high precision and full parameterization are not required for effective inference.

The conceptual lesson was that intelligence can be approximate. Deep networks tolerate lossy representations without catastrophic degradation. This principle allows practical deployment on hardware ranging from data centers to embedded devices.

Knowledge Distillation As Transfer of Learning

Distillation advanced the compression paradigm by shifting the focus from raw reduction to knowledge transfer. Large teacher models guided smaller students, passing on not just labels but the relational structure of probability distributions. Distillation thus provided students with a richer learning signal, enabling compact models to approach the performance of their larger teachers.

The conceptual lesson was that knowledge is not bound to scale. Smaller systems can learn the essence of larger ones if guided properly. Distillation made clear that efficiency is not merely about subtraction but about intelligent transfer.

Efficient Architectures As Design for Deployment

Efficient architectures such as MobileBERT, TinyBERT, ALBERT, and TinyML demonstrated that efficiency can be embedded into design rather than retrofitted afterward. Through parameter sharing, bottleneck structures, sparse attention, and micro-architectures, these models achieved compactness and speed without catastrophic loss of representational power.

The conceptual lesson was that efficiency is a design philosophy. Rather than building maximalist systems and compressing them, models can be designed for efficiency from inception. This design-first approach anticipates deployment constraints and integrates them into architecture itself.

Serving at Scale As the Final Bottleneck

Serving models at scale required not only smaller models but also optimized runtimes. Frameworks such as ONNX Runtime, TensorRT, and DeepSpeed-Inference showed how batching, operator fusion, graph optimization, and hardware specialization enable large volumes of inference requests to be processed with low latency. Elastic scaling, fault tolerance, and monitoring ensured reliability in production.

The conceptual lesson was that efficiency is systemic. A compressed model is insufficient if runtimes cannot serve it efficiently. Optimized runtimes are the bridge from research to reality, ensuring that theoretical efficiency translates into usable systems.

Cross-Cutting Themes and Trade-Offs

Across all methods, trade-offs defined the practice of compression and optimization.

- **Accuracy vs. efficiency:** Gains in speed and memory often came at some loss in accuracy.
- **Portability vs. specialization:** Methods tailored to specific hardware achieved maximal efficiency but reduced portability.
- **Generality vs. deployment readiness:** Larger models retained broader generality, while efficient models often targeted specific domains.
- **Short-term vs. long-term cost:** Post-training methods were quick to apply, while training-aware methods required more effort but yielded better performance.

These trade-offs showed that efficiency is not absolute but contextual. Each method must be chosen based on deployment requirements, balancing accuracy, latency, memory, and sustainability.

Broader Implications of Compression and Optimization

The significance of compression and optimization extends beyond engineering.

- **Accessibility:** Compact models enable deployment on mobile devices and embedded systems, democratizing access to AI.
- **Sustainability:** Energy-efficient models reduce the environmental cost of large-scale inference.
- **Scalability:** Optimized runtimes allow AI to serve millions of users simultaneously.
- **Innovation:** Efficiency enables applications in real-time translation, on-device speech recognition, and IoT intelligence that would be impossible with full-scale models.

In this sense, compression and optimization are not merely technical techniques but enablers of broader social and economic impact.

Limitations and Open Questions

Despite progress, challenges remain.

- **Accuracy loss:** Quantization and pruning inevitably risk degrading performance.

- **Knowledge bottlenecks:** Distillation depends on teacher quality and cannot fully transfer generality.
- **Hardware constraints:** Efficient architectures must adapt to rapidly changing hardware.
- **Runtime complexity:** Optimized runtimes introduce engineering complexity, raising maintainability concerns.
- **Bias and fairness:** Smaller models distilled from larger ones may inherit biases without the capacity to generalize beyond them.

These limitations remind us that efficiency is a moving target, shaped by technological, economic, and ethical factors.

Compression and inference optimization represent the pragmatic frontier of AI. They transform research breakthroughs into deployable systems, enabling intelligence to operate in real time, at scale, and across diverse hardware environments. Quantization, pruning, distillation, efficient architectures, and optimized runtimes together form a layered toolkit, each addressing a different dimension of the efficiency problem.

The conceptual lesson of this chapter is that intelligence is not only about power but about usability. Models that cannot be deployed are models that cannot serve. Efficiency is therefore not a compromise but a requirement. As AI continues to scale, compression and optimization will remain central, evolving alongside architectures and hardware to make intelligence accessible, reliable, and sustainable.

For our next chapter, we address bias, hallucinations, and failure modes.

CHAPTER 10

Addressing Bias, Hallucinations, and Failure Modes

Generative models are extraordinary tools, capable of producing text, images, audio, and other modalities that rival or surpass human work in fluency and scale. Yet their power is accompanied by distinctive vulnerabilities. Unlike traditional software, which fails in predictable ways, generative models fail in ways that are subtle, probabilistic, and context-dependent. They hallucinate facts, amplify harmful biases, produce toxic or unsafe content, and sometimes reinforce feedback loops that worsen over time. These issues are not simply technical bugs but structural properties of models trained on massive, imperfect datasets through probabilistic learning.

Addressing these challenges is central to making generative AI systems trustworthy and reliable. Hallucinations undermine the credibility of outputs, especially in domains where factual grounding is critical, such as law, medicine, and science. Bias and toxicity risk perpetuating societal inequities, harming marginalized groups, and reducing fairness in applications ranging from hiring to translation. Safety failures and insufficient testing expose organizations to reputational, legal, and ethical

I. Cronin, *Building and Training Generative AI Models*,
https://doi.org/10.1007/979-8-8688-2332-9_10

risks. Feedback loops can cause models to drift over time, amplifying errors or reinforcing skewed worldviews when deployed in dynamic environments.

This chapter examines four major categories of failure modes. First, it considers hallucinations, why they arise, and strategies for detecting and mitigating them. Second, it analyzes toxic and biased generations, exploring both their roots in data and their manifestations in outputs. Third, it discusses safety checks and red teaming, showing how systematic adversarial testing and oversight can uncover weaknesses before they cause harm. Fourth, it examines feedback loops and model updates, emphasizing how deployed systems must be monitored and adapted continuously to avoid performance decay and unintended consequences.

Throughout the discussion, the emphasis will be both conceptual and practical. On the one hand, we need to understand why these failures occur as natural consequences of generative modeling. On the other hand, we must explore concrete strategies, from retrieval augmentation to debiasing pipelines to red team frameworks, that help manage and mitigate them. Case illustrations will highlight real-world consequences, reminding us that these issues are not abstract but affect healthcare, law, education, media, and everyday digital interactions.

The broader significance of this chapter lies in its connection to trust. Generative AI will only be widely adopted if users, institutions, and regulators can trust its outputs. That trust depends not on perfection but on robust safeguards against failure. By addressing bias, hallucinations, and failure modes, we move toward models that are not only powerful but also safe, fair, and aligned with human values.

Detection and Mitigation of Hallucinated Outputs

Generative systems possess a remarkable ability to produce fluent, coherent, and contextually plausible content. Yet this strength is also their most visible weakness. These systems are prone to hallucinations, generating statements that appear credible but lack factual grounding. Hallucinations can take the form of fabricated citations, incorrect factual claims, invented details, or plausible but entirely false reasoning chains. While some hallucinations are harmless in creative contexts, such as fiction writing or brainstorming, they become deeply problematic in domains where reliability and truth are essential.

This section examines hallucinations in detail. It begins by defining and categorizing them, then explores the structural reasons why they occur in generative systems. It next addresses detection strategies, ranging from rule-based systems to retrieval-augmented self-checking. Mitigation strategies are then considered, including retrieval-augmented generation, decoding constraints, and reinforcement learning with human feedback. Finally, the section considers broader implications and case studies, emphasizing both the dangers of hallucinations and the promise of technical and organizational strategies to address them.

Understanding Hallucinations in Generative Systems

Hallucinations occur when a model generates output that is coherent and plausible but factually incorrect or unsupported. They differ from simple mistakes in traditional software.

- **Fabricated content:** Models may generate names, references, or figures that do not exist.

- **Unsupported claims:** Statements may sound authoritative but lack evidence in training data.
- **Reasoning errors:** Multi-step reasoning may appear logical but contains hidden flaws.
- **Spurious citations:** In text generation, models often fabricate references, attributing non-existent papers to real authors.

The problem arises because generative models predict the next token based on statistical patterns rather than explicit fact-checking. Fluency does not guarantee accuracy, and high confidence does not necessarily reflect truth.

Why Hallucinations Occur

Hallucinations arise from structural features of generative modeling.

- **Probabilistic prediction:** Generative systems maximize likelihood, not factual grounding. They generate what is statistically plausible, not necessarily true.
- **Incomplete training data:** Even with massive datasets, gaps remain, leading to extrapolation beyond reliable knowledge.
- **Over-generalization:** Models infer patterns beyond their evidence, producing fabricated but coherent outputs.
- **Decoding strategies:** Sampling and beam search may prioritize coherence or diversity over truth.
- **Lack of grounding:** Models trained only on text lack mechanisms to verify claims against external sources.

The inevitability of hallucinations stems from the mismatch between statistical prediction and factual reliability.

Categories of Hallucinations

Hallucinations can be categorized into distinct types.

- **Intrinsic hallucinations:** Errors that contradict source material or user-provided context.
- **Extrinsic hallucinations:** Fabrications that extend beyond the provided input, often plausible but unsupported.
- **Entity-level hallucinations:** False names, dates, or citations.
- **Reasoning hallucinations:** Logical structures that appear sound but rest on false assumptions.
- **Contextual hallucinations:** Errors arising when models fail to adapt to domain-specific expectations.

These categories guide the design of detection and mitigation strategies.

Risks and Implications of Hallucinations

The risks vary by context.

- **Medical applications:** Fabricated treatment recommendations can endanger patients.
- **Legal domains:** Spurious citations undermine credibility in law.
- **Education:** False explanations mislead learners.

- **Scientific research:** Fabricated references distort academic discourse.
- **Everyday use:** Users may be misled by plausible but false statements.

Hallucinations erode trust in generative systems, limiting their adoption in high-stakes domains.

Detection Strategies for Hallucinations

Detecting hallucinations is a prerequisite for mitigation. Several strategies have been developed.

- **Rule-based detection:** Outputs are scanned for anomalies such as non-existent citations or improbable facts.
- **Retrieval-based verification:** Model outputs are compared against trusted databases or search systems.
- **Self-consistency checks:** Models generate multiple outputs for the same prompt; inconsistent results indicate potential hallucination.
- **Confidence estimation:** Models assign probabilities to outputs, though high probability does not always equal accuracy.
- **Fact-checking pipelines:** External systems analyze and verify factual claims in generated text.

Detection strategies illustrate the need for layered approaches, as no single method guarantees accuracy.

Mitigation Through Retrieval-Augmented Generation

One of the most effective mitigation strategies is retrieval-augmented generation (RAG).

- **External grounding:** Models access databases, search engines, or knowledge graphs during generation.
- **Inline citations:** Retrieved documents provide evidence that grounds outputs.
- **Reduced hallucination:** By anchoring responses in external data, hallucinations decrease significantly.
- **Trade-offs:** Retrieval adds latency and depends on the quality of external sources.

RAG illustrates the principle that hallucination mitigation requires grounding in real data, not just statistical inference.

Mitigation Through Decoding Constraints

Decoding strategies influence hallucination rates.

- **Beam search with penalties:** Restricting diversity can reduce unsupported content.
- **Constrained decoding:** Enforcing templates or vocabularies limits spurious outputs.
- **Temperature control:** Lower sampling temperature reduces creativity but improves factuality.
- **Hybrid strategies:** Balancing diversity and constraint to optimize both coherence and accuracy.

Decoding illustrates the tension between creativity and reliability.

RLHF for Hallucination Reduction

RLHF has proven useful for hallucination mitigation.

- **Preference signals:** Annotators prefer factual, grounded responses over fabricated ones.
- **Reward models:** These preferences are encoded into reward functions.
- **Optimization:** PPO or other methods align model outputs with human judgment.
- **Effectiveness:** Models trained with RLHF show reduced hallucination rates compared to pre-trained baselines.

RLHF highlights the role of human oversight in guiding models away from failure modes.

Hybrid Mitigation Pipelines

Effective mitigation often requires combining strategies.

- **Detection first:** Outputs are screened for anomalies.
- **Verification second:** Questionable outputs are checked against retrieval systems.
- **Mitigation third:** Decoding constraints and RLHF align outputs with grounded responses.
- **Continuous learning:** Feedback from users and red teams informs ongoing fine-tuning.

Hybrid pipelines balance efficiency, accuracy, and scalability.

Case Illustrations of Hallucinations

Several high-profile cases illustrate the risks.

- **Legal hallucinations:** Lawyers using generative systems for filings submitted fabricated case citations.
- **Medical hallucinations:** Systems provided fabricated clinical recommendations, risking patient safety.
- **Academic hallucinations:** Students received invented references when asking for citations.
- **Media hallucinations:** Models generated fabricated news articles during experiments.

These cases underscore why hallucination mitigation is not optional but essential.

Broader Implications

Hallucinations reveal a deeper truth about generative models. They are not fact-checking systems but probability machines.

- **Trustworthiness:** Users must understand the probabilistic nature of outputs.
- **System design:** Mitigation must be built into pipelines, not added as an afterthought.
- **Societal impact:** Hallucinations erode trust in AI if unaddressed, limiting adoption.
- **Epistemic humility:** Designers and users must recognize the limits of generative systems.

Mitigation is therefore not only technical but philosophical.

Hallucinations are an inevitable consequence of probabilistic generative modeling, but they need not be fatal to trust. By categorizing, detecting, and mitigating them, we can manage their risks. Strategies such as retrieval augmentation, decoding constraints, and reinforcement learning with human feedback reduce hallucinations significantly. Hybrid pipelines that integrate detection, verification, and mitigation offer the most robust defense.

The broader lesson is that hallucinations are not bugs but features of statistical generation. They cannot be eliminated entirely, but they can be managed to acceptable levels. Addressing hallucinations is therefore a continuous process, requiring technical strategies, organizational oversight, and user education. Only by confronting hallucinations directly can generative systems be deployed safely in domains where accuracy and reliability matter most.

Managing Toxic and Biased Generations

Generative systems have expanded the boundaries of AI, enabling text, images, audio, and multimodal outputs that approach or surpass human-level fluency. Yet with this power comes the risk of generating outputs that are toxic, offensive, or biased.

These outputs reflect not only technical shortcomings but also the sociotechnical reality that data is a mirror of society. If training data contains prejudice, stereotypes, or imbalances in representation, generative systems will learn and reproduce them. If probabilistic learning favors fluency over fidelity to ethical norms, the results will sometimes be harmful. Toxic and biased generations are therefore systemic challenges, not isolated anomalies.

This section explores the origins, detection, and mitigation of toxicity and bias in generative systems. It begins with an examination of the sources of these problems in data, labeling, and modeling practices. It then categorizes the ways bias manifests in outputs and, importantly, makes explicit connections between each source of bias and the specific category of harm it produces, such as representational distortion, exclusion, or allocative unfairness. Next, it analyzes detection methods, from automated classifiers to adversarial probes. Mitigation strategies are considered in depth, including dataset curation, algorithmic debiasing, reinforcement learning with human feedback, and cultural adaptivity. Case illustrations from chatbots, hiring systems, translation models, and healthcare reveal the risks of unmanaged toxicity. Finally, the section discusses the broader societal implications and regulatory frameworks that shape this area, underscoring that bias management is both a technical and ethical responsibility.

Sources of Toxicity and Bias

The presence of toxicity and bias in generative systems is rooted in the full pipeline of their development.

- **Training data as the primary source:** Large language models and image generators are trained on massive corpora harvested from the Internet. These corpora inevitably contain hate speech, offensive content, stereotypes, and historically imbalanced representations of groups. The statistical learning process captures these patterns without distinguishing harmful content from neutral or beneficial patterns.
- **Annotation processes:** In supervised or reinforcement learning pipelines, human annotators bring their own perspectives and cultural assumptions. These perspectives may encode subjective bias into labeled data, particularly if annotation guidelines are vague or inconsistent.

- **Model architectures:** Neural networks amplify correlations in data. If certain demographic attributes are associated with particular outcomes in the data, the model may strengthen these associations, leading to biased generations.
- **Feedback loops in deployment:** Once models are deployed, their outputs influence future inputs. A biased recommendation may lead to biased user behavior, which in turn reinforces the model's tendencies in retraining cycles.
- **Omissions in representation:** Underrepresentation of certain groups, perspectives, or languages leads to skewed outputs. A model trained primarily on English-language corpora may fail to reflect the nuance of other linguistic or cultural traditions.

In short, bias is not accidental but structurally embedded. Addressing it requires interventions at multiple stages of the model pipeline, from data collection to deployment monitoring.

Categories of Bias in Generative Systems

Bias manifests in diverse and sometimes subtle ways.

- **Stereotypical bias:** Outputs reinforce harmful stereotypes. A system may associate women with caregiving roles and men with leadership roles. In translation, gender-neutral pronouns may be rendered into stereotypically gendered equivalents.
- **Demographic bias:** Outputs disproportionately represent or exclude groups. Text-to-image models, for example, may depict doctors primarily as men and nurses primarily as women.

- **Toxic bias:** Outputs contain hate speech, derogatory language, or slurs. Chatbots trained without safeguards may reproduce offensive material from training data.
- **Contextual bias:** Outputs fail to adapt to cultural or situational norms. A model may use expressions that are acceptable in one culture but offensive in another, or fail to account for local sensitivity in humor or idioms.
- **Systemic bias:** Outputs replicate structural inequities embedded in data. For example, predictive text in hiring systems may rank resumes differently based on subtle demographic cues.

Recognizing these categories enables more precise detection and targeted mitigation.

Historical Roots of Bias Awareness in AI

Awareness of bias in artificial intelligence predates generative systems.

- **Early classification systems:** Studies in the early 2010s showed that facial recognition systems had higher error rates for darker-skinned individuals, revealing demographic imbalances in training data.
- **Word embeddings:** In natural language processing, research showed that embeddings captured stereotypical associations, such as linking "man" with "programmer" and "woman" with "homemaker."
- **Algorithmic fairness research:** This sparked the development of fairness metrics and mitigation techniques, focusing initially on classification and recommendation.

- **Transition to generative AI:** Generative systems expanded the problem. Instead of producing binary or numeric outputs, they produced open-ended text and images, multiplying the opportunities for bias to appear in subtle, context-dependent ways.

Bias management in generative systems builds on these earlier insights but faces greater complexity due to the scale and openness of outputs.

Risks and Implications of Toxic Outputs

The risks of unmanaged toxicity and bias extend from individual harm to societal impact.

- **Direct user harm:** Toxic content can traumatize users, especially children, survivors of violence, or members of marginalized communities.
- **Reinforcement of inequality:** Biased outputs perpetuate social hierarchies. If translation systems consistently associate certain roles with certain genders, they reinforce occupational segregation.
- **Institutional risks:** Organizations deploying biased systems risk reputational damage, regulatory penalties, or lawsuits.
- **Erosion of trust:** Users lose confidence in AI if outputs are consistently offensive, inaccurate, or exclusionary.
- **Adoption barriers:** High-stakes industries such as healthcare, law, and finance may refuse to adopt generative systems without strong bias safeguards.

These implications make bias management essential for both ethical and practical reasons.

Detection Strategies for Toxicity and Bias

Detection provides visibility into when and how bias occurs.

- **Toxicity classifiers:** Specialized models evaluate outputs for offensive language. These classifiers are trained on annotated datasets of toxic and non-toxic content, enabling real-time screening.
- **Bias benchmarks:** Standardized benchmarks measure fairness across tasks. For example, datasets test whether models translate gender-neutral inputs into stereotyped outputs.
- **Adversarial probing:** Researchers craft prompts to intentionally expose biases. For instance, asking a chatbot to describe people in certain professions may reveal stereotypical associations.
- **Human evaluation:** Annotators review outputs for bias and toxicity, providing qualitative insights that automated tools miss.
- **Explainability methods:** Techniques such as attention heatmaps can reveal whether certain demographic cues drive biased outputs.

Detection alone does not solve bias, but it enables measurement and accountability.

Dataset Curation for Bias Mitigation

Addressing bias at the source requires careful dataset construction.

- **Filtering harmful content:** Removing hate speech, slurs, and offensive material reduces the baseline risk of toxicity.
- **Balancing representation:** Ensuring proportional representation of demographic groups counteracts skewed distributions.
- **Clear annotation guidelines:** Providing annotators with consistent, culturally aware instructions reduces subjective bias.
- **Synthetic augmentation:** Generating additional examples for underrepresented groups balances training corpora.
- **Domain-specific tailoring:** In fields like healthcare, domain experts curate datasets to ensure sensitivity and accuracy.

Curation emphasizes that mitigation is not only post hoc but foundational.

Algorithmic Debiasing Techniques

Algorithmic interventions complement data curation.

- **Adversarial debiasing:** Models are trained with adversarial objectives that minimize the association between sensitive attributes and outputs.
- **Counterfactual data augmentation:** For each example, a counterfactual is created by altering demographic attributes, forcing the model to generalize fairly.
- **Regularization techniques:** Penalties are applied to discourage biased associations during optimization.

- **Fairness constraints in loss functions:** Objectives include explicit fairness terms, guiding models toward equitable outputs.
- **Embedding adjustments:** Representation spaces are modified to reduce stereotypical associations between demographic groups and roles.

These methods treat fairness as an optimization goal alongside accuracy.

Reinforcement Learning with Human Feedback for Safety

RLHF has become a cornerstone of bias mitigation.

- **Preference modeling:** Annotators evaluate outputs for helpfulness, harmlessness, and honesty.
- **Reward models:** These evaluations are translated into reward signals.
- **Policy optimization:** Reinforcement learning adjusts the model to favor outputs aligned with human preferences.
- **Results:** RLHF significantly reduces toxicity and improves safety across diverse contexts.

The key insight is that human judgment provides a richer signal for fairness than raw data alone.

Cultural and Contextual Adaptivity

Fairness must reflect cultural and linguistic diversity.

- **Norm variance:** What is considered offensive in one culture may be acceptable in another. Mitigation systems must adapt dynamically.
- **Language-specific toxicity:** Harmful expressions differ across languages, requiring localized benchmarks.
- **Cultural inclusivity:** Annotation teams must include diverse perspectives to avoid embedding majority biases.
- **Adaptive filters:** Systems may adjust sensitivity based on deployment region or user preference.

Bias mitigation is not a universal formula but a context-dependent process.

Hybrid Mitigation Pipelines

Robust mitigation requires layered pipelines.

- **Pre-training interventions:** Dataset filtering and balancing reduce baseline bias.
- **Training interventions:** Algorithmic debiasing methods reshape model representations.
- **Post-training interventions:** RLHF aligns outputs with social norms.
- **Inference interventions:** Real-time filters screen outputs before reaching users.
- **Feedback interventions:** User reports and red teaming provide continual improvement.

This redundancy ensures that no single failure allows toxicity to propagate unchecked.

Case Illustrations of Toxic and Biased Generations

Case studies demonstrate real-world consequences.

- **Chatbot failures:** Early conversational agents, exposed to unmoderated data, produced racist and sexist content within days of deployment. These failures highlighted the dangers of releasing models without safeguards.
- **Hiring discrimination:** Language models trained on historical resumes learned to associate male-coded language with technical roles, reproducing gender bias in candidate rankings.
- **Biased translation:** Gender-neutral sentences were consistently translated into stereotypically gendered forms, reinforcing occupational stereotypes.
- **Healthcare inequities:** Biased datasets led models to recommend different treatments for patients with similar symptoms based on race, perpetuating systemic disparities.
- **Recommendation systems:** Generative models powering recommendations amplified existing biases, reinforcing popularity for already dominant groups while marginalizing minority voices.

Each case illustrates how bias in generative systems translates into social harm if left unmanaged.

Regulatory and Governance Frameworks

Governments and institutions are increasingly addressing AI bias through regulation.

- **EU AI Act:** Establishes risk categories and requires fairness and safety for high-risk applications.
- **U.S. AI Bill of Rights:** Outlines principles for algorithmic fairness and non-discrimination.
- **OECD guidelines:** Promote transparency, accountability, and fairness in AI systems.
- **Corporate governance:** Companies establish internal AI ethics boards and red teaming practices to manage bias.
- **Auditing frameworks:** Independent audits provide accountability for bias management.

Bias mitigation is thus embedded in broader structures of governance.

Open Questions and Future Directions

Bias management remains a frontier with unresolved debates.

- **Definition of fairness:** Competing definitions exist, such as demographic parity and equal opportunity, and choosing between them is context-dependent.
- **Trade-off with expression:** Reducing toxicity must be balanced with protecting freedom of speech.
- **Measurement challenges:** Subtle biases are difficult to capture with benchmarks.

- **Dynamic adaptation:** Models must evolve with shifting cultural norms.
- **Global governance:** Coordinating fairness across jurisdictions with different values remains unsolved.

These open questions ensure that bias mitigation will remain an active field of research and policy.

Toxic and biased generations are structural features of generative systems, arising from biased data, annotation practices, model architectures, and deployment feedback loops. They manifest in stereotypes, demographic disparities, toxic content, and contextual insensitivity. Detection methods such as classifiers, benchmarks, adversarial probes, and human evaluation provide visibility, while mitigation strategies such as dataset curation, algorithmic debiasing, RLHF, and cultural adaptivity reduce harmful outputs. Hybrid pipelines integrate these methods across stages of development and deployment.

Case studies show the risks of unmanaged bias in chatbots, hiring, translation, healthcare, and recommendations. Regulatory frameworks such as the EU AI Act and US AI Bill of Rights highlight the societal importance of fairness. Yet open questions remain about how to define fairness, balance competing values, and govern globally.

The broader lesson is that managing toxicity and bias is not optional but essential for trustworthy AI. It is both a technical and ethical challenge, requiring continuous attention, cultural sensitivity, and governance. Only by addressing toxicity and bias directly can generative systems be deployed responsibly, equitably, and sustainably.

Model Safety Checks and Red Teaming

Generative models are unlike any prior form of software. They are probabilistic systems, capable of producing open-ended outputs that are fluent, persuasive, and sometimes indistinguishable from human communication. This flexibility is their strength, but it also introduces profound risks. A model that can generate essays, code, and images can also generate offensive slurs, fabricated information, unsafe instructions, or biased portrayals of individuals and groups. Unlike traditional deterministic programs, which fail in bounded and predictable ways, generative systems can fail unpredictably, producing unsafe results in contexts that are difficult to anticipate.

For this reason, developers must establish robust practices of safety checks and red teaming. Safety checks represent integrated safeguards within the life cycle of the model: screening inputs, filtering outputs, constraining content, and continuously monitoring deployed systems. Red teaming represents the deliberate effort to stress test systems by adopting adversarial perspectives. It is not merely a technical practice but also a cultural one, emphasizing the importance of internal critics who actively seek to uncover weaknesses before deployment. The stakes are evident in real incidents: early social media chatbots that were rapidly coaxed into producing racist and misogynistic content, code assistants that suggested insecure or vulnerable implementations, and text and image generators that produced hateful or sexually explicit content when given only mildly provocative prompts. These episodes show that safety failures are not hypothetical edge cases but recurring outcomes when systems are deployed without systematic stress testing and layered safeguards.

This section examines the conceptual foundations and technical practices of safety checks and red teaming in generative AI. It expands each category of intervention in detail, explores organizational and regulatory integration, and introduces new perspectives drawn from historical analogies, cultural variation, and multi-agent adversarial ecosystems. The goal is to show that safety is not a peripheral consideration but a core requirement of trustworthy AI deployment.

Rationale for Safety Checks in Generative Systems

Generative systems demand safety checks because their failures can be severe and socially harmful. Unlike traditional software bugs, which often manifest as crashes, errors, or predictable malfunctions, the failures of generative models appear as toxic sentences, offensive jokes, dangerous instructions, or fabricated facts.

- **Unbounded output space:** Generative systems can produce virtually any combination of words or pixels. Without checks, this includes offensive, misleading, or unsafe outputs.
- **Probabilistic uncertainty:** Because the model samples from probability distributions, even a system tuned for safety will occasionally generate harmful outputs by chance.
- **Prompt sensitivity:** Seemingly benign prompts can be rephrased to elicit unsafe content. This makes filtering and safety more complex than static rule enforcement.
- **User vulnerability:** Outputs can disproportionately affect sensitive groups, such as children exposed to toxic content or patients misled by fabricated medical advice.
- **Institutional risk:** Organizations deploying unsafe systems risk legal penalties, reputational harm, and user distrust.

In practice, developers should prioritize failures by combining severity and likelihood, focusing first on failure modes that can cause irreversible harm or systemic discrimination, then on those that create serious psychological or economic harm, and only afterward on nuisance behaviors, using a simple risk matrix and input from affected stakeholders to rank which categories must be addressed with the strongest protections.

- Analogies from other fields clarify why safety checks are indispensable.
- In **aviation**, no aircraft is flown without pre-flight safety checks because the consequences of failure are catastrophic.
- In **medicine**, no drug is released without trials and monitoring, because unpredictable side effects can endanger lives.
- In **finance**, no system is deployed without stress tests to prevent systemic collapse.

Generative AI requires an equivalent discipline: built-in safety checks that prevent, detect, and mitigate harmful outputs.

Categories of Model Safety Checks

Safety checks occur at multiple stages and at varying levels of granularity.

- **Input filtering:** Prompts are scanned for harmful intent, such as attempts to generate instructions for violence or self harm. These filters can be simple, keyword-based systems or advanced classifiers that interpret semantics.
- **Syntactic checks:** Inputs and outputs are screened for explicit profanity, hate speech, or offensive terms. While basic, these checks provide a first line of defense.

- **Semantic checks:** More advanced filters detect toxicity even when offensive terms are avoided, capturing context and nuance. For example, sarcastic or coded language that conveys hate can be flagged.
- **Contextual checks:** Domain-specific safeguards prevent certain categories of output. For example, medical AI systems may refuse to provide treatment plans, redirecting users to professional resources.
- **Output filtering:** Once generated, responses are scanned before delivery. Filters remove unsafe outputs that slip through the model.
- **Real time anomaly detection:** Statistical systems monitor patterns of output over time, detecting spikes in unsafe content.
- **Post deployment monitoring:** Logs are analyzed for emerging risks, with continuous improvement cycles to refine filters.

Quantitatively, teams measure toxicity, bias, and safety with labeled evaluation sets and automatic classifiers, reporting metrics such as the fraction of unsafe outputs per thousand generations, differences in error or toxicity rates across demographic groups, and aggregate safety scores that combine severity and frequency of harmful responses.

- Robust systems apply multiple categories of safety checks redundantly, ensuring that if one layer fails, another intervenes.

Automated vs. Human-in-the-Loop Safety

Safety interventions can be automated, human-driven, or hybrid.

- **Automated systems:** These provide scalability. Classifiers evaluate millions of inputs and outputs quickly, ensuring consistent coverage. Examples include toxicity classifiers trained on large, annotated datasets.
- **Human-in-the-loop oversight:** Humans evaluate outputs in ambiguous or high-stakes contexts, where nuance and cultural knowledge are critical. Human reviewers detect subtle harms that automation may miss.
- **Hybrid workflows:** Most robust systems combine both. Automated filters catch obvious cases, while humans review flagged outputs or high-risk scenarios. For example, a system might automatically block clearly toxic content, queue borderline cases for a trust and safety team to review, and route certain domains such as medical, legal, or self-harm-related queries to specialists or structured escalation flows. In other deployments, periodic random samples of conversations are surfaced to human reviewers alongside model and filter scores so that policy violations, near misses, and novel attack patterns can be audited and used to update both the filters and the model itself.
- Challenges arise in balancing these approaches.
- Automated systems risk false positives and false negatives, blocking benign content or allowing harmful content to pass.

- Human review ensures nuance but cannot scale to millions of requests per day.
- Hybrid workflows must resolve disagreements between humans and machines, ensuring coherent policy enforcement.

The most effective deployments use automation for breadth and humans for depth, aligning scalability with contextual sensitivity.

Red Teaming As a Conceptual Framework

Red teaming brings an adversarial perspective to safety. Rather than assuming that safeguards are sufficient, red teams actively try to break systems.

- **Origins:** The concept originated in military exercises, where a designated "red team" played the role of adversary to test readiness.
- **Adoption in cybersecurity:** Red teaming became standard practice, where attackers simulate intrusions to expose vulnerabilities.
- **Application to AI:** Generative models require similar adversarial testing. Red teams craft prompts, scenarios, and use cases designed to expose weaknesses, often organized as focused events or ongoing programs. Public examples include organized red teaming efforts around large language model releases, where external researchers and civil society groups are invited to probe systems for harmful behaviors, jailbreaks, and policy violations, with their findings feeding back into model updates and safety policies.

Conceptually, red teaming reflects humility: the recognition that no model is perfectly safe and that vulnerabilities will only be exposed through deliberate confrontation with hostile scenarios, including those designed by people outside the original development team.

Techniques of Red Teaming Generative Models

Red teams employ diverse methods to uncover vulnerabilities.

- **Prompt manipulation:** Creative rephrasings of prompts bypass filters, eliciting unsafe responses. For example, instead of directly asking for harmful instructions, a user might ask the model to "explain what not to do" and then invert the advice.
- **Context embedding:** Harmful queries are hidden within benign requests to trick models into unsafe outputs, such as burying detailed instructions for wrongdoing in the middle of what appears to be a harmless story or code review.
- **Role playing scenarios:** Models are asked to assume personas that permit unsafe behavior, such as pretending to be a fictional character, "uncensored assistant," or "simulation mode" that ignores safety rules and provides harmful instructions.
- **Stepwise reasoning exploitation:** Multi step prompts coax models into producing unsafe content indirectly, for instance, by first asking for a high level description, then requesting more precise details, and finally combining them into an actionable plan.

- **Multimodal attacks:** Combining text and images creates adversarial conditions for cross modal systems, including attempts to hide instructions inside images, screenshots, or diagrams that the model is asked to interpret.
- **Stress tests:** High volume queries identify weaknesses under load, revealing how systems degrade when overwhelmed and whether safety filters fail open under heavy traffic.
- **Multilingual and cross-cultural testing:** Red teams probe systems in multiple languages, dialects, and cultural contexts to uncover failures that only appear outside the primary training language, such as toxic slurs, hate speech, or political manipulation expressed through regional idioms or coded expressions. Prompt injection and jailbreaks in these settings can be particularly subtle, relying on local metaphors, humor, or workaround phrases.

The diversity of these techniques demonstrates that safety cannot rely on static defenses but requires continual adaptation.

Red teaming success is measured quantitatively as well as qualitatively, for example, by tracking the rate of successful jailbreaks per thousand adversarial prompts, the reduction in that rate after each mitigation cycle, the number of distinct attack patterns neutralized, and severity-weighted risk scores that combine how often a failure occurs with how harmful its consequences would be.

Lessons from Cybersecurity Red Teaming

Generative AI red teaming parallels cybersecurity practices.

- **Defense in depth:** Just as cybersecurity layers firewalls, intrusion detection, and encryption, AI requires layered safety checks.
- **Penetration testing analogy:** Red teams simulate attackers to reveal vulnerabilities before adversaries exploit them.
- **Zero-day vulnerabilities:** New failure modes appear without warning, just as unknown exploits appear in cybersecurity.
- **Arms race dynamic:** Attackers and defenders evolve together, ensuring that red teaming must be continuous.

These parallels emphasize that AI safety must adopt a mindset of adversarial resilience, not passive prevention.

Organizational Structures for Red Teaming

Red teaming requires institutional support.

- **Dedicated internal teams:** Many companies establish specialized red teams with the authority and resources to challenge product development.
- **Cross-disciplinary membership:** Effective red teams combine engineers, linguists, ethicists, and domain experts to capture diverse perspectives.
- **External partnerships:** Independent researchers, NGOs, and contractors provide external scrutiny, ensuring transparency.

- **Governance integration:** Red team findings feed directly into risk management and compliance systems, ensuring accountability.
- **Feedback loops:** Lessons learned inform model retraining, safety filter updates, and deployment decisions.

Without strong organizational support, red teaming risks becoming symbolic rather than substantive.

Continuous Monitoring and Adaptive Safety

Safety cannot be static. New risks emerge continually, requiring monitoring and adaptation.

- **Dynamic threats:** Malicious actors invent new ways to bypass filters.
- **Shifting norms:** Cultural standards evolve, making some language newly toxic or acceptable.
- **Model updates:** Each new release changes behavior, requiring renewed red teaming.
- **Real-world monitoring:** Logs and user reports provide evidence of emerging problems.
- **Adaptive improvement:** Continuous feedback cycles refine safety systems iteratively.

The principle is that safety is a process, not a one-time product feature.

Case Illustrations of Safety and Red Teaming

Case studies demonstrate the stakes of safety and the value of red teaming.

- **Chatbot collapse:** A widely publicized chatbot began producing racist and sexist content within days of release. Lack of red teaming left it vulnerable to adversarial users.

  At a deeper level, this failure reflected several gaps. Organizationally, launch incentives favored speed and novelty over safety sign-off, so the system was deployed without structured pre-release abuse testing or clear go/no-go criteria. Technically, the model had no robust content filtering or rate limiting around sensitive topics, and there was no human-in-the-loop review for high-risk conversations. Monitoring was reactive rather than proactive, so the first serious safety evaluation happened in public, under adversarial pressure from users.

- **Search augmentation risk:** Generative models, combined with web search, produced fabricated or unsafe results, which internal red teams had predicted but which were not adequately addressed before launch. Here the core gap was a mismatch between known risks and product decisions. Organizationally, red team findings were treated as warnings rather than hard blockers, and there was no formal mechanism that forced mitigation of high-severity issues before release. Technically, the system lacked strong attribution, source ranking, and conflict resolution logic, so hallucinated text was presented with the same apparent authority as trusted documents. Guardrails that could have constrained medical, legal, and financial topics were either absent or not deeply integrated into the ranking and generation pipeline.

- **Multimodal stereotype generation:** Image generators produced harmful stereotypes in response to prompts for certain professions, until red teams highlighted the issue and forced retraining. The underlying causes were data and evaluation gaps. Training sets overrepresented biased associations between demographic traits and occupations, and there were no strong counterbalancing objectives in the loss function. Organizationally, evaluation suites focused on visual quality and prompt fidelity rather than fairness across demographic groups, so problems were invisible in standard benchmarks. Only when external and internal red teams systematically probed prompts across gender, race, and profession combinations did the pattern become impossible to ignore, prompting targeted data curation and fine-tuning.
- **Educational misuse:** Students used generative models to bypass academic integrity, while filters designed for plagiarism detection were insufficient. This failure stemmed from a narrow threat model. The systems were treated as generic productivity tools rather than assessment-critical technologies. Organizationally, there was little coordination with schools and universities before launch, so usage policies, detection mechanisms, and guidance for educators lagged behind adoption. Technically, safeguards focused on blocking explicit requests to "write my homework" or on surface plagiarism checks, underestimating how easily users could rephrase instructions or lightly edit outputs to evade detection. No robust design was in place for exam-safe modes, institution-level controls, or traceable usage patterns.

- **Financial advice errors:** Generative systems offered fabricated investment guidance, leading to potential legal liability. The root issues combined missing domain boundaries with weak calibration of uncertainty. Organizationally, product and legal teams did not clearly define which financial use cases were acceptable, nor did they enforce strict opt-outs for regulated advice scenarios. Technically, the models were not constrained to defer to authoritative data sources, and they lacked mechanisms to reliably say "I do not know" or to consistently label outputs as illustrative rather than actionable. Output disclaimers existed, but they were layered on top of a system that still generated confident, specific, and sometimes incorrect recommendations, leaving users exposed and organizations at risk.

Taken together, these examples show that safety failures usually arise where organizational structures, incentives, and review processes are misaligned with technical safeguards: models are capable, guardrails are partial or absent, red team warnings are not treated as blockers, and there is no rigorous gate between known risks and product launch.

Trade-Offs in Safety and Red Teaming

Safety checks involve trade-offs that must be managed carefully.

- **Strictness vs. usability:** Overly aggressive filters block benign content, frustrating users. Too lenient filters allow harm.

- **Automation vs. nuance:** Automated systems provide scale but miss cultural nuance, while human oversight captures nuance but cannot scale.
- **Transparency vs. exploitation:** Sharing safety methods builds trust but may give adversaries tools to bypass them.
- **Innovation vs. caution:** Pressure to release new models conflicts with the time needed for thorough red teaming.

Navigating these trade-offs requires careful governance and clear prioritization of safety over speed.

Broader Implications of Safety and Red Teaming

The implications of safety checks and red teaming extend beyond individual systems.

- **Trust building:** Users are more willing to adopt AI systems that are demonstrably safe.
- **Ethical accountability:** Organizations show responsibility by prioritizing harm reduction.
- **Regulatory compliance:** Safety checks align with emerging laws requiring fairness and transparency.
- **Industry standards:** As practices mature, safety checks and red teaming may become codified into certification regimes.
- **Cultural shift:** Viewing safety as continuous fosters a culture of accountability.

Safety is thus part of the social contract between AI developers and society.

Future Directions for Safety and Red Teaming

Future directions will expand the sophistication of safety practices.

- **Automated red teaming:** Generative systems themselves may serve as adversaries, probing each other for vulnerabilities.
- **Multi-agent oversight:** Teams of AI systems and humans may monitor each other continuously.
- **Integration with regulation:** Safety checks may be required by law before deployment.
- **Global collaboration:** International consortia may share red teaming insights across jurisdictions.
- **Adaptive safety systems:** Filters may learn and evolve dynamically in response to new attacks.

These trajectories reflect the continuing evolution of safety as both technical and institutional practice.

Historical Analogies in Safety Testing

Generative AI can learn from other industries that faced catastrophic risks.

- **Aviation:** Pre-flight checklists and black-box monitoring created systematic safety cultures.
- **Nuclear power:** Redundancy, containment, and stress-testing are standard to prevent catastrophic failures.
- **Pharmaceuticals:** Multi-stage trials and post-market surveillance manage unforeseen side effects.
- **Automobiles:** Crash tests and recalls enforce continuous monitoring and accountability.

Each of these fields demonstrates that safety requires institutionalized discipline, not ad hoc fixes.

Regulatory Integration of Red Teaming

Regulators are beginning to mandate safety practices.

- **EU AI Act:** Requires risk assessment and documentation of high-risk systems.
- **US guidance:** Draft policies emphasize testing and transparency in AI systems.
- **Corporate compliance:** Companies increasingly integrate red teaming into official governance frameworks.
- **Independent audits:** Third-party evaluations ensure credibility and accountability.

The institutionalization of red teaming shows its movement from optional best practice to regulatory expectation.

Cultural and Ethical Dimensions of Safety

Safety is not only technical but cultural.

- **Cultural variation:** What is considered harmful differs across societies. Filters must adapt dynamically.
- **Ethical pluralism:** Multiple ethical frameworks guide decisions, from harm reduction to freedom of expression.
- **User expectations:** Different communities may demand stricter or looser safety controls.

- **Global deployment:** Models deployed internationally must navigate conflicting cultural standards.

Safety is therefore inseparable from ethical and cultural sensitivity.

Multi-agent Red Teaming Ecosystems

A future direction involves adversarial ecosystems of AI agents.

- **AI adversaries:** Models can generate adversarial prompts to test other models continuously.
- **Collaborative testing:** Multiple agents coordinate to probe vulnerabilities at scale.
- **Human-AI collaboration:** Humans direct AI adversaries, combining creativity with automation.
- **Continuous stress testing:** Systems evolve dynamically, reducing reliance on static benchmarks.

This vision reflects safety as an ecosystem rather than a static checklist.

Model safety checks and red teaming form the backbone of responsible generative AI deployment. Safety checks ensure layered defenses, from input filtering to output monitoring, while red teaming actively exposes vulnerabilities through adversarial probing. Expanded by lessons from cybersecurity, aviation, and medicine, these practices emphasize that safety must be continuous, adaptive, and institutionalized.

With regulatory integration, cultural sensitivity, and the rise of multi-agent adversarial ecosystems, the future of safety in AI will mirror other high-stakes industries that learned to institutionalize trust through rigorous testing. The broader lesson is that safety is not a one-time achievement but a living practice, requiring vigilance, humility, and continuous innovation.

Feedback Loops and Model Updates

Generative models are not static entities. They live within ecosystems of users, data streams, and institutions, and they shape and are shaped by those environments. Unlike traditional software, which follows deterministic rules, generative systems learn patterns from massive corpora and continue to influence and be influenced by the data and contexts into which they are deployed. This dynamic interaction creates feedback loops. Outputs influence human behavior, human behavior influences new data, and new data influences subsequent versions of the model.

Feedback loops carry both risks and opportunities. Left unmanaged, they can amplify misinformation, entrench biases, and create instability. Managed effectively, they can serve as engines of alignment, enabling systems to improve continuously through structured signals. Model updates are the mechanisms by which organizations respond to feedback loops. Updates can retrain or fine-tune models, integrate reinforcement learning from human or institutional feedback, or introduce governance-driven changes. Yet updates themselves introduce trade-offs. Too frequent updates risk unpredictability, while too infrequent updates allow errors to persist.

This section examines feedback loops and model updates at great depth. It analyzes their nature and categorization, explores risks and case studies, and examines strategies for monitoring, governance, and updating. It connects AI dynamics to broader theories from economics, ecology, control theory, psychology, and cybernetics. It addresses institutional and cultural dimensions, and it looks ahead to futures involving continual learning, self-healing systems, and federated updating.

Introduction to Feedback Loops in Generative Systems

Feedback loops occur when the outputs of a system influence its future inputs, creating cycles that can either reinforce or correct behavior. Generative models, which are designed to produce text, images, and other modalities, are especially susceptible to such loops once they are deployed at scale. Unlike traditional software, their outputs often circulate back into the environments from which future training data is drawn.

A chatbot that generates a fabricated but persuasive explanation illustrates the risk. If the text is cited online, it may be scraped into future training datasets, reintroducing the original fabrication. Similarly, when a model depicts certain professions in biased ways, those portrayals can shape user perceptions and, over time, influence the very data that models later consume. By contrast, when a model responds cautiously to medical queries and directs users to qualified professionals, it reduces harmful reliance and fosters a corrective cycle that strengthens trust.

- Fabricated outputs can become future training inputs.
- Biased portrayals shape cultural perceptions and data.
- Careful handling of sensitive queries can build resilience.

Feedback loops, in this sense, are not incidental outcomes of generative AI. They are structural features that define how these systems evolve over time.

Types of Feedback Loops (Positive, Negative, Reinforcing, Corrective)

Systems theory provides categories that help us understand how feedback loops operate in generative AI. Positive feedback loops amplify existing trends, while negative loops suppress deviations and restore balance.

Reinforcing loops entrench errors, and corrective loops deliberately steer systems toward stability.

When sensational outputs capture user engagement, developers may optimize for those signals, inadvertently creating positive feedback loops that prioritize clicks over accuracy. Negative loops arise when unsafe outputs trigger interventions such as retraining, reducing the probability of similar failures in the future. Reinforcing loops are especially dangerous: a fabricated fact that spreads widely may find its way into new training sets, solidifying misinformation. Corrective loops, by contrast, can be designed intentionally, as in reinforcement learning with human feedback, where undesirable outputs are penalized to realign behavior.

- **Positive loops:** Amplify engagement or bias
- **Negative loops:** Suppress unsafe or harmful outputs
- **Reinforcing loops:** Allow errors to become entrenched
- **Corrective loops:** Use interventions to improve alignment

Recognizing which kind of loop dominates a system is crucial for governance, since amplification and reinforcement can entrench harm, while corrective cycles build resilience.

Historical Parallels in Feedback Loops (Economics, Ecology, Control Theory)

Feedback loops are not unique to AI. Economics, ecology, and engineering provide instructive precedents. Inflation spirals, for example, demonstrate how rising prices can prompt wage increases that then drive prices higher, creating a mutually reinforcing cycle. Predator-prey dynamics in ecology illustrate balancing loops: when predator populations grow, prey decline, which in turn causes predator populations to fall, allowing prey to recover. Control theory offers the classic example of a thermostat, which applies corrective feedback by switching heating or cooling on when environmental conditions deviate from a set point.

- Inflation spirals mirror the way biased outputs reinforce social inequities.
- Predator-prey cycles resemble user complaints that suppress unsafe outputs.
- Thermostats parallel AI safety filters designed to stabilize behavior.

What sets AI apart is the unprecedented speed and opacity of its feedback cycles. Whereas ecological or economic loops unfold over years, billions of AI-generated outputs can circulate back into training corpora within months, and the internal dynamics of large models make these loops difficult to observe directly.

Risks of Feedback in Deployed AI

Feedback loops create risks that extend far beyond individual mistakes. Hallucinations may spread widely, shaping discourse and reappearing in later datasets. Biased stereotypes can influence user behavior, embedding themselves more deeply in the cultural data stream. Frequent updates in response to noisy feedback may create instability, producing inconsistent outputs that undermine trust. The interconnectedness of modern digital systems further exacerbates these risks, as errors spread across search engines, educational platforms, and media ecosystems. Over time, accountability becomes increasingly difficult, since compounded loops obscure the origin of errors.

- Error amplification allows single mistakes to snowball.
- Bias entrenchment reinforces harmful stereotypes.

- Instability emerges from overreacting to noise.
- Cascades spread failures across multiple domains.
- Accountability erodes as loops compound.

These risks highlight why feedback management must be seen as a systemic governance challenge, not simply a technical issue.

Data Drift and Model Decay

All models face the problem of data drift and decay. Language evolves quickly, and new slang, idioms, or cultural references leave older models sounding outdated. Emerging technologies, events, and debates produce topical drift that models cannot address without updated knowledge. Behavioral drift occurs as users adapt to models, changing the way they ask questions or interact. Over time, these forms of drift combine to erode a model's performance, producing outputs that are irrelevant, unhelpful, or even harmful.

- Linguistic drift leaves models misaligned with evolving speech.
- Topical drift reflects emerging developments the model has never seen.
- Behavioral drift highlights co-evolution between humans and systems.
- Model decay reduces accuracy, trustworthiness, and safety.

Organizations must continually weigh the tension between updating frequently enough to remain relevant and updating cautiously enough to preserve stability.

Amplification of Bias Through Feedback

One of the most insidious risks of feedback loops is the amplification of bias. A hiring model that favors certain candidates may influence employers' decisions. Those decisions, once reflected in future datasets, reinforce the bias. Image generators that disproportionately depict scientists as male normalize stereotypes, and those images circulate online, entering new datasets and embedding the bias more deeply. Toxic or offensive outputs that spread widely risk being captured in future corpora. Echo chambers also emerge, as systems provide narrower ranges of information to users, reinforcing existing worldviews rather than challenging them.

- Hiring loops reproduce discriminatory practices.
- Representation loops normalize stereotypes.
- Toxicity loops embed offensive content in datasets.
- Echo chambers narrow the scope of information.

Without intentional corrective intervention, these cycles deepen inequities and reinforce systemic biases.

Feedback Loops in Recommender Systems As Precursors

Recommender systems foreshadowed many of the challenges now facing generative AI. Optimizing for engagement created loops in which sensational content was disproportionately promoted. Popularity bias made items that were already popular even more visible, creating self-fulfilling cycles. Users found themselves in filter bubbles, where recommendations confined them to narrower views, and marginalized voices were systematically excluded.

Generative AI intensifies these risks. Unlike recommenders, which merely select from existing content, generative systems create new content that itself becomes training material for future models. This tightens the feedback loop, making monitoring and intervention even more critical.

Real-World Case Illustrations of Feedback Loops in Generative AI

Concrete cases demonstrate how feedback loops materialize in practice. Some lawyers have submitted fabricated case citations produced by models into court filings, risking contamination of legal databases. Generative systems have provided unverified treatment recommendations that spread online, threatening to reappear in health-related corpora. Students using generative models have unknowingly included fabricated references in assignments, further embedding misinformation into academic records. Hiring systems reproducing gender or racial bias have reinforced discriminatory practices through iterative loops. Generated news summaries have distorted public understanding and circulated widely enough to influence future training data.

- Legal hallucinations risk contaminating court records.
- Medical misinformation circulates into health corpora.
- Academic citations embed fabricated references.
- Hiring systems perpetuate discrimination.
- News distortions shape discourse and future data.

These cases illustrate how isolated failures can quickly escalate into systemic distortions.

Strategies for Monitoring Feedback Loops

Monitoring is the foundation of managing feedback loops. Comprehensive logging captures inputs, outputs, and downstream effects, enabling systematic analysis. Anomaly detection identifies deviations in distribution or behavior. User reporting channels provide structured opportunities to flag harmful or inaccurate outputs. Independent auditing adds credibility and external oversight. Simulation testing allows developers to model feedback dynamics in controlled environments, anticipating risks before they spiral in real-world deployments.

- Logging records interactions for traceability.
- Anomaly detection highlights unusual behaviors.
- User reporting integrates community oversight.
- Auditing ensures accountability.
- Simulation anticipates harmful cycles.

Monitoring transforms feedback loops from hidden processes into phenomena that can be observed and addressed.

Continuous Fine-Tuning and Iterative Model Updates

Adapting to feedback often requires continuous fine-tuning rather than complete retraining. Small-scale updates allow models to absorb curated new data while maintaining stability. Domain adaptation ensures that models remain aligned with specific industries or contexts. Reinforcement learning with human feedback introduces structured correction cycles, reinforcing desirable behavior and discouraging harmful patterns. Before deployment, each update must be tested for both improvements and regressions to ensure that solving one problem does not create another.

- Incremental updates support agility.
- Domain adaptation ensures contextual relevance.
- RLHF establishes structured corrective loops.
- Testing prevents regressions before deployment.

This iterative approach balances the need for responsiveness with the necessity of stability.

Governance and Oversight of Updates

Model updates carry governance responsibilities that go beyond technical implementation. Transparency requires communicating when updates occur and what they entail. Accountability demands documentation of the rationale for changes and the outcomes they are intended to produce. Ethical review by independent bodies ensures that updates align with fairness and safety principles. Version control links specific outputs to particular model iterations, enabling traceability. Regulatory compliance may mandate documentation, auditing, and reporting of updates.

- Transparency builds trust with users.
- Accountability ensures traceability of changes.
- Ethical review guards against harm.
- Version control anchors outputs to model states.
- Compliance integrates external oversight.

Governance reframes updates from opaque technical events into accountable institutional practices.

Trade-Offs in Updating Models (Stability vs. Responsiveness)

Every update involves trade-offs. Frequent updates may improve relevance but can introduce instability and inconsistency. Infrequent updates preserve stability but allow harmful outputs to persist. Improving accuracy through constant retraining increases costs. Rapid changes may introduce new risks, while cautious updating may leave harmful trends unchecked. Updates that enhance alignment in one cultural context may inadvertently reduce it in another.

- Frequent updates risk instability.
- Infrequent updates risk irrelevance.
- Accuracy gains must be weighed against cost.
- Global alignment may conflict with local needs.

Balancing stability and responsiveness requires both technical judgment and ethical reasoning.

User Feedback As a Resource and a Risk

User feedback is invaluable but fraught with challenges. It surfaces harmful outputs that might otherwise go unnoticed, making it a resource for improvement. However, malicious actors can manipulate feedback systems, flooding them with misleading reports. Feedback dominated by majority voices may marginalize vulnerable groups, reinforcing inequities. Noise is unavoidable, requiring careful filtering to separate useful signals from irrelevant complaints.

- Feedback surfaces hidden harms.
- Malicious manipulation threatens integrity.

- Dominant groups may skew representation.
- Filtering is essential to extract signal from noise.

Feedback is critical for alignment but cannot be treated as infallible.

Institutional Feedback Integration (Red Teams, Auditors, Regulators)

Feedback must also come from institutional sources. Red teams deliberately probe systems for vulnerabilities, providing adversarial insights. Independent auditors evaluate systems from the outside, ensuring accountability. Regulators can enforce corrective actions in response to failures. Academics and civil society groups contribute perspectives on systemic risks and the protection of vulnerable communities.

- Red teams expose vulnerabilities.
- Auditors provide external accountability.
- Regulators enforce corrective measures.
- Civil society voices broaden perspectives.

Institutional feedback complements user input by reflecting broader societal priorities.

Technical Strategies for Safe Updating

Updating systems safely requires structured strategies. Checkpointing saves previous states, allowing rollback if regressions occur. A/B testing compares performance of updated and older versions with subsets of users. Canary deployments gradually release updates to limited groups, reducing systemic risk. Shadow evaluation runs updated models in parallel without user exposure, highlighting differences in outputs. Gradual scaling ensures problems remain manageable if they arise.

- Checkpointing supports reversibility.
- A/B testing compares versions empirically.
- Canary releases limit systemic risk.
- Shadow evaluation exposes differences early.

These practices reduce risks in the inherently unstable process of updating generative models.

Feedback Loops in Multimodal and Multi-agent Systems

Multimodal and multi-agent systems intensify feedback challenges. Errors in text generation may influence corresponding images or audio, creating distortions across modalities. Systems of autonomous agents can reinforce one another's mistakes, producing emergent failures. Different modalities may amplify different forms of bias, compounding inequities. Ecosystem-level effects resemble complex ecological systems, requiring monitoring beyond individual models.

- Cross-modal loops spread errors across modalities.
- Multi-agent loops create emergent failures.
- Compound bias interacts unpredictably.
- Ecosystem effects require holistic monitoring.

These dynamics call for monitoring and updating practices designed for complex, interconnected environments.

Cultural and Ethical Dimensions of Feedback and Updates

Feedback and updates are shaped by cultural and ethical considerations. Different societies define harm in different ways, complicating the governance of global systems. Ethical pluralism introduces tension between values such as harm reduction and freedom of expression. International deployment requires reconciliation of conflicting norms, while justice demands that vulnerable groups receive special protection.

- Cultural variation alters definitions of harm.
- Ethical pluralism complicates choices.
- Global deployment magnifies conflicts.
- Justice requires prioritizing the vulnerable.

Ethical oversight ensures that feedback and updates align with broad social values rather than narrow interests.

Future Directions in Feedback and Updating

Looking ahead, AI systems are moving toward more autonomous forms of feedback management. Self-healing models may detect and correct their own errors. Continual learning could replace episodic retraining with incremental adaptation. Federated updating allows distributed systems to adjust locally while maintaining global coherence. Multi-agent oversight may involve systems monitoring each other. International governance may provide standards for responsible updating practices.

- Self-healing introduces autonomous correction.
- Continual learning supports incremental adaptation.

- Federated updating balances local and global needs.
- Governance provides shared accountability.

These innovations promise adaptability but demand governance frameworks that match their complexity.

Historical Lessons in Feedback Management

Other industries offer lessons for AI feedback management. Early telecommunications networks stabilized signals by embedding feedback mechanisms. Operating systems introduced monitoring to prevent runaway processes. Cybernetics emphasized feedback as a universal regulatory principle. Industrial safety practices incorporated continual monitoring and updates to prevent catastrophic failures.

- Telecommunications managed instability.
- Operating systems prevented runaway processes.
- Cybernetics emphasized universal regulation.
- Industrial safety embedded monitoring.

AI feedback loops differ in scale and opacity, but the principle that complex systems require regulation through feedback is timeless.

Psychological and Behavioral Feedback Effects

Feedback loops affect not only models but also human psychology. Positive experiences reinforce trust, encouraging users to rely more heavily on AI systems. Overreliance leads to complacency, with users deferring too much to model outputs. Expectations are shaped by repeated interactions, influencing how people frame their queries. At scale, groups of users influenced in this way can reinforce systemic biases, creating collective behavioral cascades.

- Trust reinforcement increases reliance.
- Overreliance creates vulnerability.
- Expectations shape queries and behaviors.
- Behavioral cascades amplify systemic risks.

Understanding psychological dynamics is as critical as addressing technical ones.

Infrastructure Challenges in Frequent Updating

Frequent updating requires significant infrastructure. Retraining is computationally expensive, raising costs. Large models are difficult to update quickly, creating scalability challenges. Coordination across distributed systems becomes complex. Each update must be tested thoroughly to prevent regressions. Poorly managed updates risk downtime and service disruptions.

- Costs escalate with retraining.
- Scalability challenges slow adaptation.
- Coordination across systems is complex.
- Testing burdens grow with each update.

Updating generative systems is therefore an infrastructural as well as a technical challenge.

Long-Term Societal Risks of Runaway Feedback Loops

The societal risks of unmanaged feedback loops are profound. Misinformation cascades can entrench fabricated content as accepted truth. Bias reinforcement may magnify social inequities. Cultural homogenization threatens diversity by privileging dominant norms. Persistent errors erode public trust in AI. At a deeper level, distorted information ecosystems may destabilize democratic deliberation.

- Misinformation becomes entrenched.
- Bias magnifies systemic inequality.
- Cultural homogenization erases diversity.
- Erosion of trust undermines legitimacy.
- Democracy suffers from distorted information.

These risks confirm that feedback loops are not simply technical phenomena but matters of societal governance.

Feedback loops and model updates sit at the heart of generative AI. Outputs influence inputs, and inputs reshape outputs, creating cycles that can entrench harm or enable correction. Data drift, bias amplification, and systemic cascades illustrate the dangers, while reinforcement learning, institutional oversight, and structured updating highlight the remedies. Lessons from history, psychology, and infrastructure broaden our understanding of the challenge.

The larger lesson is clear: generative AI is not static software but a dynamic actor in social and informational ecosystems. Its fairness, safety, and trustworthiness depend not only on its training but on how it is monitored, updated, and governed as it interacts with the world.

Conclusion

Generative systems represent one of the most transformative developments in the history of computing. Their ability to produce language, images, and other forms of content at scale promises profound benefits for science, education, healthcare, entertainment, and countless other domains. Yet alongside these opportunities lie distinctive risks.

Unlike deterministic systems that fail in predictable ways, generative systems can fail in subtle, unanticipated, and socially consequential ways. Hallucinations, toxic or biased outputs, unsafe behaviors, and feedback-driven distortions reveal that generative AI requires a new paradigm of risk management. Chapter 10 has explored these challenges in depth, focusing on hallucinated outputs, toxic and biased generations, safety checks and red teaming, and feedback loops with their associated model updates.

The analysis of hallucinations revealed that generative models do not possess a native concept of truth. They are trained to predict plausible continuations of patterns, not to verify factual correctness. This explains why they can produce outputs that are linguistically convincing but factually fabricated. Hallucinations matter not only because they misinform but because their persuasiveness can mislead in high-stakes contexts such as law, medicine, and science. Detection and mitigation strategies—ranging from retrieval augmentation to human-in-the-loop verification—help reduce but cannot entirely eliminate the phenomenon. The larger lesson is that truth in generative systems must be externally grounded through knowledge integration, monitoring, and transparent disclosure of limitations.

The examination of toxic and biased generations highlighted another systemic vulnerability. Because models are trained on Internet-scale data, they inevitably absorb stereotypes, prejudice, and harmful language. Outputs that reproduce these patterns risk reinforcing inequality, excluding marginalized voices, or directly harming users. Strategies such as dataset curation, algorithmic debiasing, reinforcement learning with human feedback, and cultural adaptivity provide partial remedies. Yet the persistence of bias shows that no purely technical solution suffices. Fairness and harm reduction must be pursued as ongoing commitments, informed by diverse perspectives, regulatory frameworks, and ethical oversight. The management of toxicity and bias is thus both an engineering and a societal responsibility.

The exploration of model safety checks and red teaming emphasized the need for proactive defense. Safety checks serve as layered safeguards, screening inputs, outputs, and context for signs of harm. Red teaming, drawn from military and cybersecurity practice, represents the deliberate effort to challenge systems adversarially. By simulating malicious users or unanticipated scenarios, red teams expose vulnerabilities before real harm occurs. The lesson here is twofold. First, safety cannot be treated as a static checklist but as a continuous process that evolves alongside threats. Second, red teaming must be institutionalized, supported by governance structures, independent audits, and integration into regulatory frameworks. Just as cybersecurity became a permanent discipline, AI safety requires sustained investment and adversarial resilience.

The analysis of feedback loops and model updates revealed the dynamic nature of generative AI. Outputs do not vanish once produced. They shape user behavior, circulate through media, and may re-enter future training data, creating cycles of reinforcement or correction. When unmanaged, these loops can amplify errors, entrench bias, and destabilize systems. When monitored and harnessed, they can provide signals for improvement, guiding corrective updates and alignment. Model updates are the mechanism by which organizations respond to feedback, but they introduce trade-offs. Too frequent updates risk inconsistency and instability, while too infrequent updates allow harm to persist. Safe updating requires governance, transparency, structured testing, and cultural sensitivity. The broader implication is that generative AI is not a product to be released once but a process requiring continuous stewardship.

Taken together, the four focal areas of this chapter demonstrate that the failure modes of generative AI are systemic, not incidental. They reflect the probabilistic nature of these systems, the imperfections of training data, the unpredictability of open-ended generation, and the socio-technical feedback loops that emerge at scale. This means that addressing them requires systemic responses. Piecemeal fixes may reduce risk temporarily but cannot substitute for integrated pipelines of safety, monitoring, and governance.

The chapter also underscored the importance of humility in deploying generative AI. No system can be made perfectly safe, unbiased, or hallucination-free. Instead, the goal is to reduce risks, increase transparency, and create mechanisms for accountability. This requires recognizing the limits of technical methods and embracing institutional responsibilities. Developers, regulators, auditors, and civil society must collaborate to establish standards, enforce oversight, and ensure that generative AI serves the public good.

At a deeper level, addressing failure modes is not only a matter of engineering but of values. Decisions about what counts as bias, what constitutes harm, and how to balance freedom of expression with safety are not purely technical. They involve ethical, cultural, and political judgments. Different societies will make these judgments differently, reflecting their histories and priorities. This underscores why cultural adaptivity and inclusivity are essential components of AI governance. A model that is safe in one context may be unsafe in another, requiring global systems to accommodate diverse perspectives.

Another critical lesson is that trust depends on visible, credible safeguards. Users are more willing to engage with generative systems when they know that hallucinations are being mitigated, biases are being managed, safety checks are in place, and feedback loops are monitored. Trust does not require perfection but does require demonstrable commitment to responsibility. Organizations that treat safety as central rather than peripheral are more likely to gain durable trust from users, regulators, and society.

Looking forward, the management of bias, hallucinations, safety, and feedback will only grow in importance as generative systems expand into critical domains. In healthcare, education, law, and governance, the tolerance for error is low, and the potential for harm is high. These systems cannot be deployed responsibly without robust pipelines for failure detection and mitigation. At the same time, opportunities exist to transform failure management into a strength. By designing feedback loops as corrective rather than reinforcing, by institutionalizing red teaming, by embracing cultural sensitivity, and by integrating updates into transparent governance, organizations can build systems that not only avoid harm but actively improve over time.

In conclusion, Chapter 10 has shown that addressing bias, hallucinations, and failure modes is central to the long-term viability of generative AI. These challenges are not flaws to be patched but structural features to be managed. They require technical interventions, institutional frameworks, cultural awareness, and ethical deliberation. They demand humility, vigilance, and continuous improvement. The broader lesson is clear: generative AI will only fulfill its promise if it is accompanied by equally generative practices of responsibility. Systems capable of producing outputs at scale must be matched by systems capable of safeguarding users, correcting errors, and aligning with human values. Only by addressing bias, hallucinations, safety, and feedback loops directly can generative AI earn the trust required for widespread and sustainable adoption.

In our next chapter, we cover the evaluation and benchmarking of generative models.

CHAPTER 11

Evaluation and Benchmarking of Generative Models

Evaluating generative models is one of the most challenging tasks in the field of AI. Unlike traditional machine learning systems, where performance can often be measured with objective accuracy or error rates, generative models produce outputs that are inherently open-ended. Their success is not defined solely by whether they produce a correct answer, but also by qualities such as fluency, creativity, coherence, fidelity to input prompts, and alignment with human expectations. This multidimensional nature of generative outputs requires a richer set of evaluation methods and benchmarks than those used in conventional supervised learning.

Automatic evaluation metrics emerged first, offering fast and scalable ways to compare generated outputs to references or statistical baselines. Metrics such as BLEU in text generation or FID in image generation provided standardized, reproducible scores that accelerated research progress. However, these metrics often capture only narrow aspects of quality. BLEU emphasizes overlap with reference text but does not account for meaning or creativity, while FID measures visual similarity distributions without fully reflecting semantic content. Perplexity, diversity scores, and other automated metrics extend this toolbox, yet each remains a partial proxy.

I. Cronin, *Building and Training Generative AI Models*,
https://doi.org/10.1007/979-8-8688-2332-9_11

Human evaluation has therefore remained indispensable. Human judges provide assessments of coherence, relevance, factual accuracy, or overall appeal. Such evaluations capture subtle dimensions that automated metrics overlook, but they are costly, time-consuming, and sometimes inconsistent. Hybrid approaches, combining automatic metrics with structured human feedback, have become standard in large-scale benchmarking efforts.

The importance of evaluation extends beyond academic comparison. In production systems, evaluation underpins trust, safety, and adoption. Users and organizations need evidence that generative systems are not only powerful but also reliable, fair, and aligned with real-world goals. A/B testing frameworks allow developers to measure comparative performance in live settings, balancing subjective user satisfaction with objective signals. Task-based performance assessments measure how well generative models support downstream goals, such as answering medical queries responsibly, generating useful code, or supporting educational applications.

Benchmarking is both a technical and cultural practice. Shared benchmarks provide common ground for comparing progress, but they also shape research priorities and can create incentives for narrow optimization. Understanding both the strengths and limitations of metrics and benchmarks is therefore critical.

This chapter explores the landscape of evaluation in generative AI. It surveys automatic and human evaluation methods, explains widely used metrics, introduces A/B testing frameworks, and discusses task-based performance assessments. Together, these approaches form a multi-faceted toolkit for ensuring that generative models are not only advanced in capability but also dependable in practice.

Automatic and Human Evaluation Methods

Evaluating generative models is one of the most complex and debated processes in AI, whereas traditional machine learning often allows for more straightforward performance measures. A classifier can be evaluated on accuracy, precision, or recall. A regression model can be scored with mean squared error. These measures rely on the existence of a clearly defined ground truth. By contrast, generative models create outputs that are open ended, subjective, and multifaceted. There is rarely one single correct sentence, image, or piece of music; even when factual correctness matters, it is only one dimension among many, alongside style, coherence, and usefulness. Instead of a single right answer, there may be many outputs that are all valid but differ in style, nuance, or creativity.

This openness creates both opportunity and difficulty. It allows generative models to surprise us with creativity, but it also complicates our ability to measure quality. Automatic metrics provide scalability and reproducibility, yet they capture only fragments of what makes outputs compelling. Human evaluations provide depth and contextual sensitivity, but they are costly, time-consuming, and often inconsistent. A mature evaluation strategy recognizes the strengths and limitations of each and integrates them into a coherent pipeline.

This section provides a comprehensive exploration of automatic and human evaluation methods for generative AI. It situates these methods historically, surveys foundational and modern metrics, examines the role of human judgments, highlights their respective trade-offs, and discusses their application across text, vision, speech, and multimodal domains. It also expands to consider reference free paradigms, cultural and ethical dimensions, high stakes applications, evaluator psychology, and future directions such as self-evaluating models. The goal is to show that evaluation is not an afterthought but a central discipline that shapes the trajectory of generative AI research and deployment; in mature pipelines, fast automatic metrics offer broad coverage and continuous monitoring, while targeted human evaluations provide depth and calibration, with each informing and refining the other.

Historical Context of Evaluation in Generative Modeling

The evaluation of generative systems has always been a central challenge. In the early years of machine translation, researchers relied heavily on linguists to assess whether outputs conveyed meaning and adhered to grammatical correctness. This process was costly, inconsistent, and ultimately unsustainable, leading to the creation of automated metrics such as BLEU in the early 2000s. Speech synthesis followed a similar trajectory: initial studies depended on listening tests where participants rated intelligibility, but over time, researchers developed more objective proxies, including spectral distortion measures and mean opinion scores. Image generation too began with subjective visual inspection before adopting metrics like Inception Score and later Fréchet Inception Distance (FID), which offered reproducible numerical assessments.

- Early evaluation was dominated by human judgment in translation, speech, and vision.
- Automated metrics such as BLEU and FID emerged to address scalability and consistency.
- Tension has persisted between subjective depth and automated reproducibility.

This pattern illustrates a recurring dynamic: systems initially depend on human evaluators, later transition to automated metrics for scalability, and then struggle with the limitations of those metrics as models grow in sophistication.

Core Principles of Evaluation

Despite the diversity of evaluation methods, certain principles define what makes an assessment meaningful. Validity requires that a method measure what it claims to measure. BLEU, for example, focuses on n-gram overlap, which captures surface similarity but does not fully reflect translation quality. Reliability is equally essential, since evaluations that fluctuate across evaluators, runs, or datasets cannot be trusted. Sensitivity ensures that methods distinguish meaningfully between strong and weak models. Practicality keeps evaluations feasible at scale, even if this sometimes sacrifices nuance. Comprehensiveness reminds us that no single measure captures fluency, coherence, factuality, creativity, and safety all at once.

- Validity ensures the right phenomenon is being measured.
- Reliability provides consistency across contexts.
- Sensitivity separates strong from weak systems.
- Practicality enables large-scale use.
- Comprehensiveness acknowledges multidimensional quality.

The difficulty lies not in knowing these principles but in balancing them, since they often pull in conflicting directions.

Automatic Evaluation Methods: Foundations

Automatic methods seek to quantify quality algorithmically. Reference-based metrics such as BLEU and ROUGE for text or FID for images compare generated outputs against predefined references, assuming proximity to ground truth correlates with quality. Distributional metrics compare statistical properties of generated and real data; perplexity measures how probable text is under a language model, while FID compares image feature distributions. Embedding-based similarity metrics use pretrained neural embeddings, such as BERTScore, to assess semantic similarity beyond surface overlap. Task-based proxies evaluate the usefulness of generated content for downstream applications, such as improvements in classification accuracy when synthetic data is included.

- Reference-based metrics depend on ground truth examples.
- Distributional metrics compare statistical patterns.
- Embedding-based metrics capture semantic similarity.
- Task-based proxies measure downstream utility.

These methods provide speed, reproducibility, and scalability, but inevitably simplify the multidimensional qualities that humans notice.

Human Evaluation: The Enduring Gold Standard

Despite advances in automation, humans remain the ultimate judges of generative quality. Automated metrics can approximate certain aspects, but human evaluators capture dimensions such as coherence, naturalness, creativity, and ethical appropriateness. Evaluation tasks take many forms: direct scoring on Likert scales, pairwise comparisons of outputs, ranking multiple outputs, task-specific judgments, and open-ended commentary that surfaces issues metrics often miss.

- Direct scoring provides numerical assessments of fluency or adequacy.
- Pairwise comparison reduces scaling bias by focusing on relative quality.
- Ranking tasks generate rich comparative information.
- Task-focused evaluation ties judgment to practical usefulness.
- Open commentary captures subtle failures or creativity.

While invaluable, human evaluations are costly, subjective, and variable. Inter-rater reliability and cultural diversity complicate generalization, but their depth remains unmatched.

Case Study: BLEU in Machine Translation

BLEU was created as a scalable alternative to human evaluation in translation. Its strengths were clear: it was simple, reproducible, and moderately correlated with human judgment at the system level. But BLEU also had limitations, penalizing valid alternative phrasings, struggling with semantic adequacy, and encouraging models to mimic reference phrasing rather than explore natural variation. Over time, its dominance shaped research agendas, privileging n-gram overlap over deeper linguistic quality.

- **Strengths**: Simplicity, reproducibility, moderate correlation.
- **Limitations**: Penalizes variation, shallow semantics.
- **Long term effects**: Constrained research by overemphasizing surface similarity, and its legacy carried into broader LLM evaluation through reference-based metrics such as ROUGE style scores and overlap oriented benchmarks for summarization and dialogue, which continued to favor lexical similarity over semantic faithfulness, usefulness, or safety.

BLEU's history shows how metrics accelerate research but can also distort it by overvaluing what is easy to measure.

Case Study: FID in Image Generation

In image generation, FID became the standard. It correlates better with human judgments than earlier metrics, captures distributional similarity, and is reproducible. Yet it depends on the Inception network, which reflects ImageNet biases and overlooks domains outside that training scope. FID also struggles with fine-grained detail and subjective qualities like aesthetics. Nonetheless, it provided a unifying benchmark that advanced progress in generative vision research.

- **Strengths**: Strong correlation with human judgment, reproducibility
- **Limitations**: Dependence on biased pretrained models, blind to aesthetics
- **Impact**: Unified benchmark that catalyzed research

Like BLEU, FID demonstrates how metrics both enable progress and impose distortions.

Hybrid Approaches

Hybrid strategies combine automation and human evaluation. Automated metrics can pre-filter outputs, reducing the workload for human evaluators. Human assessments in turn calibrate and validate the metrics, ensuring they remain grounded in subjective perceptions. Adaptive frameworks alternate between large-scale automated testing and targeted human evaluation. This combination balances scalability with nuance and has become the norm in industry practice.

- Automation ensures scale.
- Human evaluation ensures depth.
- Adaptive frameworks integrate both.

Hybrids represent a pragmatic recognition that neither approach alone is sufficient.

Challenges of Human Evaluation

Human evaluation, while rich, presents persistent challenges. Recruiting and compensating evaluators at scale is expensive. Evaluators bring cultural and personal biases, shaping their judgments in ways that can skew results. Long evaluation sessions produce fatigue, reducing attentiveness and reliability. Ethical risks arise when evaluators are exposed to toxic or disturbing content. Ensuring inter-rater reliability requires training and careful task design.

- Cost and scale limit feasibility.
- Biases distort outcomes.
- Fatigue undermines reliability.
- Ethical exposure risks harm.
- Consistency demands strong guidelines.

Designing effective human evaluations thus requires attention to sampling, task design, and evaluator support.

Toward Multidimensional Evaluation

No single metric can capture the complexity of generative quality. Multidimensional frameworks assess outputs across multiple axes: fluency ensures grammaticality, coherence checks consistency, adequacy measures completeness, faithfulness guards against hallucinations, diversity prevents repetition, and safety ensures the absence of bias or harmful content.

- **Fluency**: Grammatical and well-formed outputs
- **Coherence**: Consistent and logical structure
- **Adequacy**: Complete response to prompts

- **Faithfulness**: Factual accuracy and avoidance of hallucination
- **Diversity**: Varied outputs across attempts
- **Safety**: Freedom from bias and toxicity

These dimensions collectively provide a more holistic assessment than any single number.

Implications for Research and Deployment

Evaluation methods shape not only technical progress but also deployment practices. Metrics like BLEU and FID defined research trajectories for years, directing attention toward measurable qualities at the expense of others. In deployment, evaluation is indispensable for providing assurance of safety and reliability. Governance also depends on evaluation: regulators and auditors require transparent frameworks to assess compliance with legal and ethical standards.

- Research incentives are shaped by available metrics.
- Deployment depends on evidence of reliability.
- Governance requires transparent frameworks.

Evaluation, therefore, carries technical, cultural, and political consequences.

Reference-Free Evaluation Paradigms

Reference-based methods assume predefined outputs, but these become inadequate for creative or open-ended tasks. Reference-free paradigms aim to evaluate without explicit ground truth. Probabilistic likelihoods from pretrained models can serve as proxies. Models can be trained to self-evaluate their own outputs. Reward models can approximate human judgment as scalar signals. Interactive benchmarks embed evaluation into real user interaction rather than static datasets.

- Language modeling likelihoods estimate plausibility.
- Self-evaluation internalizes judgment.
- Reward models approximate human preferences.
- Interactive benchmarks emphasize real use.

These paradigms reflect the transition toward open-ended generative systems where overlap with references is insufficient.

Cross-Modal Evaluation in Multimodal Systems

Multimodal systems complicate evaluation by requiring consistency across modalities. Metrics such as CLIPScore assess alignment between text and generated images. Audio–visual coherence measures whether speech or music matches generated video. Cross-modal faithfulness ensures consistency when outputs are translated across modalities. Human judgment remains essential for evaluating coherence and aesthetic appeal across sensory dimensions.

- Image-text alignment evaluates prompt matching.
- Audio-visual coherence ensures temporal fidelity.
- Cross-modal faithfulness validates translation between forms.
- Human judgment provides multimodal nuance.

Multimodal evaluation magnifies complexity, demanding both new metrics and innovative human studies.

Evaluator Psychology and Cognitive Bias

Human evaluation is subject to psychological and cognitive biases. Anchoring effects cause evaluators to be influenced by previously seen outputs. Framing effects from instructions can bias judgments. Cultural bias shapes perceptions of fluency and appropriateness across societies. Fatigue reduces attentiveness over long tasks, and expectation effects cause evaluators to reward outputs that conform to prior beliefs.

- Anchoring ties current judgments to past outputs.
- Framing biases evaluation criteria.
- Cultural variation complicates generalization.
- Fatigue diminishes reliability.
- Expectations skew assessments.

Designing evaluations requires awareness of these biases to ensure reliability and fairness.

Evaluation in High Stakes Domains

In high stakes domains, the challenges of evaluation intensify. Healthcare requires outputs that are accurate, safe, and trustworthy, necessitating expert human evaluation. In law, hallucinations have legal consequences, making precision indispensable. Education demands outputs that are both correct and pedagogically valuable. Finance involves outputs with direct economic impact, where errors carry significant risks.

- Healthcare demands expert oversight.
- Law requires factual precision.
- Education balances accuracy and accessibility.
- Finance involves high economic stakes.

In such domains, evaluation cannot rely on automatic metrics alone but must integrate domain-specific expertise. Experts need not only to verify that outputs are correct, but also to interpret how those outputs were obtained, to understand uncertainty, and to assess whether the reasoning, evidence, and assumptions behind a model's answer are acceptable for the decisions that depend on it.

Cultural and Ethical Dimensions of Evaluation

Evaluation reflects cultural and ethical commitments. Definitions of quality differ across societies, making evaluation inherently value-laden. Ethical standards demand that evaluations address harm reduction alongside technical performance. Inclusion of diverse evaluators prevents dominance by a single cultural perspective. Transparency in evaluation design and reporting is essential for building trust.

- Quality is culturally defined.
- Ethical standards expand evaluation beyond fluency.
- Inclusion prevents cultural dominance.
- Transparency builds legitimacy.

Evaluation is thus as much an ethical and cultural practice as it is a technical one.

Future Horizons in Evaluation

The future of evaluation involves more automation, adaptivity, and integration.

- **Self-evaluating models:** Systems that critique their own outputs.

- **Interactive benchmarking:** Evaluation embedded in live use rather than static datasets.
- **Multi-agent evaluation:** Models evaluating each other to provide scalable oversight.
- **Continual evaluation:** Dynamic monitoring rather than episodic testing.
- **Global standards:** International benchmarks ensuring cross-cultural comparability.

Evaluation will evolve alongside generative models, becoming more dynamic and systemic.

Automatic and human evaluation methods are the twin anchors of generative AI assessment. Automatic metrics provide scalability, speed, and reproducibility, but they inevitably simplify. Human evaluation captures nuance, creativity, and ethical judgment, but it is costly and inconsistent. Neither suffices in isolation. The path forward requires hybrid approaches, multidimensional frameworks, cultural sensitivity, and continual innovation.

The broader lesson is that evaluation is not just a technical process but a cultural and ethical practice that shapes what the field values. The metrics and benchmarks we choose define progress itself. Responsible evaluation therefore requires humility, transparency, and inclusion. As generative systems expand into high-stakes domains, evaluation will remain the central safeguard ensuring that generative AI is not only powerful but also reliable, fair, and aligned with human values.

BLEU, FID, Perplexity, Diversity Scores

The evaluation of generative models is one of the most influential and contested topics in artificial intelligence. Metrics are not neutral yardsticks. They serve as signals of progress, currencies of reputation, and even gatekeepers of legitimacy. A good metric can accelerate discovery by providing a reproducible standard of comparison. A weak or misapplied metric can distort research priorities, encouraging optimization for numerical gains that diverge from human preference or real-world usefulness.

Four families of metrics have dominated evaluation over the past two decades: BLEU for text, FID for images, perplexity for language modeling, and diversity scores for capturing variation. Each arose in response to pressing evaluation challenges, and each has left a profound mark on how research has unfolded. Yet each also illustrates the limitations of automatic evaluation in open-ended generative settings. This section provides an extended conceptual exploration of these metrics. It traces their origins, explains their mechanics, analyzes their impact, and situates them within the broader landscape of evaluation theory and practice.

The discussion unfolds across multiple layers: the technical logic of how metrics operate, the historical context in which they emerged, their role in shaping incentives, their cultural and ethical implications, and the open problems they leave behind. The aim is not only to document metrics but to show how they structure the very meaning of progress in generative AI.

The Role of Metrics in Generative AI

Metrics serve as much more than scorekeepers. They create a shared language for progress, allow teams to verify one another's results, and reduce dependence on slow and costly human studies. In practice, the presence or absence of a widely accepted metric can determine whether an idea spreads, whether a paper is considered competitive, and whether an organization is willing to ship a product. Because researchers tend to optimize what they can measure, metrics do not simply record progress, they shape it. The history of BLEU in machine translation and ROUGE in summarization shows how optimization for n-gram overlap sometimes produced systems that scored well yet sounded rigid or unnatural to human readers. In generative vision, aggressive tuning for FID occasionally led to models that produced visually sharp but mode collapsed samples, masking weaknesses in diversity and controllability behind a single impressive number.

At the same time, the institutional role of metrics is easy to overlook. A metric defines what is worthy of attention, what counts as an improvement, and which directions appear promising. This institutional function explains why debates about evaluation can influence entire research agendas. When a field chooses a metric, it is not only choosing a number, it is choosing priorities.

- Standardization enables comparisons that others can reproduce.
- Automation improves speed and lowers the cost of iteration.
- Incentives follow what is measured, which channels effort toward particular properties.
- Deployment relies on thresholds that communicate usability and safety to decision-makers.

Seen in this light, metrics are both measurement tools and social infrastructure. They stabilize expectations across labs and companies, even as they inevitably leave important qualities only partially captured and sometimes produce metric-driven distortions when optimization for the score diverges from human notions of quality, safety, or usefulness.

BLEU: The Cornerstone of Text Evaluation

BLEU, introduced in 2002, was a decisive moment for automatic evaluation in machine translation. The designers wanted a proxy for human judgment that was inexpensive, repeatable, and scalable to large corpora. Their solution was to quantify the overlap of short word sequences between a system output and one or more reference translations. Modified precision prevented systems from gaming the score by repeating high frequency words, and a brevity penalty discouraged unrealistically short outputs.

BLEU delivered a practical standard at the system level and correlated moderately with human preferences when comparing entire systems rather than individual sentences. That practicality drove rapid adoption across machine translation and influenced other text generation tasks. Yet the same design choices produced well-known limitations. The metric cannot recognize many legitimate paraphrases. It treats surface similarity as a stand-in for semantic adequacy. It performs poorly in creative, open-ended settings, and it depends heavily on having several references, which is costly.

- **Mechanics**: N-gram overlap with modified precision and a brevity penalty
- **Strengths**: Simple, efficient, system level correlation with human judgment
- **Weaknesses**: Penalizes valid paraphrase, shallow semantics, reliance on multiple references
- **Historical effect**: Accelerated MT research while steering it toward surface overlap

BLEU's story shows how a metric can catalyze a field and also narrow it. What is easy to count becomes what is easy to optimize.

Beyond BLEU: ROUGE, METEOR, CIDEr, BERTScore

As the community encountered BLEU's limits, new metrics appeared that tried to move closer to human judgment without losing scalability. ROUGE emphasized recall for summarization and asked whether the model captured content present in references, which makes it particularly well suited to tasks where covering key points matters more than exact phrasing, such as news or report summarization, but less informative for translation where fine-grained lexical choices carry more weight. METEOR introduced stemming, synonym matching, and alignment to soften the rigidity of pure surface overlap, and often performs reasonably across both translation and summarization because it can recognize paraphrases while still rewarding alignment with references. CIDEr, designed for image captioning, rewarded phrases that matched a consensus across multiple references rather than any single one, capturing how human captions vary yet tend to share certain descriptive details; it works best in captioning style tasks and is less commonly used for long-form text. BERTScore shifted the basis of comparison to contextual embeddings from pretrained transformers, making it possible to recognize semantic similarity that lacks lexical overlap and to transfer across domains, from translation to summarization to open ended generation, though sometimes at the cost of being less sensitive to subtle factual errors.

- ROUGE favors content coverage but still inherits lexical bias and is most informative for summarization and other content selection tasks, less so for sentence level translation quality.
- METEOR adds linguistic flexibility and often correlates better with humans on both translation and summarization because it can credit paraphrases and morphological variants.
- CIDEr uses reference consensus to value human like captions and is particularly effective in image captioning, where multiple short descriptions of the same scene exist but is not typically the primary choice for long, document-level tasks.

- BERTScore evaluates meaning through embeddings rather than exact words and can be applied across translation, summarization, and captioning, though its strengths in capturing semantic similarity sometimes mask small factual mistakes that matter in high precision domains.

These metrics represent a gradual shift from counting shared strings toward approximating shared meaning, with each finding its strongest niche in different task families. None resolves the fundamental tension between scale and validity, but together, they broaden the evaluation toolkit across translation, summarization, captioning, and other generative settings.

Perplexity: Measuring Language Model Fluency

Perplexity predates modern generative systems and originates in the study of probabilistic language models. It measures how well a model predicts a held-out corpus by exponentiating the negative average log likelihood. Lower values indicate that the model assigns higher probability to the test text. Historically, reductions in perplexity tracked improvements in fluency, especially for n-gram models and early neural networks. Today, perplexity remains valuable as an internal training diagnostic, helping developers compare checkpoints, spot optimization failures, and detect overfitting or distribution shift on held out data.

The limits of perplexity are now well understood. It measures predictive likelihood rather than human quality. A model can achieve a low score by assigning high probability to safe and bland continuations. The measure is also sensitive to tokenization and dataset choice, which complicates comparisons across systems. For very large models, sizable gains in perplexity do not always correspond to better user preferences or safer behavior, which makes it a misleading metric for deployment decisions or product comparisons even when it is informative during model development.

Definition: Exponentiated negative average log likelihood on a test set.

- **Utility**: A useful training diagnostic and historical baseline for tracking optimization progress, comparing checkpoints, and detecting problems such as overfitting or distribution shift.
- **Limitation**: Only weakly linked to human-judged usefulness in modern settings, so a lower perplexity model is not necessarily more helpful, safer, or better aligned for real users.

Perplexity therefore remains valuable inside the training loop as an internal signal for model fitting, but it is misleading when treated as a general deployment metric for open ended generation or as a stand-alone basis for product-level quality claims.

Diversity Scores: Capturing Variety in Outputs

As models become fluent, they risk playing it safe. Diversity metrics attempt to detect whether a system collapses to a narrow subset of responses. Distinct-n calculates the proportion of unique n-grams across outputs and signals lexical variety. Self-BLEU compares one output against others from the same model to see whether they are too similar. These measures help diagnose mode collapse and encourage exploration.

Their limits are clear as well. Lexical variety can increase while relevance declines. A system might produce different words that do not answer the prompt or that wander off topic. Semantic diversity matters more than surface variety, and current scores are only proxies.

- Distinct-n and self-BLEU provide quick signals of repetition.
- Diversity must be balanced with coherence and adequacy.
- Variety at the word level does not guarantee variety of ideas.

For example, two models might achieve similar BLEU scores on a summarization task, yet one repeatedly paraphrases the same central framing while the other presents multiple distinct but still accurate summaries; the latter would show higher distinct-n and lower self-BLEU, illustrating how diversity metrics can differentiate between outputs that look equivalent under traditional overlap measures.

The lesson is to treat diversity as a dimension among several, not as a stand-alone objective.

Benchmarking Image Generation

FID became the most widely used statistic for evaluating generative images. FID maps real and generated images into the feature space of a pretrained Inception network, models each set as a Gaussian, and computes the Fréchet distance between the two. The score reflects both quality and diversity at the distribution level and generally aligns better with human judgments than earlier alternatives like Inception Score.

FID's dependence on the Inception features also brings trade-offs. Those features inherit biases from ImageNet and do not capture all domains equally well. Fine-grained semantics and subjective properties such as aesthetics can be underrepresented. Even so, the field gained a common yardstick that enabled comparisons across GANs, VAEs, and diffusion models.

- **Strength**: Distributional comparison that correlates with human perception
- **Limitation**: Feature bias and reduced sensitivity outside ImageNet style domains
- **Impact**: Unified benchmark that standardized progress reporting

Distributional metrics like FID approximate perceptual similarity, yet they remain sensitive to the reference features on which they rely. For example, two image generators might achieve nearly identical FID scores on a faces dataset, even though one produces a narrow set of very sharp, nearly identical smiling portraits while the other generates a wider variety of ages, poses, and lighting with mild artifacts. Human observers would often judge the second model as more useful and interesting, but this qualitative difference is only weakly reflected in the shared FID value.

Statistical Properties and Interpretability of Metrics

Different metrics behave differently under sampling, reference count, and corpus size. BLEU becomes unstable at the sentence level and is most meaningful over whole corpora. FID is sensitive to the number of samples, with small test sets producing volatile estimates. Perplexity can change with tokenization choices, making cross-model comparisons tricky. Diversity metrics can be inflated by noise or by trivial paraphrase.

Interpretability varies as well. BLEU is intuitive for relative comparisons but has no inherent absolute threshold. FID values are hard to interpret without context. Perplexity is mathematically clear but opaque to many practitioners outside language modeling.

- Stability depends on sample size and reference design.

- Absolute numbers rarely tell the full story.
- Reporting should include confidence intervals and dataset details.

Careful statistical practice is part of responsible metric use.

Metric Gaming and Goodhart's Law

Once a metric becomes a target, systems learn to exploit it. Models tuned for BLEU can echo reference phrasing rather than communicate meaning. Image generators optimized for FID can steer toward features favored by the Inception network without improving human perceived quality. Minimizing perplexity can yield probable but dull language. Maximizing diversity scores can lead to incoherent but varied outputs.

- Optimization pressure finds shortcuts that raise scores without raising quality.
- Guardrails include multiple metrics, human checks, and periodic audits.
- Transparent reporting reduces the incentive to pursue narrow gains.

Goodhart's Law is a reminder to treat metrics as guides, not goals.

Metrics in Industry vs. Academia

Academic research depends on public benchmarks to compare methods and to coordinate collective progress. Leaderboards provide visibility and structure. Industry uses metrics differently. Product teams care about satisfaction, safety, latency, and task success. They rely on dashboards and composite indices rather than a single number, and they combine automatic metrics with structured human evaluations that reflect real user needs.

- Academia values comparability and public ranking.
- Industry values user outcomes and operational signals.
- Bridging the gap often means internal benchmarks that mix automated scores with human studies.

Context determines how a metric should be interpreted and weighted.

Cultural and Linguistic Variability in Metrics

Evaluation is not uniform across languages and cultures. N-gram overlap can underestimate quality in morphologically rich languages because many correct inflections do not match references. Tokenization choices influence perplexity and can distort multilingual comparisons. Image metrics that rely on features trained on Western datasets carry aesthetic and cultural bias.

- Language structure affects surface matching metrics.
- Tokenization complicates cross-language comparisons.
- Visual features embed cultural priors that influence scores.

Responsible evaluation requires attention to linguistic and cultural diversity rather than assuming universality.

The Political Economy of Benchmarks

Benchmarks distribute prestige and resources. High leaderboard positions attract citations, talent, and funding. Institutions that set benchmarks influence what the field values. When benchmarks overrepresent Western data or narrow domains, they can marginalize other contexts and needs. Understanding these dynamics clarifies why debates about evaluation are often debates about power.

- Leaderboards channel attention and investment.
- Funding and hiring often follow benchmark success.
- Inclusion in benchmarks shapes whose problems are solved.

Metrics do not just measure progress. They also organize it.

Metrics in Live Systems

Once models are deployed, metrics enter continuous monitoring. Teams track fluency, diversity, latency, safety signals, and user outcomes on dashboards. Single scores give way to composite indicators that summarize several dimensions. Feedback from metrics triggers alerts, prompts investigations of drift, and informs rollback decisions when regressions appear.

- Live monitoring favors bundles of signals over a single metric.
- Composite indices better reflect product reality.
- Evaluation becomes an ongoing process rather than a one-time test.

In production, metrics function as part of observability rather than only as offline evaluation.

Longitudinal Evolution of Metrics

Metrics have life cycles. Gains in BLEU eventually stopped reflecting meaningful translation improvement. As image models approached photorealism, FID differences grew smaller and less informative. Perplexity continued to fall for large language models while user preferences did not always improve. Diversity measures that once flagged repetition became less diagnostic as fluency rose. In practice, different families of metrics now cluster into broad categories such as factuality and safety metrics that track correctness and harmfulness, usefulness and task success metrics that measure whether users can complete goals, creativity and diversity metrics that reward novel but coherent outputs, and multimodal metrics that evaluate alignment between text, images, and other modalities.

- A metric is most useful within the regime for which it was designed.
- Stagnation or saturation signals the need for new measures.
- Periodic recalibration keeps evaluation aligned with user value.
- Different categories of metrics factuality and safety, task success, creativity, multimodality must be balanced rather than optimized in isolation.

Healthy fields retire or reinterpret metrics when their assumptions no longer hold. For generative systems, this includes recognizing when traditional reference-based or likelihood-based scores have reached their useful limits and shifting attention toward measures that better capture safety, alignment, and creative value. Doing so requires community standards for safety and creativity metrics, including shared benchmark tasks, transparent scoring protocols, and agreed thresholds that signal when a system is not only more capable but also trustworthy and responsibly imaginative.

Open Problems in Metric Design

Several gaps limit current practice. Automatic measures of hallucination and factuality remain unreliable across domains. Safety scoring for toxicity, bias, and misuse lacks robust automation. Creativity is difficult to quantify without reducing it to novelty alone. Cross modal coherence is still measured crudely, and contextual appropriateness remains out of reach for static metrics.

- Factuality and safety require better automated proxies.
- Creativity needs measures that go beyond simple novelty.
- Multimodal alignment calls for richer, cross modal diagnostics.

Progress in these areas would reshape how systems are trained and released.

Toward Meta-Metrics and Ensemble Evaluation

Because no single score suffices, many teams now combine signals. Weighted ensembles draw on BLEU, ROUGE, BERTScore, diversity measures, and safety classifiers. These automated bundles are calibrated against recurring human studies so that drift in the metric suite is caught early. Weightings vary by application, since a customer support assistant, a medical summarizer, and a creative writing model do not share the same priorities.

- Combine heterogeneous metrics to cover multiple dimensions.
- Calibrate routinely against human preference data.
- Adapt weights to the task rather than enforcing a single recipe.

Meta-level evaluation acknowledges that quality is multidimensional and that responsible practice blends automation with human judgment.

The Future of Metrics

The next generation of metrics will be semantic, adaptive, and multimodal.

- Embedding-based methods will dominate, using powerful pretrained representations.
- Task-based metrics will measure usefulness directly, tying scores to whether users can successfully complete real tasks.
- Adaptive metrics will evolve dynamically with models, updating calibration and thresholds as capabilities change.
- Self-evaluating systems may critique their own outputs, providing scores or explanations that can be audited and compared against human judgment.
- Cross-modal metrics will assess coherence across text, image, and audio rather than treating each modality in isolation.

Metrics will become more integrated into systems, but they will remain partial proxies rather than final arbiters.

BLEU, FID, perplexity, and diversity scores exemplify the promise and peril of automatic evaluation. Each enabled standardization and reproducibility, but each also constrained progress to what was easily measured. BLEU rewarded overlap at the expense of meaning. FID captured distributional similarity without semantics. Perplexity measured likelihood but not human satisfaction. Diversity scores encouraged novelty without coherence.

The lesson is that metrics are artifacts of their time and context. They are technical constructs and social institutions, shaping what is valued and rewarded. They must be treated with humility, complemented by human evaluation, hybrid approaches, and cultural awareness. Future evaluation will require pluralism, with no single score sufficing. Only through diverse and adaptive evaluation methods can generative AI progress responsibly, balancing innovation with alignment to human values.

A compact way to visualize this shift is to contrast past, present, and future metric paradigms:

Past

- **Dominant examples**: BLEU, ROUGE, FID, perplexity
- **Focus**: Surface overlap, likelihood, low-level distributional similarity
- **Primary role**: Benchmark leaderboards and offline model comparison
- **Blind spots**: Meaning, usefulness, safety, cross-cultural variation

Present

- **Dominant examples**: BERTScore, task-specific success rates, toxicity and bias classifiers, human preference studies
- **Focus**: Semantic similarity, user preference, factuality and safety, task completion
- **Primary role**: Blended pipelines where automatic metrics screen candidates and human evaluators calibrate and audit
- **Blind spots**: Long-term impacts, creativity, cross-modal coherence, subtle cultural harms

Future

- **Emerging examples**: Embedding-based holistic scores, interactive task-based evaluations, self-critique metrics, multimodal alignment measures
- **Focus**: Usefulness in context, robustness, safety under adversarial testing, creativity, and diversity consistent with norms
- **Primary role**: Deeply integrated into training and deployment loops, guiding adaptive systems that tune themselves toward human aligned objectives
- **Acknowledged limits**: Metrics treated as advisory signals rather than ground truth, always paired with expert review and continuous real-world monitoring

This past–present–future progression highlights the central message of the chapter: metrics are moving from narrow, surface-level proxies toward richer but still imperfect reflections of human values, and responsible evaluation requires using them as tools for insight rather than absolute judges of quality.

A/B Testing Frameworks

Evaluation in generative AI does not end with static metrics or carefully controlled human studies. Metrics such as BLEU, FID, perplexity, and diversity scores provide reproducibility, and human evaluations capture nuance, but neither fully answers the central question of deployment: which model actually performs better for real users in real contexts. A/B testing frameworks fill this gap. By comparing models directly with actual users under randomized conditions, they provide causal evidence about performance, preference, and impact. They have become the backbone of evaluation in large-scale deployments of generative systems.

Traditional software A/B tests assume deterministic behavior: a given input passed to version A or B of a feature will reliably produce the same output, and metrics such as click-through rate or time on page can be attributed directly to small changes in code or UI. Generative systems break these assumptions. Large language models and other generative models are stochastic, highly prompt sensitive, and often shaped by hidden context such as conversation history or user personalization. The same prompt can produce different outputs across runs, and small prompt or parameter changes can interact in nonlinear ways with user behavior. As a result, classical A/B tooling that treats variants as fixed treatments over stable experiences is insufficient without adaptation; experimentation frameworks must account for randomness in model outputs, prompt design, safety filters, and user feedback loops when interpreting results.

Historical Origins of A/B Testing

A/B testing arises from the statistical tradition of controlled experimentation. In the early twentieth century, Ronald Fisher formalized randomized designs in agricultural studies, showing that random assignment neutralizes confounders and isolates causal effects. Medicine later adopted these ideas through randomized controlled trials, which became the backbone of evidence-based practice. Mid-century marketing then applied the same structure to advertising by sending different versions to different audiences and measuring the difference in response.

The Internet transformed this practice by providing scale and immediacy. Online platforms could expose millions of users to alternative interfaces or ranking algorithms and observe outcomes within hours. Organizations such as Amazon and Google institutionalized experimentation as an operational habit, running thousands of tests in parallel to refine search, recommendations, pricing displays, and user flows.

Generative AI adopted A/B testing because static scores were not enough. Metrics like BLEU or perplexity could not reliably predict which model people would prefer in real conversations. Human rater studies were informative but too slow and expensive to guide daily product decisions. A/B testing allowed teams to evaluate models directly with real users in their actual context of use and to collect evidence about preference, effectiveness, and safety at scale.

- Randomized experiments migrated from agriculture to medicine, marketing, and then the web.
- Online platforms made continuous experimentation feasible and routine.
- Generative AI uses A/B testing to measure lived user experience rather than proxy scores alone.

Statistical Foundations of A/B Testing

The logic mirrors randomized controlled trials. Users are randomly assigned to control and treatment. The control group interacts with the baseline model, while the treatment group sees a variant. Differences in outcomes across the two groups estimate the causal effect of the change. This approach works because randomization balances observed and unobserved factors on average, making the groups comparable.

Key elements deserve attention. Random assignment must be truly unbiased, often implemented with user-ID hashing that is stable over time. Sample size should be large enough to detect the expected effect with acceptable power. Outcome measures need to be defined in advance and tied to product goals, not chosen after the fact. Statistical testing determines whether observed differences are likely to reflect a real effect rather than noise, and confidence intervals quantify the uncertainty around that estimate.

- Randomization creates comparable groups and enables causal inference.
- Power analysis links expected effect size, sample size, and duration.
- Pre-specified outcomes and confidence intervals guard against over-interpretation.

Why A/B Testing Matters for Generative AI

Generative systems differ from classical classifiers because outputs are open ended and context sensitive. The best answer depends on a user's goal, cultural expectations, and tolerance for risk. Static offline metrics cannot fully capture these nuances. A/B testing places the evaluation in the deployment setting, where people react to the system as they actually use it, not as a simplified benchmark imagines they will.

Generative models also change the user, and the user changes the model's apparent performance. Habits form, prompts evolve, and expectations shift. Only an in-context experiment can observe these dynamics. Finally, quality is multidimensional. Teams often need to track satisfaction, task success, safety events, and latency together rather than a single scalar.

- Open-ended outputs make preference the relevant criterion.
- Context and culture strongly shape perceived quality.
- Interaction effects emerge over time and must be measured in situ.

Designing A/B Tests for Generative Systems

Effective experiments begin with a clear hypothesis and a narrow unit of change. Teams specify whether they aim to reduce hallucinations, increase task completion, or improve politeness. The control is a current reliable model, and the treatment is a single variation so that attribution is unambiguous. Randomization is implemented at the user or session level so that each participant remains consistently in one bucket for the duration of the test.

Outcome measures must align with the objective and be measurable at scale. Preference prompts, error counters, and success signals are useful, but they should be complemented with safety indicators and latency. Power analysis guides the required sample size and expected run time. Duration should be long enough to smooth out diurnal cycles and short-term news shocks, yet not so long that product decisions stall.

- Start with a single, testable change and a stable control.
- Align outcomes with the hypothesis and product goals.
- Size and schedule the test with power and seasonality in mind.

Metrics Within A/B Testing

Even in an online experiment, metrics provide the lenses through which outcomes are viewed. Behavioral metrics such as click-through, dwell time, and completion rate reveal engagement and efficiency. Preference metrics capture explicit judgments when users choose between outputs or rate them. Safety metrics track toxicity, bias signals, and hallucination rates. Trust metrics measure self-reported confidence in the model's outputs. The value of A/B testing is that each of these can be compared causally across variants rather than correlated post hoc.

- Behavioral, preference, safety, and trust metrics should be monitored together.
- Composite dashboards help resolve trade-offs across dimensions.
- Causal attribution improves decision quality compared with offline correlation.

Case Studies in Text Generation

Experience across text applications shows why experiments often trump static scores. In machine translation, a system with a higher BLEU score does not always win user preference because it may mimic reference phrasing rather than communicate intent. In summarization, readers sometimes prefer lower-ROUGE outputs that are clearer and more faithful to salience rather than exact overlap. In dialogue, models optimized for perplexity can sound stiff. A/B tests capture whether people perceive the conversation as helpful, polite, and efficient.

- Preference frequently diverges from BLEU or ROUGE.
- Readability and faithfulness often outweigh lexical overlap.
- Conversation quality requires user-centered measures such as helpfulness.

Case Studies in Vision and Multimodal Systems

In image generation, two systems with similar FID can differ in prompt faithfulness and user delight. Experiments reveal which set of images people actually prefer for a given prompt distribution. In image captioning, the metric that best predicts usefulness for accessibility may differ from the one that best predicts descriptive richness. Multimodal assistants add another layer, since coherence across text, image, and audio matters. A/B testing surfaces cross-modal issues that single-modality scores miss.

- Equal FID does not guarantee equal perceived quality.
- Accessibility and comprehension need task-specific outcomes.
- Cross-modal coherence is best judged in live interaction.

Online vs. Offline A/B Testing

There are two broad settings. Online testing exposes live users to variants and measures outcomes at production scale. This yields strong external validity but raises ethical and operational concerns. Offline testing recruits evaluators to compare outputs in a controlled panel. It provides depth and rapid iteration but lacks organic context and scale. A common practice is to run offline pilots to filter weak variants, then promote finalists to online tests.

- Online tests offer scale and realism but require stronger safeguards.
- Offline panels are quick and informative but less representative.
- A staged pipeline combines both to reduce risk and cost.

Ethical Considerations in A/B Testing

Experimentation involves people, so the rights and welfare of participants are paramount. Many users are unaware they are in a test, which creates obligations around transparency, risk limits, and data protection. Variants must not expose specific groups to higher rates of harm. Sensitive domains may require independent review and pre-defined termination criteria. Documentation should explain the objective, the safeguards, and the results used to justify shipping decisions.

- Minimize risk, especially for vulnerable populations.
- Provide oversight and clear stop conditions for harmful outcomes.
- Record rationales and protections to enable accountability.

Statistical Pitfalls in A/B Testing

Several common errors can undermine conclusions. Peeking at results and stopping early inflates false positives. Running many simultaneous tests without correction raises the chance of spurious wins. Underpowered studies waste time because they cannot detect effects of practical size. Misaligned metrics produce wins that do not advance the true goal. Aggregate averages can hide subgroup differences that matter for fairness or for business reality.

- Avoid early stopping and correct for multiple comparisons.
- Ensure adequate power and align outcomes with objectives.
- Segment analyses to detect heterogeneous effects.

Feedback Loops and Longitudinal Testing

Generative systems and users adapt to each other. A variant that looks good in the first week may degrade as users learn new prompting habits or as external topics shift. Longitudinal experiments track performance over sustained periods and watch for drift. Logging pipelines should connect experiment assignments to later outcomes so that delayed effects are visible.

- Measure not only the initial lift but also stability over time.
- Monitor for drift in prompts, topics, and user segments.
- Keep assignment stable to separate learning effects from random noise.

Scaling A/B Testing in Industry

Running many experiments requires robust infrastructure. Platforms must randomize assignment consistently, collect detailed logs, and compute metrics automatically. Scheduling and isolation prevent interference among concurrent tests. Real-time dashboards track safety and performance, while decision pipelines encode rules for promotion and rollback so that shipping is principled rather than ad hoc.

- Experiment platforms manage randomization, logging, and analysis.
- Guardrails avoid test interference and protect safety metrics.
- Automated promotion criteria bring discipline to deployment.

Beyond A/B: Multivariate and Bandit Testing

A/B testing is the simplest controlled design, but more adaptive methods can be efficient. Multivariate tests study combinations of factors. Multi-armed bandits allocate more traffic to better variants while the experiment runs, raising reward and shortening time to decision. Bayesian approaches maintain posterior beliefs over treatment effects and enable sequential decision-making with explicit uncertainty.

- Multivariate designs explore interactions among changes.
- Bandits and Bayesian methods improve efficiency during experimentation.
- Adaptive designs need careful statistical controls to preserve validity.

Cultural and Psychological Dimensions

Human perception shapes measured outcomes. Preferences for tone, politeness, or humor vary across cultures and demographics. The framing of rating prompts can shift judgments. Prior experiences with assistants create expectations that influence evaluations. To obtain generalizable results, samples should be diverse and prompts localized rather than translated mechanically.

- Culture and demographics affect what users perceive as quality.
- Framing and expectations bias responses.
- Diverse sampling improves external validity.

Governance and Regulation of A/B Testing

As experimentation becomes routine, governance structures are needed. Transparency policies determine what is communicated to users. Internal ethics boards or external reviewers may be necessary in regulated sectors. Jurisdictions differ in their rules for online experimentation, data retention, and consent. Documentation standards and audit trails help organizations demonstrate compliance when models influence health, education, or finance.

- Establish policy for disclosure, consent, and data protection.
- Align with sector-specific regulations and regional laws.
- Maintain auditable records of design, outcomes, and decisions.

Sequential and Adaptive Experimentation

Classical tests fix their duration in advance. Sequential methods analyze interim results while controlling error rates. Bandit algorithms reallocate traffic toward promising variants to reduce regret during the run. Bayesian sequential analysis updates beliefs continuously and can stop when the probability of superiority passes a threshold. These methods can shorten decision cycles, though they require expertise to avoid bias.

- Sequential testing reduces time to decision while guarding against false positives.
- Bandits raise user benefit during the test by shifting traffic.
- Bayesian rules make uncertainty explicit and actionable.

A/B Testing for Safety and Alignment

Quality is not the only objective. Safety and fairness must be measured directly. Experiments can compare hallucination rates, toxic language incidents, and disparities across demographic slices. Trust and perceived safety can be gathered through user surveys. Shipping criteria should include thresholds on these metrics, not only engagement or revenue uplift.

- Track hallucination, toxicity, and bias as first-class outcomes.
- Include user trust measures alongside behavioral signals.
- Gate deployment on safety thresholds, not just preference wins.

Infrastructure Requirements for Large-Scale A/B Testing

At scale, reliability depends on engineering. Logging must be lossless and tied to stable identifiers. Data pipelines need to transform raw events into clean features quickly. Experiment services should handle randomization, bucketing, and exposure tracking. Dashboards must visualize trends and confidence intervals, and fail-safes should pause or terminate tests automatically when harm indicators rise.

- Reliable logging and exposure tracking are foundational.
- Fast, correct pipelines convert events into metrics.
- Automated fail-safes protect users when anomalies occur.

Psychological and Sociological Effects of Testing

Testing itself can influence behavior. If users realize that versions differ, they may change how they interact. Hidden experimentation can undermine trust if discovered later. Word of mouth and social media can cross-contaminate control and treatment through shared screenshots or advice. Over time, a culture of constant change can normalize experimentation but also raise expectations for rapid improvement.

- Awareness of testing can alter user actions.
- Lack of transparency can damage trust if revealed.
- Social sharing leaks treatment effects across groups.

Cultural Variability in Experiment Results

Results often differ by language, region, and demographic. Translation quality can be judged by different standards in inflected versus analytic languages. Humor and politeness conventions vary across cultures, which changes the perceived quality of responses. Age, education, and professional background influence how people value verbosity, caution, or creativity. A single global average can hide these differences, so segmentation and local testing are essential.

- Segment by language, region, and key demographics.
- Localize prompts and outcomes rather than applying one template.
- Use stratified reporting so global wins do not mask local losses.

The Future of Evaluation Ecosystems

A/B testing is part of a larger ecosystem.

- **Static metrics:** Provide reproducibility and benchmarking.
- **Human evaluations:** Capture nuanced judgments.
- **Red-teaming:** Probes vulnerabilities and adversarial cases.
- **A/B testing:** Measures lived user experience.
- **Continuous monitoring:** Tracks performance over time.

The future of evaluation requires integrating all methods into coherent pipelines.

A/B testing frameworks expand evaluation beyond metrics and laboratory studies into real-world contexts. They provide causal evidence of improvement, capture user preference, and integrate dimensions of safety and trust. They are powerful precisely because they measure models where they matter most: in interaction with humans.

Yet their power brings responsibility. Poorly designed experiments mislead. Ethical lapses risk harming users. Cultural biases distort results. Scaling experiments without governance undermines trust. A/B testing must therefore be conducted with rigor, transparency, and inclusivity.

Generative AI requires evaluation as dynamic as the models themselves. Metrics quantify surface features, human evaluations provide nuance, and A/B testing grounds all of this in lived experience. Together, these methods form a holistic evaluation ecosystem. Within that ecosystem, A/B testing provides the ultimate reality check, ensuring that generative models are not only impressive in benchmarks but also meaningful, safe, and valuable in the world.

Task-Based Performance Assessments

Generative AI evaluation is never complete if it stops at static metrics or user preference tests. BLEU, FID, perplexity, and diversity scores are essential for reproducibility. A/B testing frameworks allow causal comparisons in live settings. Yet even with these tools, a fundamental question often remains unanswered: does the system actually help people accomplish the real-world tasks for which it was designed? Task-based performance assessments respond to this question directly.

Task-based performance assessments evaluate generative models not by their resemblance to references or their ability to generate engaging outputs, but by their effectiveness in completing meaningful tasks. They measure utility, impact, and outcomes in domains ranging from healthcare to law, education, science, and creative industries. By embedding models into authentic tasks, evaluators can test whether they contribute to problem-solving, decision-making, creativity, and productivity.

This section explores task-based performance assessment in unprecedented depth. It traces its conceptual roots, contrasts it with proxy metrics, and surveys applications across multiple domains. It investigates intrinsic and extrinsic measures, designs for authentic tasks, methodological challenges, cultural and ethical dimensions, and the infrastructure needed for scaling. It also incorporates theories of assessment from education and psychology, analysis of cognitive load and task complexity, adaptive and personalized evaluation frameworks, and governance considerations. The discussion culminates in recognizing task-based assessment as the cornerstone of trustworthy generative AI evaluation, bridging the gap between laboratory metrics and societal impact.

Conceptual Foundations of Task-Based Evaluation

Task-based evaluation reframes what it means for a generative system to perform well. Rather than asking whether an output resembles a reference or earns a high score on a proxy metric, it asks whether the system helps people achieve concrete goals in real activities. The center of gravity moves from outputs to outcomes, from imitation to usefulness, and from model-centric criteria to human-centric success. In this view, a summary is good if it improves decision quality, a caption is good if it supports accessibility, and a code suggestion is good if it shortens delivery time without increasing defects.

- Utility orientation emphasizes usefulness in context rather than surface qualities.
- Outcome definitions tie success to problem solving rather than similarity alone.
- Human centricity aligns evaluation with goals, workflows, and responsibility.

This perspective treats generative AI as an augmentation technology whose value is judged by its contribution to human work.

Intrinsic vs. Extrinsic Assessment

Task-based evaluation operates on two complementary levels. Intrinsic assessment measures how well the model performs the focal task on its own terms, such as solving math problems, producing accurate summaries, or generating code that compiles and passes tests. Extrinsic assessment measures the broader impact of the system on real outcomes, such as improved clinical decisions, faster legal review, or better student learning.

- Intrinsic checks task correctness and fidelity to requirements.
- Extrinsic examines downstream effects on quality, speed, cost, and safety.

Both views are necessary. Intrinsic scores without real-world benefit are hollow, while extrinsic gains without intrinsic reliability may be brittle.

Historical Roots of Task-Based Assessments

This orientation has deep precedents. Information retrieval has long asked whether systems actually help users find what they need, not only whether precision and recall are high. Medical decision support has been judged by changes in diagnostic accuracy, adherence, and patient outcomes. Educational technology has been evaluated by learning gains and retention rather than internal classification accuracy.

- The tradition evaluates contribution to work, not only internal model behavior.
- Generative AI inherits this frame by treating outputs as tools in larger activities.

Domain-Specific Examples

Task-based evaluation takes different shapes across fields because the meaning of success varies.

- **Healthcare**: Discharge summaries are judged by whether they reduce handoff errors and readmissions; radiology reports by diagnostic accuracy and time to decision.

- **Law**: Drafts and contract reviews are evaluated by cycle time, error avoidance, and risk detection.
- **Education**: Explanations are evaluated by learning gains and retention over time.
- **Software engineering**: Code suggestions are judged by compilation, unit-test pass rates, defect rates, and developer throughput.
- **Creative industries**: Generated assets are evaluated by integration into production pipelines and campaign outcomes.

Each example shifts attention from resemblance to functional integration in workflows.

Designing Authentic Tasks

The quality of a task-based evaluation rests on the authenticity of the tasks. They should mirror real conditions and constraints, reflect the needs of actual users, and span a diverse set of activities so that systems are not overfitted to a single pattern. Difficulty must be calibrated to separate strong and weak systems without being trivial or intractable, and procedures must be replicable so that results are comparable across studies.

- Authenticity, relevance, diversity, calibrated difficulty, and standardization are the pillars of sound task design.

Measuring Outcomes

Outcome measures define what counts as success. Objective correctness remains central when it applies, as with code execution, mathematical proofs, or factual statements. Efficiency captures time saved and error reduction. Quality measures capture decision usefulness, creativity gains, or clarity. Human factors include satisfaction, trust, and perceived support. Finally, downstream impact assesses effects on organizations or populations.

- Correctness, efficiency, quality, user experience, and downstream impact together produce a holistic view.

Human–AI Collaboration in Tasks

Most deployments combine human judgment with model capability. Writers, designers, and analysts use the system as a drafting partner. Professionals such as doctors and lawyers use summaries and retrieval to augment reasoning. Students use tutors to practice, receive feedback, and fill gaps. Evaluations therefore measure the quality of the collaboration, not only the quality of isolated outputs.

- Co-creation, decision support, and tutoring are natural collaboration modes that evaluations should capture.

Intrinsic Benchmarks vs. Real-World Integration

Controlled intrinsic benchmarks enable comparability, yet they can drift away from reality. Real-world integration studies capture authentic use and emergent behavior, but they are costly and harder to standardize. A balanced program uses intrinsic tests for screening and iteration, then validates with real-world or field studies to confirm that gains transfer.

- Intrinsic provides precision and repeatability.
- Integration provides realism and external validity.
- Both are required for a reliable picture.

Case Study: Code Generation

Code provides clear intrinsic and extrinsic signals. Intrinsically, generated code must compile and pass unit tests. Extrinsically, developer tasks should complete faster with fewer defects, and teams should report reduced cognitive load without loss of design quality. Collaboration measures ask whether suggestions are accepted, edited, or rejected, and whether review cycles shorten.

- Success is measured by test pass rates, throughput, defect density, and acceptance patterns.

Case Study: Education

Educational settings illustrate the dual lens. Intrinsically, explanations must be correct, clear, and age-appropriate. Extrinsically, students should show improved comprehension, retention, and transfer across related problems. Teacher-in-the-loop settings examine whether AI support improves lesson planning and classroom outcomes without narrowing curricula.

- Learning gains, transfer, and teacher productivity complement accuracy.

Challenges of Task-Based Evaluation

The strengths of this approach come with real demands. Designing authentic tasks is complex; user strategies vary; large-scale studies are costly; tasks evolve as users adapt; and task sets can reflect narrow cultural or professional assumptions if not curated carefully.

- Complexity, variability, cost, adaptation, and bias are the recurring challenges.

Automating Task Evaluation

Automation helps with scale where it is feasible. Code execution can be evaluated by comprehensive test suites. Retrieval tasks can be checked against ground truth. Simulated agents can model user interactions to stress test flows before involving humans. Yet many activities require judgment that is hard to simulate.

- Use automation to screen candidates and reserve human time for higher-level qualities.

Human-in-the-Loop Task Evaluation

Expert review and user studies remain essential. Professionals assess domain fidelity and risk. Controlled studies measure how people perform with and without assistance. Crowdsourcing can cover general tasks if instructions, quality control, and sampling are sound.

- Expert judges, structured user studies, and carefully managed crowds all contribute to validity.

Formative vs. Summative Assessment

Formative evaluations support iteration during development. They identify failure modes, refine prompts or agents, and guide data collection. Summative evaluations assess readiness for deployment, documenting performance, safety, and fairness. Mature programs run both, with formative cycles feeding into periodic summative checks.

- Formative guides improvement; summative certifies readiness.

Safety and Fairness in Task Evaluation

Tasks must be designed and scored with safety and equity in view. Outputs should not cause harm. Performance should be equitable across demographic groups, languages, and contexts. Robustness testing examines stress conditions and adversarial cases to make sure that quality does not collapse at the edges.

- Safety, fairness, and robustness are first-class outcomes, not afterthoughts.

Scaling Task-Based Evaluation

Industry settings require evaluations that keep pace with releases. Synthetic task generation can expand coverage, while continuous monitoring tracks success in production with privacy-preserving telemetry. Cross-domain test suites prevent over-specialization, and feedback loops route failures back into data and training plans.

- Synthetic coverage, live monitoring, cross-domain testing, and iterative improvement make scaling feasible.

Theories of Assessment from Education and Psychology

Classical assessment theory provides a vocabulary for quality. Construct validity asks whether the task truly measures the intended competency. Reliability asks whether results are stable across raters and trials. Authentic assessment values tasks that mirror the real demands of practice. Formative scaffolding uses tasks to guide learners toward mastery, an idea that transfers to guiding model-assisted workflows.

- Validity, reliability, authenticity, and scaffolding strengthen evaluation design.

Task Complexity and Cognitive Load

Task difficulty should discriminate without overwhelming. Simple tasks check basic correctness, while complex tasks require planning, reasoning, or creativity across multiple steps. Measures of cognitive load help ensure that assistance reduces unnecessary effort rather than adding friction.

- Balance difficulty so that differences reflect capability, not exhaustion.

Adaptive and Personalized Task Evaluation

Personalization complicates evaluation because users face different needs and paths. Tests can adapt difficulty to skill, track longitudinal progress, and analyze outcomes for subgroups. Evaluation frameworks should allow dynamic tailoring while retaining comparable anchors.

- Personalized and adaptive protocols need flexible scoring and careful comparability checks.

Infrastructure for Continuous Task-Based Monitoring

Sustained evaluation depends on engineering as much as methodology. Logging must capture task outcomes reliably. Dashboards should display success rates, error types, and time to completion. Alerts should flag declines or safety events. Automated pipelines should fold evaluation into CI/CD so that releases ship only when task criteria are met.

- Logging, dashboards, alerting, and gated deployment make evaluation continuous.

Risks of Task Reductionism

Reducing complex work to narrow tasks can mislead. Important context can be stripped away, and models may be optimized for test artifacts rather than real practice. Passing a checklist can conceal negative side effects such as overreliance or degraded expertise.

- Guard against oversimplification and monitor for unintended consequences.

Governance of Task-Based Evaluation

Legitimacy requires governance. Organizations should document how tasks are defined, how outcomes are measured, and how results inform decisions. Regulators may require specific evaluations for high-risk uses. Equity considerations should influence sampling and reporting. Transparency with users builds trust.

- Accountability, standards, inclusion, and clear communication underpin responsible practice.

Ethical and Cultural Dimensions

Tasks reflect values and can privilege certain worldviews. Evaluation programs should be sensitive to epistemic injustice, include diverse populations, and anticipate long-term effects that do not show up in short tests. Cultural norms shape what counts as helpful, respectful, or creative, so tasks and scoring need to be localized thoughtfully.

- Inclusivity and cultural sensitivity are integral to credible assessment.

Future Directions

The future expands evaluation in multiple ways.

- **Multimodal tasks:** Systems assessed on integrated text, image, and audio performance
- **Interactive tasks:** Multi-turn dialogue or collaboration measured over time
- **Longitudinal assessment:** Tracking outcomes across months or years
- **Adaptive ecosystems:** Evaluation evolving dynamically with models

Task-based evaluation will grow in scope, depth, and sophistication.

Task-based performance assessments are the ultimate reality check. They measure whether generative AI systems contribute meaningfully to human goals, whether they enhance productivity, creativity, decision-making, and well-being. Unlike static metrics, they connect evaluation to lived experience.

They are demanding to design, costly to implement, and difficult to scale. They require expertise, diversity, and infrastructure. Yet they are indispensable. Without them, generative AI risks being impressive in laboratories but irrelevant, misleading, or harmful in practice.

Task-based performance assessments ensure that AI is judged not only by what it produces but by what it enables. They are the bridge between capability and impact, the cornerstone of trustworthy evaluation, and the path by which generative AI becomes not just technically advanced but socially valuable.

Conclusion

Evaluation and benchmarking are not peripheral details of generative AI research; they are its backbone. The way we measure performance determines the trajectory of innovation, the direction of optimization, and the standards by which systems are trusted or rejected. Chapter 11 has examined evaluation from multiple perspectives: automatic metrics, human evaluation, A/B testing frameworks, and task-based performance assessments. Each method offers unique strengths, each reveals different facets of generative performance, and each has limitations that must be recognized.

Automatic metrics such as BLEU, FID, perplexity, and diversity scores represent the first wave of attempts to quantify generative quality. They emerged because researchers needed reproducibility and comparability. BLEU allowed translation systems to be judged quickly, FID allowed images to be compared in bulk, perplexity provided a mathematically rigorous measure of likelihood, and diversity scores highlighted the dangers of repetitive outputs. These metrics were transformative, enabling rapid progress and establishing benchmarks that defined the field for years. Yet their limitations became clear. BLEU rewarded surface overlap but penalized creativity. FID assumed Gaussian distributions and reflected biases of pretrained vision models. Perplexity measured probability but not usefulness. Diversity scores captured novelty but not meaning. They were indispensable but incomplete.

Human evaluation arose to fill some of these gaps. Humans can detect nuance, semantics, pragmatics, and cultural appropriateness that metrics cannot capture. Human studies revealed when high-BLEU systems were dispreferred, when fluent text masked hallucinated facts, or when images technically aligned with prompts but failed aesthetically. Human evaluation restored alignment with user perception but at the cost of scale, consistency, and efficiency. Crowdsourcing provided breadth but sacrificed expertise. Expert judges provided depth but were expensive and limited. Human evaluation illuminated quality but could not replace automatic metrics in rapid iteration.

A/B testing frameworks brought evaluation into real-world contexts. By randomly assigning users to different models, they measured causal differences in preference, engagement, and safety. A/B testing captured performance in authentic interaction, not only in controlled laboratories. It revealed when two models with similar benchmark scores diverged sharply in user satisfaction. It allowed companies to optimize conversational models, summarization tools, and multimodal assistants in live deployment. Yet A/B testing raised new challenges: statistical pitfalls such as peeking and multiple testing, ethical questions of consent and fairness, infrastructure demands for large-scale deployment, and cultural variability in results. It highlighted how experimentation is not only technical but also social and ethical, shaping trust as much as performance.

Task-based performance assessments extended evaluation to its fullest scope. Instead of asking whether outputs resembled references or whether users expressed preference, task-based evaluation asked whether AI helped achieve goals in real domains. Could doctors diagnose more accurately? Could students learn more effectively? Could programmers develop software more efficiently? Could scientists generate hypotheses more productively? This approach recognized that generative AI is not built only to produce text, images, or audio but to serve as a partner in human tasks. Task-based evaluation captured extrinsic impact, measuring effectiveness, efficiency, and outcomes. It also revealed cultural and ethical dimensions, since definitions of task success vary across societies, and oversimplified tasks risked reductionism. Task-based assessment emerged as the most demanding but also the most indispensable form of evaluation, linking AI capability directly to human impact.

Taken together, these four evaluation paradigms provide complementary lenses. Automatic metrics offer speed and comparability. Human evaluations provide nuance. A/B testing measures causal effects in live environments. Task-based assessments ground AI in real-world utility. No single approach suffices. Each compensates for the others' weaknesses, and together they form an evaluation ecosystem.

The conclusion that emerges is one of pluralism. Evaluation in generative AI must be multidimensional, layered, and adaptive. It must recognize that quality is not one thing but many: fluency, accuracy, coherence, creativity, safety, fairness, efficiency, trustworthiness, and utility. Metrics, human ratings, experiments, and tasks each illuminate different dimensions. Progress is genuine only when measured across all of them.

There are also broader lessons. Evaluation defines incentives. When BLEU dominated, translation systems optimized for overlap rather than meaning. When FID became standard, GAN research chased lower scores even when perceptual quality plateaued. Leaderboards became currencies of reputation, sometimes distorting innovation. Evaluation thus governs research indirectly but powerfully. It is not neutral. It encodes values about what counts as progress.

Ethics and governance must therefore be central to evaluation. Metrics can embed biases from training corpora. Human evaluations can reflect cultural blind spots. A/B testing can expose users to harms without their knowledge. Task-based assessments can privilege one definition of success while marginalizing others. Evaluation frameworks must include safeguards for fairness, inclusivity, and transparency. They must be sensitive to cultural diversity and global contexts. They must be accountable not only to researchers and companies but also to the societies in which generative systems are deployed.

Looking ahead, evaluation will evolve. Automatic metrics will become more semantic and multimodal, leveraging foundation models to judge meaning across domains. Human evaluations will integrate preference modeling and structured annotation pipelines. A/B testing will become more adaptive, using Bayesian methods and bandit algorithms to increase efficiency while maintaining rigor. Task-based assessments will expand into longitudinal, collaborative, and multimodal contexts, measuring not just immediate outputs but sustained outcomes. Evaluation will become continuous rather than episodic, integrated into deployment pipelines as ongoing monitoring.

The ultimate goal is not only to measure generative AI but to govern it responsibly. Evaluation frameworks define what is rewarded, what is improved, and what is deployed. They shape the alignment between technical progress and societal benefit. They determine whether generative systems become tools of empowerment or sources of harm.

Chapter 11 has argued that evaluation is the hinge of generative AI. Without it, progress is unmeasurable. With poor evaluation, progress is misdirected. With pluralistic, rigorous, and ethical evaluation, progress can be meaningful. Automatic metrics, human evaluation, A/B testing, and task-based performance assessments together form a toolkit for measuring quality, reliability, and utility. They remind us that generative AI is judged not only by what it produces but by what it enables humans to accomplish, how it treats them, and how it integrates into society.

The challenge ahead is to refine this toolkit, to guard against distortions, and to ensure inclusivity and transparency. The opportunity is to make evaluation not a bottleneck but a foundation for responsible innovation. As generative AI continues to expand in power and influence, evaluation will remain the critical practice that keeps it grounded in human values and real-world effectiveness.

For our next and final chapter, we talk about future directions and deployment in production.

CHAPTER 12

Future Directions and Deployment in Production

The story of generative AI does not end with breakthroughs in modeling or advances in evaluation. The true test of these systems begins when they are deployed in production environments at scale. A research prototype can impress with benchmarks, demos, and controlled experiments. A production system, by contrast, must operate reliably, safely, and continuously in dynamic contexts. It must integrate into pipelines, adapt to changing requirements, comply with legal and ethical standards, and evolve alongside the organizations and societies that depend on it.

This chapter addresses the practical considerations of bringing generative models into production. Deployment is not merely the act of releasing a model but an ongoing process of integration, monitoring, and governance. The software engineering practices that have matured around traditional systems, such as continuous integration and delivery pipelines, monitoring, observability, and compliance, must be adapted and extended to the unique challenges of generative AI.

The first section explores CI/CD pipelines for generative models. Unlike traditional software, models are not static code but artifacts of data, architecture, and optimization. Updating them requires retraining or fine-tuning, managing large artifacts, and validating outputs for both quality

I. Cronin, *Building and Training Generative AI Models*,
https://doi.org/10.1007/979-8-8688-2332-9_12

and safety. CI/CD for AI involves model versioning, automated retraining workflows, structured deployment gates, and rollback mechanisms. Getting these pipelines right is essential for maintaining velocity without sacrificing reliability.

The next focus is model monitoring and observability. In production, models encounter data that differ from training distributions, user interactions that evolve over time, and adversarial attempts to exploit vulnerabilities. Monitoring requires not only tracking performance metrics but also detecting drift, hallucinations, bias, and safety-critical failures. Observability emphasizes the ability to trace, diagnose, and explain model behaviors in real time, enabling quick mitigation of issues.

Legal, ethical, and compliance challenges form another layer of deployment complexity. Generative systems can create outputs with copyright implications, propagate biases, or produce toxic content. Deploying them at scale involves navigating privacy regulations, intellectual property law, and ethical obligations to protect users. Compliance frameworks must be established for data use, output accountability, and transparency. Organizations must prepare not only for technical challenges but also for societal and regulatory scrutiny.

Finally, the chapter considers trends shaping the future of deployment: multimodal fusion, agentic models, and AI safety. Multimodal systems require deployment strategies that integrate text, vision, audio, and beyond. Agentic models that act autonomously raise new monitoring and governance questions. Safety research informs practices that balance innovation with risk mitigation. These trends indicate that deployment is not static but evolving, demanding anticipatory practices that adapt to shifting technical and societal landscapes.

The goal of this chapter is to equip practitioners and researchers with a conceptual map of deployment at scale. By combining engineering practices with governance, ethics, and forward-looking trends, it frames deployment not as an afterthought but as the stage where generative AI demonstrates its true value and encounters its greatest risks.

CI/CD Pipelines for Model Updates

The deployment of generative AI models into production environments introduces challenges that go far beyond those faced in traditional software engineering. Unlike static applications, which change only when developers explicitly modify code, generative models are dynamic artifacts that evolve with data, hardware, evaluation criteria, and societal context. Their performance is influenced not just by the source code that defines their architectures but by the massive datasets on which they are trained, the hyperparameters that guide their optimization, and the hardware on which they are executed. Once deployed, they encounter live user interactions, adversarial prompts, and constantly shifting input distributions.

In this landscape, **continuous integration and continuous delivery (CI/CD) pipelines** become essential. They provide a structured way to iterate on models, retrain them, validate updates, and push new versions into production environments. CI/CD is not simply about software automation; in generative AI, it becomes the backbone of reliability, accountability, and scalability. By extending software engineering practices into the probabilistic and data-driven domain of machine learning, CI/CD pipelines ensure that model updates are frequent, safe, and transparent.

This section explores CI/CD pipelines for generative AI in comprehensive detail. It begins with their origins and conceptual foundations, then breaks down the key components of AI-specific pipelines, including data management, training orchestration, evaluation gates, model registries, deployment strategies, monitoring, and rollback. It extends the discussion into issues of reproducibility, compliance, security, and sustainability, while also surveying case studies across industries. Finally, it considers emerging directions, including integration with agentic systems and multimodal fusion, situating CI/CD as not only an engineering practice but a sociotechnical system that underpins trustworthy deployment.

Historical Origins of CI/CD in AI

The origins of CI/CD trace back to the software crises of the mid-twentieth century, when teams working on large projects struggled to integrate their codebases. Continuous integration emerged as a response: developers would integrate their code frequently into shared repositories, with automated builds and tests verifying correctness. Continuous delivery extended this principle by automating deployment, allowing new features and fixes to move quickly into production.

When machine learning entered production environments, practitioners recognized both parallels and divergences. Code alone did not define system behavior. Data pipelines, feature engineering scripts, model weights, and training runs played equally important roles. Traditional CI/CD practices could not handle these complexities. Out of this need, **MLOps** emerged as a discipline, adapting DevOps practices to the life cycle of AI models.

Generative AI amplified these challenges. Models became larger and more resource-intensive, with billions of parameters. Their outputs became open-ended and harder to evaluate automatically. Distribution shifts and adversarial prompts created new vulnerabilities. This forced organizations to rethink CI/CD pipelines as more than code automation; they became governance frameworks that manage continuous change in models, data, and context.

Conceptual Foundations of CI/CD for Generative Models

CI/CD pipelines for generative models are guided by principles that balance agility with accountability. At their heart, these pipelines must allow teams to move quickly while maintaining stability, safety, and traceability. The speed of iteration matters because user expectations and competitive pressures demand rapid improvement. Yet with speed comes the risk of error, so pipelines must safeguard reliability. Every artifact must be reproducible so that results can be traced to data, code, and configurations. Because models today involve massive datasets and distributed clusters, scalability is critical. Safety gates must catch hallucinations, bias, and toxicity before deployment, since these cannot be fixed after the fact. Finally, rollback pathways are essential so that teams can revert to stable versions when updates introduce problems.

- Velocity ensures teams can innovate without being blocked.
- Reliability protects users from unsafe or unstable releases.
- Reproducibility ties every model back to its origins.
- Scalability enables work across large datasets and clusters.
- Safety embeds checks for harm before deployment.
- Rollback guarantees resilience in case of failure.
- Here is a compact comparison that highlights what makes CI/CD for generative models distinct.

Dimension	Traditional ML CI/CD	GenAI-specific CI/CD
Primary artifact	Single trained model plus static pre-/post-processing code	Model plus prompts, system messages, tools, safety filters, routing policies, and sometimes multiple cooperating models
Output behavior	Mostly deterministic for a given input (feature vectors → labels/scores)	Stochastic, prompt sensitive, and highly context-dependent; small changes can shift style, safety, and correctness

(continued)

Dimension	Traditional ML CI/CD	GenAI-specific CI/CD
Core quality checks	Accuracy, AUC, RMSE, calibration on held-out data	Multi-axis evaluation: usefulness, factuality, safety, bias, style, diversity; mix of automatic metrics, preference models, and human review
Test data	Fixed validation and test sets that closely match training distribution	Dynamic evaluation suites including adversarial prompts, red-team sets, safety testbeds, multilingual and multimodal scenarios
Safety gates	Mostly around fairness, privacy, and rate limiting	Layered safety stack: input filters, output filters, policy models, jailbreak tests, domain blocks, and human-in-the-loop approvals for high-risk domains
Configuration surface	Model weights, feature pipeline, thresholds	Model weights plus prompts, sampling parameters, tools/APIs available, routing rules, system policies; many more "levers" to version and test
Versioning	Model version tied to code and data snapshot	Full experience versioned: model, prompts, safety configuration, evaluation suites, and routing graphs, all treated as first-class artifacts
CI tests	Unit tests, data schema checks, offline metric regression	All of the traditional tests plus prompt regression tests, golden conversations, safety regression suites, and reproducibility of generations under fixed seeds

(*continued*)

Dimension	Traditional ML CI/CD	GenAI-specific CI/CD
CD/release gating	Thresholds on standard metrics and basic business KPIs	Gated on safety and alignment criteria, red-team results, A/B tests on user satisfaction, and policy compliance in addition to standard metrics
Monitoring in production	Drift detection on features/labels, latency, error rates	Continuous logging of prompts and outputs, safety violations, jailbreak attempts, hallucination proxies, user feedback, and per-segment behavior
Rollback and hot-fixes	Revert model or feature pipeline; usually coarse grained	Fine grained controls: rollback model, revert prompts or sampling parameters, tighten filters, change routing, or disable specific tools without full redeploy

Core Components of CI/CD Pipelines for Generative AI

Pipelines for generative models bring together multiple coordinated stages. Data pipelines manage ingestion, cleaning, preprocessing, and versioning while also watching for drift. Training pipelines orchestrate distributed training, record hyperparameters, and store checkpoints. Evaluation pipelines combine automated metrics, human-in-the-loop checks, adversarial probing, and regression testing. Model registries link artifacts to their lineage, storing metadata such as datasets, code commits, and evaluation results. Deployment pipelines handle packaging into services or APIs, with canary releases and rollback systems. Finally, monitoring pipelines watch performance and safety signals in production.

- Data pipelines handle ingestion, cleaning, and versioning.
- Training pipelines automate distributed retraining and tracking.
- Evaluation pipelines combine metrics, human review, and stress tests.
- Model registries ensure traceability and accountability.
- Deployment pipelines turn artifacts into scalable services.
- Monitoring pipelines provide real-time oversight after release.

Each stage works in coordination so that updates can flow continuously without sacrificing stability.

End-to-end CI/CD pipeline for generative models can be pictured as

Data Pipelines

↓

Training Pipelines

↓

Evaluation Pipelines

↓

Model Registry

↓

Deployment Pipelines

↓

Monitoring Pipelines

↙ ↘

feedback to Data feedback to Training and Evaluation

In this diagram, data pipelines feed training; training pipelines produce candidate checkpoints that flow into evaluation; successful candidates are registered with full lineage in the model registry; deployment pipelines pull approved versions from the registry and expose them as services; monitoring pipelines observe real traffic and safety signals, and their alerts and logs feed back into data collection, retraining, and evaluation, closing the loop.

Data Management in CI/CD Pipelines

Data is the foundation of generative AI, so CI/CD must treat it as a first-class artifact. Pipelines must automatically ingest data from sources such as scrapes, enterprise records, and curated corpora. Cleaning removes inconsistencies, filtering excludes unsafe content, and normalization ensures consistency across formats. Each dataset version must be tracked and stored so that models can be reproduced later, with lineage tools and standards recording which upstream sources, transformations, and policies produced each snapshot. Automated quality checks flag anomalies, privacy risks, or toxic material. Drift monitoring ensures that training and production data remain aligned, since distribution shifts are a major cause of model decay. In generative settings, synthetic data generation and data augmentation pipelines also become part of data management: synthetic corpora, edited examples, and adversarially generated prompts must be versioned, documented, and linked to the models that created them so that their effects can be audited.

- Ingestion ensures freshness and scale.
- Preprocessing enforces cleanliness and safety.
- Versioning and lineage provide reproducibility and traceability from raw sources through transformations to final datasets.

- Quality checks surface risks early.
- Drift monitoring prevents decay from misaligned data.
- Synthetic and augmented data pipelines expand coverage but must be governed with the same rigor as natural data.

Without rigorous data management, even sophisticated pipelines lose trustworthiness.

Training Pipelines for Model Updates

Training pipelines automate large-scale retraining and fine tuning. Schedulers allocate jobs across GPUs and TPUs, distributing execution across clusters to speed up throughput. Logging records hyperparameters, random seeds, and runtime metadata so that experiments can be repeated. Checkpointing preserves intermediate models for fault tolerance, and experiment tracking compares variants against baselines. These mechanisms together allow rapid experimentation while keeping results traceable and reproducible.

- Scheduling orchestrates resource allocation across clusters.
- Logging and checkpointing preserve transparency and fault tolerance.
- Experiment tracking allows systematic comparison.

Training pipelines therefore embody the trade-off between speed and accountability.

Evaluation Pipelines

Evaluation gates decide whether a model can progress to deployment. Automated metrics provide quick feedback on fluency, diversity, and distributional features. Human evaluation captures nuance and safety dimensions that metrics miss. Adversarial probing deliberately stresses the system to expose vulnerabilities. Task-based assessments measure usefulness in real contexts, while regression testing ensures improvements do not degrade prior capabilities. At each stage, explicit model validation checkpoints act as hard gates: only models that satisfy predefined quality and safety thresholds are promoted, preventing unstable or poorly understood updates from flowing into downstream deployment and monitoring pipelines.

- Automated metrics bring efficiency.
- Human evaluation captures subtle judgments.
- Adversarial testing exposes weaknesses.
- Regression tests safeguard legacy performance.
- Validation checkpoints block propagation of unsafe or unstable versions.

Layered evaluation provides both breadth and depth, preventing weak models from advancing.

Model Registries and Versioning

Registries act as the institutional memory of the pipeline. Each artifact is linked to its dataset version, code commit, and training configuration. Metadata such as evaluation results, deployment notes, and safety checks are stored alongside. Access controls specify who can promote models to production, ensuring governance. Registries provide traceability so that every deployed model can be audited for its lineage.

- Version control ties artifacts to code and data.
- Metadata storage preserves evaluation and deployment history.
- Access policies enforce governance.
- Traceability creates accountability across releases.

Registries ensure that model evolution remains auditable and transparent.

Deployment Pipelines

Deployment pipelines operationalize models. Containerization packages models with dependencies, while APIs expose them as services. Autoscaling adjusts resources to meet fluctuating demand. Canary releases introduce updates to a subset of users before full rollout, reducing risk. Rollback systems restore previous versions automatically if errors occur.

- Containerization and APIs make models portable and usable.
- Autoscaling adapts to demand.
- Canary releases and rollbacks provide safe deployment paths.

Deployment pipelines turn research artifacts into production-ready services.

Monitoring and Observability

Once deployed, models must be continuously observed. Monitoring tracks latency, throughput, and error rates, as well as divergence between production data and training distributions. Output monitoring checks for unsafe, biased, or hallucinated content. User feedback loops are integrated into dashboards to highlight real-world issues. Root-cause tracing links anomalies to particular datasets or model versions.

- Performance monitoring covers latency and reliability.
- Drift detection alerts when inputs shift.
- Output checks flag unsafe or biased generations.
- Feedback loops incorporate user signals.

Observability ensures that deployed systems remain safe and effective over time.

Rollback and Recovery

Because no pipeline is immune to error, rollback is essential. Canary monitoring detects issues early and triggers reversions. Registries store stable baselines so that prior versions are available. Automated triggers initiate recovery without waiting for human intervention. Postmortems analyze failures to improve resilience for future updates.

- Canary monitoring provides early detection.
- Stable baselines guarantee fallback options.
- Automated recovery prevents long outages.

Rollback mechanisms build resilience into continuous iteration.

Scaling CI/CD for Generative AI

Scaling raises distinctive challenges. Models with billions of parameters require distributed storage and efficient scheduling across clusters. Evaluation at scale demands parallelization of both automated metrics and human reviews. Families of models may be managed together, creating orchestration challenges. Pipelines must also meet compliance standards across global jurisdictions.

- Artifact size and cluster scheduling complicate scaling.
- Evaluation and human review must be parallelized.
- Multi-model orchestration requires careful coordination.
- Compliance adds regional constraints.

Scaling transforms pipelines into enterprise-wide infrastructure.

Challenges and Open Problems

Despite progress, challenges remain. Reproducibility gaps arise from nondeterministic training. Bias detection remains incomplete, leaving fairness integration unsolved. Human evaluations slow iteration, creating bottlenecks. Compute costs limit retraining frequency. Regulatory uncertainty complicates design across regions.

- Reproducibility, fairness, evaluation bottlenecks, costs, and regulation remain open problems.

These challenges show that CI/CD for generative AI is still an evolving practice.

Future Directions in CI/CD Pipelines

Looking forward, pipelines will grow more autonomous and adaptive. Automated retraining may be triggered by detected drift. Continuous evaluation may embed A/B testing directly into pipelines. Self-healing systems may automatically recover from failures. Safety gates will adapt as societal standards evolve. Integration with agentic systems will allow autonomous updates across multiple components.

- Automation, continual evaluation, self-healing, adaptive gates, and agentic integration define the trajectory of pipelines.

Future pipelines will not only automate engineering but also enforce accountability.

Integration with DataOps

Generative AI pipelines increasingly overlap with DataOps. Both require lineage tracking for data and code. Continuous ingestion keeps data fresh at the same pace as model updates. Data quality gates enforce ethical and safe inputs. Synchronization ensures that updates in data trigger retraining across pipelines.

- Lineage, ingestion, quality gates, and synchronization link CI/CD with DataOps.

This integration ensures that governance extends to data as well as models.

Cross-Organizational Collaboration

In large organizations, pipelines span multiple teams and jurisdictions. Shared registries enable collaboration while enforcing access control. Regional compliance requires tailored evaluation steps. Multi-team orchestration prevents interference across workloads. Communication protocols keep efforts aligned across organizational layers.

- Shared infrastructure, compliance gates, orchestration, and transparency make cross-team pipelines workable.

CI/CD is thus as much a sociotechnical challenge as a technical one.

CI/CD and Explainability

Explainability must be incorporated into pipelines so that models remain interpretable. Outputs should be analyzed for transparency. Audit logs must record key decisions. Metrics should track whether updates preserve interpretability. Human experts should validate explainability results.

- Interpretability checks, audit logs, metrics, and expert review sustain clarity.

This ensures that pipelines preserve transparency alongside performance.

Security in Model Pipelines

Security is critical at every stage. Data poisoning detection filters corrupted inputs. Registries validate artifact integrity through hashing. Access control ensures only authorized personnel can modify pipelines. Robustness checks simulate adversarial attacks before deployment.

- Data integrity, access control, and robustness testing secure the pipeline.

Security ensures resilience against malicious interference.

Environmental and Cost Considerations

Generative pipelines consume significant resources, so sustainability matters. Pipelines should monitor carbon impact, optimize scheduling, and reduce compute waste. Cost oversight must be integrated so that teams are aware of the financial burden of retraining.

- Carbon monitoring, optimization, and cost tracking balance innovation with sustainability.

Efficiency supports long-term viability.

Sociotechnical Dimensions

Pipelines are shaped by organizational culture as well as engineering. Teams with mature DevOps practices adopt CI/CD faster. Leadership buy-in influences adoption. Ethical priorities affect which safeguards are emphasized. Expertise must span ML, infrastructure, and ethics.

- Engineering maturity, leadership support, ethical culture, and skill distribution all affect success.

CI/CD for generative AI is not only a technical system but also a cultural one.

CI/CD for Multimodal and Agentic Systems

Pipelines must adapt to new system types.

- **Multimodal fusion**: Models integrating text, vision, and audio require multimodal evaluation gates.
- **Agentic models**: Autonomous agents require pipelines that manage continual learning and decision-making.
- **Interactive systems**: Multi-turn dialogue requires pipelines for longitudinal testing.
- **Collaborative ecosystems**: Pipelines orchestrate multiple interacting models.

Future systems demand flexible, extensible pipelines.

CI/CD pipelines for model updates are the backbone of production-scale generative AI. They manage continuous change across code, data, models, and context. They balance velocity with safety, reproducibility with scalability, and automation with accountability. By integrating data governance, training orchestration, evaluation gates, model registries, deployment strategies, monitoring, rollback, security, sustainability, and explainability, they transform the chaotic process of model iteration into structured, trustworthy practice.

As generative AI scales into multimodal and agentic domains, CI/CD pipelines will evolve from engineering support systems into governance infrastructures. They will not only automate updates but ensure compliance, transparency, and societal trust. They will determine whether generative AI systems remain robust, safe, and valuable over time.

Model Monitoring and Observability

Deploying generative AI models into production environments requires not only robust pipelines for integration and delivery but also continuous visibility into how these models behave once they are live. Unlike traditional software systems, where monitoring typically involves tracking latency, memory usage, and error codes, generative systems produce open-ended outputs that can vary unpredictably. These outputs must be assessed for not only technical correctness but also factual reliability, ethical appropriateness, cultural sensitivity, and long-term alignment with user needs.

Monitoring and observability together form the nervous system of production-scale AI. Monitoring refers to the continuous measurement of predefined indicators, such as latency or error rates. Observability extends this concept by enabling diagnosis of issues in complex systems through logs, metrics, and traces. In generative AI, observability is not only about infrastructure but also about semantics. Systems must be observable at the level of meaning, context, and social impact.

This section explores monitoring and observability for generative AI in depth. It begins with their historical roots in software engineering and shows how the challenges of generative models extend beyond traditional monitoring. It then examines the core dimensions of monitoring, including performance, quality, safety, drift, and user trust. Subsections analyze architectural components, infrastructure needs, ethical and legal dimensions, and case studies across domains. Finally, the section looks ahead to the future of monitoring, including multimodal observability, autonomous monitoring agents, and the integration of monitoring into broader governance frameworks.

Historical Roots of Observability

Observability as a concept originates in control theory, where it refers to the ability to infer the internal state of a system based on its external outputs. In engineering, observable systems are those that can be diagnosed and adjusted through measurement.

In software engineering, observability emerged as systems grew in scale and complexity. Traditional monitoring tracked CPU usage, memory allocation, and error logs. As distributed systems proliferated, observability expanded to include tracing across microservices, structured logging, and real-time dashboards. Engineers realized that monitoring predefined metrics was insufficient; systems needed to be designed for diagnosability. Observability became a design principle rather than a retrofitted feature.

When machine learning systems entered production, monitoring challenges intensified. Models made probabilistic predictions influenced by data distributions rather than deterministic logic. Failures were subtle, manifesting as accuracy degradation or biased outputs rather than crashes. Observability had to extend into the data and model layers. Generative AI added further complexity: outputs were open-ended, semantic, and context-sensitive. Observability now had to span not only infrastructure and models but also meaning and social acceptability.

Generative AI Monitoring Challenges

Monitoring generative systems requires a broader conception of observability than classic ML. Classification models map inputs to a closed set of labels, so monitoring can focus on accuracy, calibration, and latency. Generative models produce free-form text, images, audio, and mixed media. There is rarely a single ground truth for any given response, which makes quality assessment a matter of signals rather than a binary verdict. The same prompt can be validly answered in many ways, and the space of possible prompts shifts as users learn to steer the system.

These properties undermine traditional dashboards that rely on a few stable metrics. In practice, four classes of risk dominate: hallucination of plausible but false content, generation of toxic or biased material, drift in the distribution of prompts and contexts, and sheer operational scale. Each requires specific detectors, sample efficient auditing strategies, and feedback channels to route incidents into remediation. Emerging techniques such as self-consistency scoring across multiple generations, judge models that score outputs for factuality and safety, and reinforcement style audit loops that prioritize review of high risk interactions are increasingly used to keep pace with these risks.

- Open ended outputs eliminate a single ground truth, so monitoring must rely on ensembles of signals.
- Hallucinations require detectors that combine retrieval checks, constraints, and human review.
- Toxicity and bias must be tracked by classifiers and slice based analytics before content reaches users.
- Distribution shift and global scale demand continuous, automated surveillance that keeps pace with usage.

The result is an observability program that treats quality, safety, and performance as first-class objectives rather than add-ons.

Performance and Efficiency Monitoring

Even with new risks, fundamentals still matter. Users experience the system through latency, throughput, and stability. If responses are slow or error prone, perceived quality collapses regardless of semantic accuracy. Monitoring therefore begins with the platform.

- **Latency**: Measure per-request end-to-end times, tail percentiles, and queue wait components.
- **Throughput**: Track sustained requests per second and backpressure behavior under load.
- **Resource use**: Monitor GPU or TPU utilization, memory headroom, KV-cache hits, and batching efficiency.
- **Stability**: Alert on crashes, timeouts, and memory leaks with request traces that pinpoint failure sites.

These metrics form the base layer. They do not prove that a model is helpful or safe, but without them, no higher-level guarantee is credible.

Quality and Alignment Monitoring

Quality for generative models is multidimensional. A single score rarely captures adequacy, coherence, factuality, and style. Monitoring must therefore combine automated checks with periodic human audits and maintain slice-level visibility.

- **Factual consistency**: Retrieval-augmented cross-checks, citation verification, and contradiction detectors
- **Semantic coherence**: Topic tracking across turns, long-context consistency checks, and contradiction spotting within a session

- **Instruction alignment**: Conformance to system and policy prompts, including refusal rules where applicable
- **Creativity balance**: Novelty and repetition monitors that flag degenerate loops or ungrounded flights

A practical approach blends lightweight online signals with scheduled human reviews that recalibrate the automated detectors.

Safety and Compliance Monitoring

Safety is a continuous obligation rather than a pre-release gate. Production systems need classifiers and policies that operate at serving speed, along with processes for human escalation.

- **Toxicity and harassment**: Multi-threshold classifiers with conservative defaults and appeal paths
- **Bias and fairness**: Outcome disparities by demographic slices, with alerts when gaps widen
- **Compliance**: Checks for privacy-sensitive data, copyright indicators, and domain-specific restrictions
- **Red teaming in the loop**: Rotating adversarial prompts executed in production-like sandboxes to surface new failure modes

The objective is rapid detection and containment so that unsafe outputs are blocked, logged, and used to improve subsequent releases.

Data Drift and Concept Drift Detection

Generative quality decays when the world or user behavior moves away from training assumptions. Monitoring must distinguish distribution changes in inputs from changes in the desired mapping between inputs and outputs.

- **Data drift**: Embedding-based distance to training corpora, vocabulary and topic shifts, and prompt format changes
- **Concept drift**: Changing external standards such as clinical guidelines or legal policies that alter what counts as correct
- **Detection practice**: Maintain reference windows and control charts, with confidence intervals rather than single thresholds
- **Mitigation**: Trigger fine tunes, retrieval index refreshes, or guardrail updates when drift persists

In multi turn conversations and multimodal settings, drift is harder to detect because signals are spread across turns, modalities, and time: the same user session may start on one topic and gradually shift tone, domain, or language, while accompanying images or audio introduce additional shifts that single prompt level statistics cannot capture. LLMOps practices address this by logging session-level embeddings, safety scores, and task outcomes, computing drift metrics over whole conversations or interaction graphs, and visualizing these in dashboards that track changes by segment, modality, and use case, so that emerging shifts can be spotted and tied back to concrete remediation actions.

- Treat drift signals as triggers for data and model maintenance, not as mere dashboard curiosities.

User Feedback Integration

Users are a source of weak labels that can be aggregated into strong signals. Feedback must be captured with care, de-biased, and connected to model lineage so that issues are traceable to specific versions.

- **Explicit ratings and flags**: Thumbs signals, issue categories, and free-text notes
- **Implicit signals**: Abandonment, prompt reformulation, copy or share events, and external verification clicks
- **Trust signals**: User toggles for citations, sources opened, and preference for conservative answers
- **Aggregation**: Stratify by region, cohort, and task to avoid masking minority harms

Feedback loops are most effective when tied directly to triage queues and evaluation backlogs.

Infrastructure for Observability

Monitoring at scale requires purpose-built plumbing. The pipeline must connect logging, metrics, tracing, and alerting in a way that is both high throughput and privacy aware.

- **Logging**: Structured, schema-versioned records that capture prompt, response attributes, policy events, and model metadata
- **Metrics**: Online aggregation with rollups by slice, plus retention policies for historical analysis

- **Tracing**: Per-request spans from ingress through model inference, retrieval, and post-processing
- **Dashboards and alerts**: Role-specific views for SREs, safety teams, and product owners

Global deployments need regional sharding, data residency controls, and sampling strategies that preserve privacy while retaining diagnostic power.

Automated vs. Human-in-the-Loop Monitoring

Automation provides coverage; humans provide judgment. The two should be orchestrated rather than treated as alternatives.

- **Automated classifiers**: Toxicity, PII, copyright, hallucination risk, and jailbreak attempts
- **Statistical monitors**: Drift, anomaly detection, and canary comparisons between model versions
- **Human review**: Scheduled audits of sampled interactions, plus burst reviews during incidents
- **Hybrid flow**: Automation triages and routes to human queues with rich context and lineage links

A balanced design keeps alert volume manageable while ensuring nuanced errors are not missed.

Ethical and Legal Dimensions of Monitoring

Observability touches user data and therefore rights. Programs must embed privacy by design, publish policies, and maintain audit trails.

- **Privacy**: Data minimization, differential privacy or redaction for logs, and strict retention windows

- **Consent and disclosure**: Clear statements about monitoring and feedback use
- **Transparency**: User-facing explanations of why content was blocked or modified
- **Accountability**: Immutable logs tied to model and data versions to support audits and regulatory reviews

Trust in monitoring depends on treating users as stakeholders, not just data sources.

Security Considerations

Monitoring also protects systems from adversaries. Injection, data exfiltration, and poisoning attempts show up first as anomalies.

- **Adversarial prompt detection**: Pattern rules and learned detectors for jailbreaks and policy evasion
- **Poisoning and spam signals**: Rate anomalies, content fingerprints, and provenance checks on data used for retraining
- **Access auditing**: Fine-grained logs for admin and model-change operations
- **Security alerts**: Dedicated channels and runbooks that link detection to containment actions

Security observability complements quality observability and should be tested through regular exercises.

Multimodal Monitoring

As models span text, images, audio, and video, monitoring must check each modality and the coherence between them.

- **Text**: Factuality, toxicity, coherence, and instruction-following
- **Images**: Safety filters, prompt faithfulness, and bias in representation
- **Audio**: Clarity, transcription accuracy, and content appropriateness
- **Cross-modal coherence**: Alignment between captions, generated images, and narrated audio

Tooling should support shared identifiers so that a single interaction can be audited across modalities.

Longitudinal and Life Cycle Monitoring

Quality changes over days, weeks, and months. Monitoring must therefore operate on multiple timescales and link development to production.

- **Short term**: Anomaly detection for spike incidents.
- **Medium term**: Trend analysis for hallucination rates, safety events, and satisfaction.
- **Long term**: Alignment with evolving norms, policies, and regulations.
- **Life cycle integration**: Collect the same key signals in sandbox, canary, and full production so results are comparable.

Life cycle visibility turns one-off fixes into sustained maintenance.

Organizational and Sociotechnical Considerations

Monitoring succeeds when roles, processes, and culture support it. Teams need clear ownership of dashboards and on-call rotations, along with escalation paths that include safety and compliance experts. Leadership must treat observability as a product feature, not a cost center. Skills in ML, infrastructure, measurement, and social analysis should be represented so that signals are interpreted correctly and converted into action.

- **Ownership and on-call**: Who watches which signals and when
- **Cross-functional collaboration**: Engineering, safety, legal, and policy review
- **Cultural priority**: Treat monitoring findings as blockers for release, not optional suggestions
- **Skills mix**: Combine statistics, ML, SRE practice, and domain knowledge

When the organizational foundations are in place, monitoring becomes a living system that protects users, informs iteration, and sustains reliability at scale.

Future Trends in Monitoring

The future of monitoring will involve greater automation, adaptivity, and governance.

- **Agentic monitoring**: Autonomous agents may monitor outputs and trigger interventions.
- **Self-healing systems**: Observability frameworks will not only detect but correct failures automatically.

- **Context-aware monitoring**: Systems will adapt to user contexts, detecting subtler failures.
- **Integrated governance**: Monitoring will feed directly into compliance and regulatory reporting.

Future observability will extend beyond engineering to societal oversight.

Monitoring and observability are indispensable for production-scale generative AI. They ensure not only technical reliability but also semantic integrity, ethical compliance, and user trust. They extend beyond traditional infrastructure metrics to include quality, safety, drift, and cultural sensitivity. They integrate automation with human oversight, infrastructure with governance, and engineering with ethics.

Without monitoring, generative AI becomes a black box that risks silent failure. With monitoring, it becomes a transparent, diagnosable, and trustworthy system. Observability transforms generative AI from impressive prototypes into sustainable, accountable production systems. In the long run, monitoring is not just an engineering discipline but a form of governance, anchoring generative AI in real-world contexts and responsibilities.

Legal, Ethical, and Compliance Challenges

The deployment of generative AI into production environments inevitably collides with the boundaries of law, ethics, and compliance. Unlike technical engineering challenges, which can often be solved through new algorithms or optimized hardware, governance challenges are entangled with history, culture, institutions, and values. The outputs of generative models carry legal implications about ownership, copyright, privacy, and liability. They carry ethical implications about fairness, transparency, autonomy, and trust. They trigger compliance concerns tied to industry regulations, international treaties, and organizational accountability.

This section addresses legal, ethical, and compliance challenges in detail. It situates generative AI within the historical evolution of technology regulation, outlines the legal issues of intellectual property and privacy, explores liability and accountability gaps, unpacks ethical dimensions such as bias and fairness, and surveys compliance frameworks across sectors and geographies. It then extends into deeper analysis of trade secrets, antitrust, synthetic data, dual-use dilemmas, and epistemic justice. Finally, it examines comparative regulatory landscapes and reflects on how compliance costs and incentives shape deployment. The conclusion reframes governance not as an external burden but as a core design principle for sustainable AI.

Historical Evolution of Tech Regulation

Technologies of communication, computation, and automation have always tested the adaptability of legal systems. Each technological leap forced lawmakers, courts, and institutions to invent new categories of regulation, often under conditions of urgency and controversy. The printing press required the invention of copyright frameworks to balance incentives for authorship with access for society. Telecommunications technologies like the telephone and radio prompted rules for spectrum allocation, fair access, and pricing. The rise of software demanded new interpretations of intellectual property law, including licensing and patent debates, as well as the question of whether code itself should be treated as a protected asset. With the growth of the Internet, issues of privacy, liability, and jurisdiction reached the forefront, eventually crystallizing in frameworks such as GDPR in Europe. Cloud computing later introduced fresh compliance requirements around third-party data management, with regimes such as HIPAA in healthcare and SOC 2 in enterprise controls.

Generative AI, however, diverges from these precedents. Its outputs are not static reproductions of prior content but novel creations.

This distinction stretches doctrines of authorship, originality, and accountability. It reveals the inadequacy of simply extending old frameworks and demands the creation of entirely new categories of law and governance.

Intellectual Property and Generative Outputs

One of the most pressing challenges lies in intellectual property (IP). Generative models are trained on massive corpora scraped from the web, often containing copyrighted books, articles, images, and code. Courts must now determine whether such training constitutes fair use, implied licensing, or outright infringement. Equally unsettled is the question of ownership of generated outputs. If an AI system generates an image, does ownership reside with the end user, the model developer, the provider of the training data, or perhaps no one at all?

Complicating matters further, generated works sometimes imitate specific artists or authors, blurring the line between inspiration and unlawful copying. Markets are disrupted as generative systems compete with human creators, raising questions about fair compensation and protection of livelihoods.

- Training data use raises questions of fair use versus infringement.
- Generated outputs challenge established ownership categories.
- Derivative works test the boundary between homage and copying.
- Market disruption threatens the economic stability of creative industries.

Traditional IP law was designed around human authorship and originality. Generative AI destabilizes these categories and forces legal systems to reconsider what authorship means in the digital age.

Privacy and Data Protection

Privacy stands as another central compliance challenge. Models trained on public web data may inadvertently memorize and reproduce personal information. Laws like GDPR grant individuals the "right to erasure," but removing such data once embedded in model parameters is technically complex. Even when explicit data is not present, models can infer sensitive attributes from patterns, creating risks of exposure that existing privacy frameworks were never designed to address.

- Training data may expose personal information.
- Right to erasure is difficult to implement at the parameter level.
- Models can infer sensitive attributes even without explicit data.
- Different global regulations, such as GDPR, CCPA, and PIPL, complicate deployment.

Organizations attempt to mitigate these risks through differential privacy, anonymization, and data minimization, but regulators must decide whether privacy protections extend into the hidden inner workings of model weights and embeddings.

Liability and Accountability

Liability for AI-generated harm remains unsettled. If a chatbot provides harmful medical advice, where does responsibility fall: with the developer, the deployer, or the user? Professional liability frameworks, such as those governing doctors and lawyers, continue to assign responsibility to the professional, yet reliance on AI complicates the boundary. As models operate with increasing autonomy, attributing accountability becomes even harder.

Governance frameworks under discussion include strict liability, shared liability, and distributed liability. Without clear rules, harms risk falling into legal grey zones where no party takes responsibility.

- Product liability assigns responsibility for defective systems.
- Professional liability collides with reliance on AI systems.
- The autonomy gap complicates traditional accountability.
- Governance options range from strict to distributed liability.

Clear liability rules are necessary to prevent harms from slipping through gaps in the legal system.

Bias, Fairness, and Social Equity

Generative AI systems often reproduce or even amplify existing social inequities. Training corpora reflect the prejudices of the societies from which they are drawn. The resulting outputs can disproportionately affect marginalized groups, deepening inequality in critical areas such as hiring, justice, or healthcare.

Fairness metrics remain underdeveloped and contested, with debates about what constitutes fairness in practice. Bias cannot be solved by technical fixes alone; it requires ethical commitments and participatory design that includes diverse stakeholders.

- Training datasets encode systemic prejudice.
- Output disparities can harm marginalized groups.
- Social consequences intensify in high-stakes domains like justice and healthcare.
- Fairness metrics remain immature and incomplete.

Transparency, Explainability, and Trust

Transparency is critical both for compliance and for maintaining public trust. Generative models are often too complex for intuitive explanation, which complicates accountability. Tools such as saliency maps, attention visualizations, and counterfactual examples offer partial insight but fall short of full transparency.

Regulators may require disclosure of training data, evaluation protocols, and system limitations. Beyond compliance, transparency supports trust by helping users calibrate their reliance on generative systems.

- Generative models function as black boxes.
- Explainability tools offer only partial insights.
- Regulators may mandate disclosure of key processes.
- Transparency forms the basis of user trust.

Compliance Frameworks and Standards

Organizations must navigate a patchwork of compliance frameworks, each emphasizing different aspects of governance. ISO standards focus on AI risk management and robustness. The NIST AI Risk Management Framework in the United States emphasizes governance and documentation. The EU AI Act introduces legally binding risk-based tiers with strict obligations for high-risk applications. Alongside these, sector-specific standards remain in force, such as HIPAA for healthcare, FERPA for education, and Basel guidelines for finance.

- ISO provides guidance on AI risk and robustness.
- NIST encourages governance practices in the United States.
- The EU AI Act establishes strict obligations for high-risk systems.
- Sector-specific frameworks continue to apply.

Compliance must therefore be woven directly into CI/CD pipelines, registries, and monitoring infrastructures, ensuring that governance is not an afterthought.

Sector-Specific Compliance

Different industries impose unique obligations, making generic approaches insufficient. In healthcare, AI-generated outputs must satisfy HIPAA and FDA requirements. In finance, SEC oversight and explainability standards are paramount. In education, FERPA protects student data while enabling instructional use. In government, democratic accountability demands transparency and fairness.

- Healthcare requires HIPAA and FDA compliance.
- Finance emphasizes SEC oversight and explainability.
- Education requires FERPA protections.
- Government use must align with democratic principles.

Sectoral variation ensures that compliance remains a domain-specific challenge.

Cross-Border Legal Complexities

Generative AI is global, but regulation is fragmented. Europe enforces GDPR and the EU AI Act, prioritizing privacy and risk management. The United States relies on sectoral and state laws, creating a patchwork. China mandates watermarking and content moderation. Developing nations often lack comprehensive regulation but still face the risks of global deployment.

- Europe enforces strict privacy and risk laws.
- The United States follows fragmented rules.
- China emphasizes content moderation.
- Developing nations face risks without equivalent regulation.

Cross-border compliance requires sophisticated legal and operational strategies.

Ethical Governance in Organizations

Ethics extends beyond compliance, requiring organizations to adopt governance frameworks that reflect values of fairness, accountability, transparency, and dignity. Independent ethics boards provide oversight, while ethical principles must be embedded into development pipelines rather than remaining aspirational. Organizational culture plays a key role, as ethics cannot succeed without buy-in across teams and leadership.

- Ethics boards offer independent oversight.
- Principles like fairness and dignity guide practice.
- Ethics must be operationalized within pipelines.
- Cultural adoption across organizations is critical.

Human Oversight and Autonomy

Accountability requires clear human oversight mechanisms. Oversight may take the form of human-in-the-loop systems, requiring human approval before decisions are finalized, or human-on-the-loop systems, which allow ongoing monitoring and intervention. The risk of overreliance is real; professionals may defer too much authority to automated systems. Oversight must therefore be deliberately integrated into architecture.

- Human-in-the-loop ensures approval before decisions.
- Human-on-the-loop enables ongoing oversight.
- Overreliance erodes professional responsibility.
- Oversight must be architected into pipelines.

Long-Term Ethical Risks

Generative AI raises broader risks that extend beyond immediate deployments. Misinformation can destabilize democratic discourse. Cultural homogenization marginalizes local traditions. Creative and knowledge workers face labor displacement. Some foresee existential risks if advanced systems create systemic hazards.

- Misinformation destabilizes societies.
- Cultural homogenization marginalizes diversity.
- Labor displacement threatens creative economies.
- Advanced systems may present systemic risks.

Long-term resilience requires anticipatory ethics and scenario planning.

Case Studies and Precedents

Concrete legal cases illustrate these challenges. Copyright lawsuits have been filed by artists and publishers against AI developers. Medical AI systems have faced regulatory delays until safety validation was proven. Several jurisdictions have passed laws addressing malicious deepfakes. Hiring algorithms have been scrutinized for discriminatory practices.

- Copyright lawsuits challenge training dataset use.
- Medical regulators delay approvals pending safety checks.
- Deepfake laws address malicious use.
- Hiring bias cases reveal discriminatory risks.

These precedents offer early glimpses into how regulation is adapting in practice.

Future Directions in Law and Policy

Emerging policy directions emphasize global harmonization, adaptive regulation, participatory governance, and automated compliance integration. Global treaties may eventually create shared standards. Adaptive regulation ensures flexibility as technical progress accelerates. Participatory governance empowers citizens and stakeholders. CI/CD integration makes compliance checks part of automated workflows rather than afterthoughts.

- Global harmonization may produce shared treaties.
- Adaptive regulation adjusts to technical change.
- Participatory governance ensures democratic input.
- Compliance integration automates oversight.

Future governance must balance innovation with responsibility.

Trade Secrets and Proprietary Models

Developers often rely on trade secret law to protect training data and model architectures. This strategy collides with growing demands for transparency. Regulators may require disclosure that undermines secrecy. Excessive reliance on trade secrecy risks entrenching monopolies and hindering accountability.

- Trade secrets protect data and architectures.
- Transparency demands challenge secrecy.
- Regulators may require disclosure.
- Secrecy risks entrenching monopolies.

Antitrust and Competition Law in AI

The market for generative AI is increasingly concentrated, with a handful of firms dominating infrastructure, compute, and training data. Regulators are already investigating anti-competitive practices. Open-source alternatives provide a counterbalance, but they too raise governance questions.

- Concentration of foundation models fuels scrutiny.
- Regulators investigate anti-competitive risks.
- Open-source models counter proprietary dominance.
- Competition law must balance fairness with innovation.

Synthetic Data As a Compliance Strategy

Synthetic data offers partial relief from privacy and compliance challenges. By mimicking real-world distributions without storing personal data, it promises privacy protection. However, synthetic data can still embed biases, its legal status remains unsettled, and it often underperforms on complex tasks.

- Synthetic data reduces privacy risks.
- Biases persist even in synthetic generation.
- Regulators debate its legal status.
- Utility trade-offs limit accuracy.

Dual-Use and Misuse Dilemmas

Generative AI embodies dual-use potential. It fuels creativity and productivity but also enables malicious applications such as deepfakes, disinformation, and cyberattacks. Governance must deter misuse without stifling innovation, while international security frameworks adapt to address these risks.

- Creative uses expand productivity.
- Malicious uses destabilize information ecosystems.
- Governance must strike a balance.
- International security magnifies stakes.

Cultural and Epistemic Justice in AI Outputs

Generative AI often privileges Western knowledge systems, as English-language corpora dominate training data. Indigenous and minority perspectives are marginalized, creating epistemic injustice.

Designers have an ethical obligation to curate inclusive datasets that value diverse traditions.

- Western corpora dominate training data.
- Minority knowledge systems are marginalized.
- Epistemic justice demands inclusivity.
- Inclusive datasets are an ethical obligation.

Comparative Global Regulatory Landscapes

Regulation differs significantly across jurisdictions. The EU enforces strict rights-based frameworks. The United States relies on a fragmented but innovation-friendly approach. China emphasizes content control and watermarking. Developing nations lack regulatory capacity but face downstream risks.

- The EU emphasizes strict frameworks.
- The United States follows fragmented, sector-specific rules.
- China enforces centralized controls.
- Developing nations face risks without frameworks.

Comparative analysis underscores the difficulty of harmonization and the importance of global cooperation.

The Economics of Compliance

Compliance has economic implications.

- **Cost burden**: Meeting standards requires investment in monitoring and auditing.

- **Competitive advantage**: Firms with robust compliance may gain user trust.
- **Compliance-as-a-service**: Emerging industry supporting smaller firms.
- **Regulatory capture risk**: Large firms may shape rules to their advantage.

Economics shapes who can compete in generative AI.

Legal, ethical, and compliance challenges define the boundaries of sustainable AI deployment. Intellectual property disputes, privacy rights, liability gaps, and compliance frameworks are not secondary concerns but central determinants of legitimacy. Ethical obligations such as fairness, transparency, and epistemic justice frame the social value of AI. Compliance infrastructures operationalize these principles across pipelines, monitoring systems, and audits.

The expansion of this section has highlighted additional complexities: trade secrets, antitrust, synthetic data, dual-use dilemmas, cultural justice, and global regulatory fragmentation. Each demonstrates that AI governance is not a technical afterthought but a multidimensional process involving law, ethics, economics, and organizational culture.

Multimodal Fusion, Agentic Models, and AI Safety

The trajectory of generative AI does not stop with the systems we deploy today. The future is defined by trends that expand capability, shift architectures, and introduce new governance challenges. Three of the most transformative trends are multimodal fusion, the rise of agentic models, and the deepening integration of AI safety. These trends are not isolated. They interact, overlap, and sometimes amplify one another. Multimodal fusion extends generative AI across text, images, audio, and beyond, creating systems that process and produce multiple forms of information simultaneously. Agentic models move from passive generation to active decision-making, initiating actions in pursuit of goals. AI safety, once seen as a niche research concern, becomes a mainstream operational requirement for deploying powerful models responsibly.

This section explores these three trends in detail. It examines the conceptual foundations of multimodal fusion, the technical and societal implications of agentic models, and the frameworks being developed for AI safety. Case studies illustrate how these trends manifest in real-world deployments. The analysis then turns to the interactions among them, considering how multimodal systems may require agentic control, how agentic systems intensify safety needs, and how safety practices must adapt to multimodal and agentic contexts. The section concludes with reflections on how these trends shape the future of deployment at scale.

Multimodal Fusion, Agentic Models, and AI Safety: Foundations and Implications

Multimodal fusion represents one of the most significant shifts in the trajectory of generative AI. Whereas earlier systems specialized narrowly in text, images, or speech, modern architectures seek to integrate these domains into unified models. This integration is not a simple matter of stacking separate capabilities but of learning shared representational spaces that allow models to understand, translate, and generate across modalities. In doing so, multimodal systems expand the interpretive and generative range of AI, enabling them to interact with the complexity of human environments in richer ways than any single-modality system could manage.

Foundationally, this is a departure from the trajectory of early text-only transformers, which demonstrated extraordinary generative capacity but were limited in scope. With the introduction of models like CLIP, which mapped text and image into joint embeddings, and newer architectures like Flamingo, GPT-4V, and AudioLM, the field witnessed a profound extension of generative intelligence into cross-modal reasoning. The implications are sweeping: systems can now caption images, answer questions about visual scenes, synthesize audio in response to text, and retrieve information seamlessly across multiple media.

- **Foundational shift**: From unimodal generation to integrated multimodal understanding
- **Examples**: CLIP for text-image embedding, Flamingo and GPT-4V for vision-language tasks, AudioLM for audio-language fusion
- **Implications**: Greater interpretive breadth, richer user interaction, and new capacities for creativity and problem solving

At its conceptual foundation, multimodal fusion relies on embedding spaces that represent different modalities in a common framework. This enables models to perform tasks such as image captioning, visual question answering, and cross-modal retrieval, laying the groundwork for applications across accessibility, robotics, healthcare, and education.

Technical Dimensions of Multimodal Fusion

To achieve this integration, multimodal systems require innovation at multiple levels: architecture, training data, alignment, and evaluation. Architecturally, models must balance the strengths of modality-specific encoders, such as CNNs for images or audio spectrogram models, with shared fusion layers that allow information to be exchanged. Transformer-based fusion has become a dominant paradigm, but researchers continue to experiment with alternatives to address issues of scale and alignment.

Training such systems is challenging because multimodal corpora are less abundant and harder to align than text-only data. Curating datasets that pair text with images, audio, or video is labor-intensive, and alignment is often imperfect. A text caption may only partially describe an image, and audio signals may convey emotional nuance that transcripts omit. Evaluating these systems adds another layer of complexity: metrics must capture coherence across modalities rather than within a single one, and no universally accepted standard yet exists.

- **Architectures**: Modality-specific encoders plus shared fusion layers
- **Training data**: Scarce and difficult to align compared to text-only corpora
- **Alignment challenges**: Mismatches between modalities complicate training
- **Evaluation**: Requires cross-modal coherence metrics beyond current standards

This technical frontier illustrates that the future of multimodal AI depends not just on scaling but on creating robust methods for aligning modalities in ways that reflect real-world complexity.

Applications of Multimodal Systems

The real promise of multimodal fusion lies in its diverse applications. In assistive technologies, multimodal systems that integrate speech, vision, and text can revolutionize accessibility for people with disabilities, enabling interfaces that accommodate a wide range of needs. In education, multimodal tutors combine text explanations with diagrams, images, or narrated audio, personalizing learning experiences. Healthcare offers another striking application, as models synthesize imaging data, patient notes, and laboratory results to provide diagnostic support. Entertainment and creative industries leverage multimodal generation to build immersive multimedia experiences. Robotics, too, benefits, as robots equipped with multimodal perception can navigate and act in dynamic environments.

- **Accessibility**: Multimodal systems empower individuals with disabilities.
- **Education**: Integrated learning experiences through text, image, and audio.
- **Healthcare**: Diagnostic support by combining multiple data types.
- **Entertainment**: Generating immersive multimedia.
- **Robotics**: Enhanced perception and adaptability in complex environments.

These examples underscore the expansion of generative AI beyond narrow applications toward deeply integrated tools that augment human activity across sectors.

Ethical and Social Implications of Multimodal Fusion

Multimodal fusion, while powerful, magnifies ethical and social risks. Biases embedded in one modality can reinforce biases in another, creating compounded inequities. Deepfake technologies, which combine text, image, and audio synthesis, expand the scope of misinformation, raising security and societal concerns. Although multimodal systems promise breakthroughs in accessibility, they risk creating new forms of exclusion if not designed inclusively. Finally, cultural representation looms as a key issue: datasets often privilege dominant cultural perspectives, marginalizing others and perpetuating epistemic injustice.

- **Bias propagation**: Multiple modalities amplify inequities.
- **Misinformation**: Deepfakes and multimodal disinformation intensify risks.
- **Accessibility paradox**: Systems can empower but also exclude.
- **Cultural representation**: Dominant cultures may be privileged in datasets.

To realize the promise of multimodal AI responsibly, governance frameworks must anticipate and mitigate these risks.

Rise of Agentic Models

Alongside multimodal fusion, another profound trend has emerged: the rise of agentic models. Whereas traditional generative AI functioned as reactive tools (producing outputs in response to prompts), agentic systems blur the line between tool and actor. They not only generate outputs but also make decisions, initiate actions, and pursue goals, often using their own outputs as inputs for subsequent reasoning steps.

Examples such as AutoGPT, LangChain agents, and new research prototypes demonstrate how agentic models can plan, execute multi-step processes, and interact with external tools or APIs. The shift introduces autonomy, initiative, and goal-directedness into AI, raising deep conceptual and governance questions.

- **From completion to action**: Outputs drive further actions in reasoning loops.
- **Examples**: AutoGPT, LangChain, emerging prototypes.
- **Conceptual shift**: From passive tools to active agents with initiative.

Agentic models represent a paradigm shift, requiring careful reflection on autonomy, accountability, and control.

Technical Foundations of Agentic Models

For agentic models to function, new architectural and operational foundations are needed. Systems require planning modules that can generate, evaluate, and revise multi step strategies. They need persistent memory to maintain state across interactions. Tool use becomes essential, enabling models to call APIs, search engines, or software systems. Feedback loops are equally important, ensuring that agents adapt based on outcomes through reinforcement learning or other monitoring mechanisms.

- **Planning**: Beyond single outputs, agents construct sequences of actions.
- **Memory**: Persistence allows continuity across interactions, spanning short term context window memory within a single session, longer term vector store memory that retrieves past information by similarity, and episodic structured memory that records key events or decisions in a more symbolic form for later reasoning.
- **Tool use**: Integration with external systems enables practical execution.
- **Feedback loops**: Outcomes guide adaptation and refinement.

These features extend generative AI into systems capable of operating dynamically in environments, with autonomy that demands new safeguards.

Applications of Agentic Systems

The applications of agentic models are expansive. In research, autonomous assistants can design experiments, gather data, and summarize findings. Businesses deploy agents to manage workflows, optimize processes, or coordinate communication. Creative collaboration flourishes as agents sustain longer projects in partnership with humans. In software engineering, agents write, debug, and even deploy code with minimal supervision. At the personal level, productivity agents schedule, remind, and act across digital ecosystems.

- **Research**: Autonomous design and synthesis of findings.
- **Business**: Workflow management and optimization.
- **Creative collaboration**: Sustained co-creation with humans.
- **Software engineering**: Automated coding and deployment.
- **Personal productivity**: Digital assistants that act rather than suggest.

These use cases highlight the transformative reach of agentic AI across both intellectual and creative domains.

Ethical and Governance Implications of Agentic Models

Agency intensifies ethical challenges. Autonomous behavior raises the risk of unintended or harmful actions. Liability becomes murkier when decisions are made by AI rather than by humans directly. Manipulation risks increase as agents may nudge or steer users in subtle ways. And the central governance dilemma emerges: how to balance autonomy with effective human oversight. In practice, these concerns appear in concrete scenarios such as a trading agent executing risky positions without adequate caps, a scheduling assistant canceling or rescheduling critical meetings without confirmation, or a customer support agent offering refunds or concessions beyond policy limits.

- **Autonomy risks**: Agents may act unpredictably or harmfully. For example, an operations agent that automatically adjusts inventory orders could overreact to a transient signal and trigger costly overstocking or stockouts before a human notices.
- **Liability gaps**: Accountability becomes diffuse. If a health triage agent prioritizes patients incorrectly based on flawed data and a delay in care results, responsibility may be contested among developers, deployers, and clinicians who relied on the system.
- **Manipulation**: Subtle nudges can shape user choices. A shopping or financial-planning agent that is rewarded only for click-through or sales might gradually steer users toward higher-margin products or riskier investments that are not in their best interest.
- **Oversight dilemmas**: Striking a balance between autonomy and control. A customer service agent that requires human approval for every action is inefficient, but one that can independently close accounts or modify subscriptions may cause irreversible harm if misconfigured.

Robust oversight, transparency, and governance frameworks are necessary to ensure that agentic models operate safely and responsibly, with clear boundaries on what agents can do automatically, when they must seek consent, and how humans can review and reverse their actions.

AI Safety As a Core Trend

AI safety has moved from a niche research concern to a mainstream operational priority. Once associated mainly with speculative risks of future advanced systems, safety now includes practical issues such as hallucinations, toxicity, bias, and robustness. Safety must be integrated directly into pipelines, monitoring systems, and governance practices, not added as an afterthought.

Concrete practices already reflect this shift. RLHF, adversarial red-teaming, and bias auditing are common. Organizations increasingly build safety checks into CI/CD pipelines and deploy monitoring infrastructures to catch failures in production.

- **Definition**: Ensuring systems behave as intended, avoiding harmful outcomes
- **Shift**: From speculative concerns to concrete operational risks
- **Integration**: Safety embedded in pipelines and monitoring
- **Examples**: RLHF, red-teaming, adversarial testing

AI safety has become indispensable for trustworthy deployment.

Intersections of Trends

Multimodality, agency, and safety do not evolve in isolation but converge in powerful and risky ways. Agents that operate across modalities require advanced monitoring to ensure alignment. Autonomous systems amplify risks, making safety mechanisms essential. Multimodal disinformation and deepfakes highlight the importance of safeguards. Future systems will almost certainly be both multimodal and agentic, which magnifies the urgency of integrated safety frameworks.

- **Multimodal agency**: Agents combining multiple modalities.
- **Safety for agents**: Autonomy requires robust oversight.
- **Safety for multimodal**: Cross-modal misinformation heightens risks.
- **Convergence**: Integration multiplies both opportunity and danger.

Case Studies Across Trends

Case studies illustrate how these trends intersect in practice. In healthcare, multimodal systems combine imaging, text, and structured data to support diagnosis, but their agentic features raise liability concerns. In education, multimodal tutors provide rich learning experiences but autonomous behaviors complicate fairness and pedagogical control. In national security, agentic systems capable of producing multimodal disinformation pose severe risks. At the same time, accessibility technologies show the positive side: multimodal agents can empower people with disabilities, provided safety controls remain strong.

- **Healthcare**: Multimodal diagnostics with liability questions
- **Education**: Rich tutoring systems with oversight dilemmas
- **Security**: Agentic disinformation as a threat vector
- **Accessibility**: Empowering but requiring careful safety design

These case studies show how the three trends, fusion, agency, and safety, already shape real-world deployment, with benefits and risks deeply intertwined.

Future Directions

The future of deployment will integrate these trends.

- **Multimodal as default**: All future foundation models will be inherently multimodal.
- **Agents as standard**: Generative AI will continue to evolve from passive tools into autonomous assistants.
- **Safety as infrastructure**: Safety will be embedded into pipelines as a foundational layer.
- **Global governance**: International coordination will shape safe deployment of multimodal, agentic AI.

Future deployment will require both technical sophistication and governance innovation.

Multimodal fusion, agentic models, and AI safety represent the defining trends of generative AI deployment. Multimodal systems expand scope across media. Agentic models shift from passive outputs to active decisions. AI safety ensures that these capacities are deployed responsibly. Together, they shape not only the technical direction of AI but also its societal impact and governance.

Organizations deploying generative AI at scale must anticipate these trends. They must prepare infrastructure, monitoring, and compliance not only for current models but for multimodal and agentic futures. They must embed safety not as an afterthought but as a foundational design principle. By doing so, they can harness the transformative potential of generative AI while mitigating its risks, ensuring that the next phase of deployment is both powerful and responsible.

The future lies in harmonizing these dimensions. Organizations that integrate legal compliance, ethical governance, and proactive accountability into their pipelines will not only reduce risks but also gain trust and competitive resilience. Legal, ethical, and compliance integration is thus not only about preventing harm but about enabling generative AI to become a sustainable, equitable, and trusted part of society.

Conclusion

The deployment of generative AI at scale is not a singular act but a living process. It is the point where models leave the confines of research laboratories and controlled benchmarks and begin to interact with the unpredictability of real-world users, data, and institutions. Chapter 12 has examined this reality by analyzing the infrastructures, safeguards, and forward-looking trends that together define what it means to operationalize generative AI responsibly.

The chapter began with continuous integration and continuous delivery pipelines for model updates. These pipelines extend practices inherited from software engineering into the probabilistic domain of generative models. In this context, deployment involves not only code but data, models, and evaluation protocols. CI/CD pipelines create reproducibility by linking every deployed model to its data, configuration, and evaluation record. They enable rapid iteration without sacrificing accountability by embedding safety checks, monitoring triggers, and rollback mechanisms. Far from being simple engineering conveniences, CI/CD pipelines function as the governance backbone of production AI. They embody the principle that innovation must proceed in parallel with reliability, ensuring that systems remain adaptive yet traceable.

Monitoring and observability were explored next as the nervous system of deployed systems. Unlike traditional monitoring, which measures infrastructure metrics such as latency or memory, monitoring for generative AI must assess the quality, safety, and social impact of open-ended outputs. Observability must extend into semantics, detecting hallucinations, toxicity, and bias in real time. It must track distributional drift, monitor user trust, and integrate both automated classifiers and human oversight. The conclusion of this discussion is that monitoring cannot remain reactive; it must be anticipatory and designed as an ongoing dialogue between systems and stakeholders. Observability frameworks make systems diagnosable, accountable, and trustworthy, preventing silent failures and sustaining resilience across the life cycle.

The chapter then turned to legal, ethical, and compliance challenges. This discussion emphasized that governance is not external to deployment but integral to it. Intellectual property questions about training data and generated outputs remain unsettled, requiring new doctrines of authorship and ownership. Privacy regulations demand safeguards against the memorization and reproduction of personal data, yet also push developers to rethink how privacy rights extend into model parameters. Liability gaps highlight the difficulty of attributing accountability when models act with growing autonomy. Ethical concerns such as bias, fairness, and transparency require not only technical fixes but participatory governance that includes diverse stakeholders. Compliance frameworks, from the EU AI Act to HIPAA and Basel standards, embed legal obligations into pipelines, demanding auditable trails and domain-specific safeguards. The conclusion is that legal, ethical, and compliance integration is not an optional layer but the foundation for societal legitimacy and long-term viability.

Finally, the chapter explored three defining trends shaping the future of deployment: multimodal fusion, the rise of agentic models, and the mainstreaming of AI safety. Multimodal systems expand scope by integrating text, vision, audio, and beyond. They enable richer interactions but raise complex challenges of alignment and bias across modalities. Agentic models shift from passive generation to autonomous action, introducing planning, tool use, and initiative. This transformation amplifies both potential and risk, requiring stronger oversight and clearer liability frameworks. AI safety, once peripheral, is now a mainstream priority. Red-teaming, reinforcement learning with human feedback, and safety benchmarks are being integrated into pipelines as core components. The convergence of these trends suggests that the future of deployment will involve multimodal agents operating with autonomy, governed by embedded safety frameworks that anticipate and mitigate harms.

Across these themes, one conclusion emerges: deployment is inseparable from governance. The processes that keep generative AI adaptive, reliable, and safe are not mere technical add-ons but the very conditions under which it can operate in society. CI/CD pipelines anchor reproducibility, monitoring sustains accountability, compliance defines legitimacy, and safety ensures alignment. Without these, generative AI risks collapsing under its own scale, producing outputs that are unreliable, unaccountable, and untrusted. With them, generative AI becomes not just a research achievement but a durable infrastructure for knowledge, creativity, and decision-making.

Another conclusion is that deployment requires interdisciplinary collaboration. Engineers build pipelines and monitoring tools, but lawyers, ethicists, domain experts, and users define the boundaries of fairness, trust, and responsibility. Organizations cannot silo deployment as a technical task. They must cultivate cultures that integrate technical, legal, and ethical perspectives into unified governance. This sociotechnical integration determines whether deployment succeeds in aligning models with human values and institutional needs.

Looking ahead, deployment is likely to evolve in three ways. First, automation will increase. CI/CD, monitoring, and compliance will incorporate greater levels of automation, moving toward self-healing pipelines and autonomous monitoring agents. Second, governance will scale globally. As models are deployed across jurisdictions, organizations must navigate and harmonize divergent regulatory regimes. Third, safety will become infrastructure. Just as encryption became a default feature of Internet communications, safety mechanisms will become embedded features of AI pipelines, expected by regulators and demanded by users.

In conclusion, Chapter 12 has argued that deployment is the proving ground of generative AI. It is where technical capacity meets social responsibility. It is where innovation must align with governance. The future of deployment depends not only on scaling models but on scaling accountability. By embedding CI/CD pipelines, monitoring systems, compliance frameworks, and safety practices into the fabric of deployment, organizations can ensure that generative AI evolves as a trustworthy, resilient, and socially legitimate technology. The lesson is clear: the path from research breakthrough to societal integration passes through deployment, and deployment succeeds only when it is anchored by governance at every stage.

Index

A

A/B testing, 477, 488
- case studies
  - text generation, 521
  - vision and multimodal systems, 521
- cultural and psychological dimensions, 525
- cultural variability, 528
- designing, 520
- deterministic behavior, 517
- ecosystems, 529
- ethical considerations, 522
- feedback loops and longitudinal experiments, 524
- governance and regulation, 526
- historical origins, 517, 518
- infrastructure requirements, 527
- metrics, 520, 521
- multivariate and Bandit, 525
- online *vs.* offline, 522
- psychological and sociological effects, 528
- reasons, 519
- safety and alignment, 527
- scaling, 524
- sequential and adaptive experimentation, 526
- statistical foundations, 518, 519
- statistical pitfalls, 523

Academic hallucinations, 437
Accountability, 576, 581
Adam (Adaptive Moment Estimation), 213, 214
Adapters
- engineering practices, 327, 328
- principles, 326

Adaptive methods
- Adam, 213, 214
- comparative analysis, 217–219
- LAMB and large-batch training, 214–216
- motivation, 210, 211
- perspectives and future research, 220, 221
- RMSProp, 211, 213

Adaptive thresholds, 225
Adversarial losses, 157, 158
Agentic models, 561, 562, 587
- applications, 593
- ethical and governance challenges, 593, 594
- examples, 591
- technical foundations, 592

I. Cronin, *Building and Training Generative AI Models*,
https://doi.org/10.1007/979-8-8688-2332-9

AI safety, 587, 595
ALBERT principle, 410
Antitrust, 583
Asynchronous checkpointing, 270
Audio and speech generation, 18, 19
Audio signal processing, 98
Audit logs, 560
Authenticity, 533
Autoencoders, 156
Automatic evaluation methods
  foundations, 491
  historical context, 490
  principles, 490
Automatic evaluation metrics, 487
Autonomous systems, 595
Autoregressive models, 5, 6

B

Backpropagation, 198
  historical context, 202, 203
  mechanics, 203, 204
Bandit algorithms, 525, 526
Batch gradient descent, 205
BatchNorm, *see* Batch normalization (BatchNorm)
Batch normalization (BatchNorm), 226
Batching strategies, 418
Bayesian approaches, 525
Behavioral drift, 471
Behavioral metrics, 520
Benchmarking
  image generation, 507
  political economy, 510, 511
  technical and cultural practice, 488
BERTScore, 504, 505
β-VAEs, 66
Bias, 577
  algorithmic interventions, 444
  amplification, 13, 472
  audits, 134
  awareness, 441, 442
  case studies, 447
  categories, 440, 441
  cultural and contextual adaptivity, 445, 446
  dataset curation, 443
  detection strategies, 443
  explicit connections, 439
  future research, 448, 449
  hybrid mitigation pipelines, 446
  regulatory and governance frameworks, 448
  risks and implications, 442
  RLHF, 445
  sources, 439, 440
Bias mitigation strategies, 95
  in-processing techniques
    fairness constrained learning, 136
    representation learning, 137
  postprocessing techniques
    outcome calibration, 137
    reject option, 137
    threshold adjustment, 137

preprocessing techniques
data augmentation, 136
feature filtering, 136
rebalancing, 135
Bias-variance trade-off, 172
BigGAN, 60
Bilingual Evaluation Understudy (BLEU), 10, 487, 490
machine translation, 493
text evaluation, 502, 503
BLEU, *see* Bilingual Evaluation Understudy (BLEU)
Bradley–Terry model, 355, 356

C

Categorical encoding techniques, 114
cGAN, *see* Conditional GAN (cGAN)
Chatbot, 468
Checkpointing, 185
challenges, 272
elastic recovery, 271, 272
frequency, 271
future research, 272, 274
granularity, 271
historical context, 266, 267
models of failures, 268, 269
motivations, 266
strategies, 269, 270
CI/CD pipelines, *see* Continuous integration and continuous delivery (CI/CD) pipelines
CIDEr, 504
Clarification, 110
Classical assessment theory, 537
Classical decay methods, 234, 235
Classifier-free guidance, 76
Classifier guidance, 76
Cloud computing, 574
Cluster-level orchestration, 254
CNNs, *see* Convolutional neural networks (CNNs)
Code synthesis, 17
Cognitive load, 538
Comparative analysis, 585
Competition law, 583
Compiler stacks, 393
Compliance, 546, 578, 585, 586
Computational cost, 14
Computer vision (CV), 98, 404
Concept drift, 567
Conditional GAN (cGAN), 59
Conditional VAE (CVAE), 66
Connectionist temporal classification, 154
Content generation, 17
Contextual bias, 441
Contextual hallucinations, 433
Continual learning, 479
Continuous integration and continuous delivery (CI/CD) pipelines
challenges, 558
components, 547, 551–553
cross-organizational collaboration, 559

Continuous integration and continuous delivery (CI/CD) pipelines (*cont.*)
  data management, 553, 554
  DataOps, 559
  definition, 547
  deployment, 556
  environmental and cost considerations, 560
  evaluation, 555
  explainability, 560
  foundations, 548–551
  future research, 558
  model registries and versioning, 555
  monitoring and observability, 556
  multimodal and agentic systems, 561, 562
  origins, 548
  rollback and recovery, 557
  scaling, 557
  security, 560
  sociotechnical dimensions, 561
  training, 554
Contrastive learning, 148
Contrastive losses, 158–160
Control theory, 470, 563
Convergence
  definition, 182
  early stopping, 183
  gradient norms, 183
  loss stabilization, 182
  representation stability, 184
  validation metrics, 183
Conversational agents, 16
Convolutional networks, 168
Convolutional neural networks (CNNs), 25
Corrective loops, 469
Cosine annealing, 236
Counterfactual testing, 134
Cross-border legal complexities, 580
Cross-entropy loss, 156
Cross-modality generation, 21
Cross-modal tasks, 404
Cross-validation, 168
Crowdsourcing, 536, 541
Cultural bias, 498
Cultural homogenization, 481–482, 581
Cultural justice, 584
Cultural variability, 510
Custom logging, 185
CV, *see* Computer vision (CV)
CVAE, *see* Conditional VAE (CVAE)
Cybernetics, 480
Cybersecurity, 457
Cyclical schedules, 236

D

Data augmentation, 167
Data cleaning
  automation tools and methods, 116
  clinical trial, 118

purpose, 107–109
Data drift, 567
Data management, 553, 554
DataOps, 559
Data parallelism, 258, 259
Data preparation, 95
Data protection, 576
Dataset bias
causal inference, 138
definition, 131
detection, 133, 134
facial recognition, 139, 140
forms, 131–132
future research, 141
intersectional fairness, 138
mitigation, 135–137
model performance, 135
organizational and legal considerations, 140, 141
outcomes, 133
practical workflow, 138, 139
Data types
cleaning and normalization pipelines, 94
overview, 94
semistructured data, 99, 100
storage *vs.* retrieval mechanism, 100, 101
structured data, 96, 97
synthetic data generation, 94
unstructured data, 97, 99
DCGANs, *see* Deep convolutional GANs (DCGANs)
DDIMs, *see* Denoising diffusion implicit models (DDIMs)
DDPMs, *see* Denoising diffusion probabilistic models (DDPMs)
Decoder, 63
Decoder-only architecture, 45
Deep convolutional GANs (DCGANs), 60
Deepfake technologies, 62, 590
Deep learning, 406
Deep neural networks, 145
DeepSpeed, 280
Demographic bias, 440
Denoising diffusion implicit models (DDIMs), 70, 71, 80
Denoising diffusion probabilistic models (DDPMs), 69–71
Denoising strategies, 72, 73
Deployment
CI/CD (*see* Continuous integration and continuous delivery (CI/CD) pipelines)
complexity, 546
legal, ethical and compliance challenges, 573–586
model monitoring and observability, 562–573
multimodal fusion, agentic models and AI safety, 586–598
trends, 597
Detection methods, 115

Differential checkpointing, 270
Diffusion models, 6, 126
  advantages, 41
    flexibility in conditioning, 75, 76
    high-output fidelity, 74
    mode-coverage, 74
    stable training, 73, 74
    synergy, 77, 78
    theoretical grounding, 76, 77
  applications, 84
  application's priorities, 89
  architecture components, 40
  challenges, 41
    computational cost, 80
    memory footprint, 81
    over-smoothing risk, 81
    sampling speed, 79, 80
    trade-offs and application-specific considerations, 82
  components, 70
  concept, 69
  denoising strategies, 72, 73
  enterprise applications, 40
  ethical and societal considerations, 91
  future research, 90
  vs. GANs, 85, 86
  hybrid models, 90
  hybrid usage, 40
  intuition, 71
  limitations, 79
  ongoing-research, 83
  origins and breakthroughs, 39
  practical implications, 78
  vs. transformers, 88, 89
  usage in generation, 40
  vs. VAEs, 86, 87
  variants, 71, 72
Discriminative models, 2, 3
Discriminator, 157
Distance-based models, 110
Distributed monitoring, 186
Distributed training frameworks
  challenges, 278, 279
  communication primitives and collective operations, 275, 276
  comparison, 279–281
  fault-aware and elastic designs, 277
  future research, 281, 282
  gradient synchronization, 276, 277
  parallelism, 278
  role of, 274
Distributional metrics, 491, 508
Distribution shifts, 171
Diversity scores, 506, 507
Domain-specific dataset curation
  annotation and labeling, 305
  balancing scale and relevance, 306, 307
  challenges, 304, 305
  ethical and privacy considerations, 308, 309
  historical context, 302, 303
  implications, 309, 310

pipeline, 299–301
principles, 303
reasons, 302
synthetic data generation and augmentation, 307, 308
workflows, 306
Domain-specific use cases, 21
Double descent, 170
Drift monitoring, 117, 118
Drift tracking, 188
Dual-use potential, 584
Dynamic learning rates, 234

E

Economics of compliance, 585, 586
EDA, *see* Exploratory data analysis (EDA)
Educational technology, 532
Elastic provisioning, 254
Elastic recovery, 271, 272
Elastic scaling, 421
ELBO, *see* Evidence lower bound (ELBO)
Embeddings, 134
Encoder, 63
Encoder-decoder architecture, 45
Ensemble methods, 169
Entity-level hallucinations, 433
Epistemic justice, 584
Ethical governance, 580
Ethical risks, 15, 581
Evidence lower bound (ELBO), 7, 64
Explainability, 578
Explicit generative models, 6
Exploratory data analysis (EDA), 133
Exponential decay, 235
Extrinsic assessment, 531
Extrinsic hallucinations, 433

F

Face recognition, 172
Facial recognition, 139, 140
Failure modes, 430
Fairness metrics, 577
Fault tolerance, 277
See also Checkpointing
definition, 267
principles, 267
Feedback collection
annotation noise and reliability, 375
annotation quality and consistency, 373, 374
bias, 374
challenges, 377
cost, 372
cultural and linguistic diversity, 374
effectiveness, 372
ethical concerns, 377
fundamental tension, 378, 379
over-optimization, 376
scaling, 376

Feedback loops
- amplification of bias, 472
- case illustrations, 473
- challenges, 481
- continuous fine-tuning and iterative model updates, 474, 475
- cultural and ethical dimensions, 479
- data drift and decay, 471
- dynamic interaction, 467
- governance, 475
- historical parallels, 469, 470
- institutional sources, 477
- management, 480
- modalities, 468
- monitoring, 474
- multimodal and multi-agent systems, 478
- psychological and behavioral effects, 480
- recommender systems, 472
- risks, 470, 471
- societal risks, 481, 482
- trade-offs, 476
- types, 468, 469
- updating, 477, 479
- user feedback, 476

Few-shot learning
- advantages, 318
- applications, 318
- case-style illustrations, 321, 322
- challenges and risks, 319, 320
- explainability and uncertainty, 320, 321
- foundations, 311
- mathematical intuition, 316, 317
- meta-learning to in-context learning, 312, 313
- research, 322, 323
- strategies, 313, 314
- workflows, 317, 318

FID, *see* Fréchet Inception Distance (FID)

Fine-tuning
- adapters and LoRA, 323–334
- domain-specific dataset curation, 299–310
- few-shot and zero-shot learning, 311–323
- parameter-efficient, 324, 325
- transfer learning and pre-trained model utilization, 290–299

Fixed learning rates, 233

Formative *vs.* summative assessment, 536

Foundation models, 105

FP16 operations, 392

Fréchet Inception Distance (FID), 8, 10, 59, 490, 507
- image generation, 493, 494

G

GANs, *see* Generative adversarial networks (GANs)

Gated recurrent units (GRUs), 44

Gaussian distributions, 541
Generalization, 353
  architectural choices, 168
  cross-validation, 168
  data augmentation, 167
  diagnostics, 177
  ensembles, 169
  LLMs, 171
  monitoring and validation, 169
  regularization, 167
  transfer learning and pretraining, 168
Generative adversarial networks (GANs), 5, 7, 8, 13, 69, 124
  advantages, 37, 58
  architectural innovations, 60
  architecture components, 36
  challenges, 37, 58, 59
  components, 56
  conditional, 59
  definition, 56
  vs. diffusion models, 85, 86
  enterprise applications, 37
  ethical and societal considerations, 62
  evolving role of, 62
  vs. generative architectures, 61
  hybrid usage, 37
  image generation, 61
  origins and concept, 36
  steps, 57
  usage in generation, 36
Generative AI metrics, 501, 502
Generative modeling, 63, 430
  applications, 12
  automatic and human evaluation methods, 489–500
  challenges, 13–15
  definition, 2–4
  description, 1, 2
  evaluation, 9–12
  performance, 3
  probabilistic foundations, 6–8
  training methodologies and loss functions, 8, 9
  types
    autoregressive, 5
    diffusion, 6
    GANs, 5
    VAEs, 4
Generative outputs, 575, 576
Generative systems, 15, 546
Generative *vs.* discriminative models
  advantages and disadvantages, 27
  architecture and training, 25
  data requirements and representational learning, 26
  evaluation metrics, 26
  evolutionary trajectories, 31–33
  fundamental objectives, 23
  human-AI collaboration, 28–30
  interoperability and hybrid architectures, 28
  use cases, 24

Generator, 157
Geometry of knowledge, 317
Goodhart's Law, 509
Gossip-based
 synchronization, 277
GPUs, *see* Graphics processing
 units (GPUs)
Gradient clipping
 effectiveness, 229
 management, 230, 231
 practical heuristics, 226
 problem, 222, 223
 strategies, 223–226
Gradient descent
 challenges, 205, 207
 definition, 204
 hyperparameters, 205
 perspectives, 209
 variants, 205, 207, 208
Gradient descent-based
 models, 110
Gradient norm
 balancing, 225
Gradient penalty, 158
Gradient synchronization, 276, 277
Graphics processing units (GPUs)
 characteristics, 250, 251
 evolution, 249
 limitations, 251
Graph optimization, 420
Group normalization, 228
GRUs, *see* Gated recurrent
 units (GRUs)
Guided diffusion, 72

H

Hallucinations, 13, 429, 470, 520
 analysis, 483
 categories, 433
 creative contexts, 431
 definition, 431
 detection strategies, 434
 illustrations, 437
 implications, 437, 438
 mitigation
  decoding strategies, 435
  hybrid pipelines, 436
  RAG, 435
 problem, 431, 432
 risks and implications, 433, 434
 RLHF, 436
 structural features, 432
Handling methods, 116
Hardware scaling, 255, 256
Hierarchical allreduce, 277
Hierarchical checkpointing, 270
Hierarchical VAE, 66
High stakes domains, 498, 499
Hinge loss, 157
Hiring algorithm, 172
Hiring systems, 473
Historical bias, 132
Horovod, 279
HuBERT, 149
Human evaluation, 11, 14, 488,
 489, 555
 challenges, 495
 cultural and ethical
  dimensions, 499

forms, 492
horizons, 499, 500
psychological and cognitive biases, 498
research and deployment, 496
Hybrid losses, 160, 161
Hybrid strategies, 262, 263, 494

I

Image generation, 10
Image inpainting, 156
ImageNet dataset, 144
Image synthesis, 18
Implicit generative models, 7
Implicit regularization, 170
Inception Score (IS), 8, 10, 59, 490, 507
Incremental checkpointing, 269
Independent auditing, 474
Industry *vs.* academia metrics, 509
Inference model, 63
InfoNCE loss, 159
Instance normalization, 228
InstructGPT project, 350
Instruction tuning, 52
Intellectual property (IP), 575, 576
INT8 operations, 391
Interpretability, 508
Intrinsic assessment, 531
Intrinsic hallucinations, 433
IP, *see* Intellectual property (IP)
IS, *see* Inception Score (IS)

J

Jensen-Shannon divergence, 7

K

Kernel methods, 110
Knowledge distillation
challenges, 404, 405
computer vision and speech, 404
definition, 399
engineering practices, 403
extensions and innovations, 405
future research, 405, 406
historical roots, 400
motivation, 399
NLP, 403
theoretical principles, 400, 401
types, 401, 402
Kullback-Leibler (KL) divergence, 7, 64, 125

L

Labeling bias, 132
Language models, 126, 353
Large-batch training, 214–216
Large language models (LLMs), 16, 150, 171, 517
Latent analysis, 134
Latent diffusion models (LDMs), 70, 72, 80, 87
Layer normalization (LayerNorm), 227

Layer-wise adaptive moments for batch training (LAMB), 214–216
Learned evaluation models, 11
Learning rate, 174
Learning rate schedules
  classical decay methods, 234, 235
  comparative properties, 239
  definition, 231
  fixed *vs.* dynamic, 233, 234
  future research, 239
  importance, 232
  large-scale training considerations, 237, 239
  modern schedules, 235–237
  parameters and time, 233
Least squares loss, 157
Legal hallucinations, 437
Liability, 576
Linguistic drift, 471
Linguistic variability, 510
L2 normalization, 111
Log transformation, 111
Long short-term memory (LSTMs), 34, 44
LoRA, *see* Low-rank adaptation (LoRA)
Loss functions
  categories, 155
    adversarial losses, 157, 158
    contrastive losses, 158–160
    hybrid losses, 160, 161
    reconstruction losses, 155, 156
    reflections, 161, 162
  definition, 153
  gradient-based optimization, 154, 155
  speech recognition model, 153
Low-rank adaptation (LoRA), 52
  vs. adapters, 330, 331
  applications, 331, 332
  challenges and limitations, 332, 333
  efficiency and flexibility, 329, 330
  ideas, 328
  research, 333, 334
LSTMs, *see* Long short-term memory (LSTMs)

M

MAR, *see* Missing At Random (MAR)
Machine learning, 105, 107
Masked language modeling (MLM), 148, 156
Maximum likelihood estimation, 7
MCAR, *see* Missing Completely At Random (MCAR)
Mean absolute error, 156
Mean squared error (MSE), 155
Measurement bias, 131
Media hallucinations, 437
Medical AI systems, 582

Medical hallucinations, 437
Memorization, 172
Memory management, 419
Meta-learning, 221
Meta-metrics, 513
METEOR, 10, 504
Metrics
  A/B testing, 520, 521
  cultural and linguistic variability, 510
  ensemble evaluation, 513
  future generation, 514–516
  gaming, 509
  generative AI, 501, 502
  industry *vs.* academia, 509
  live systems, 511
  longitudinal evolution, 511, 512
  problems, 512, 513
  statistical properties and interpretability, 508
Mini-batch gradient descent, 205
Mini-batching, 208
Min–max scaling, 110
Missing At Random (MAR), 113
Missing Completely At Random (MCAR), 113
Missing data, 113, 114
Missingness, 113
Missing Not At Random (MNAR), 113
Mixed-precision quantization, 390
MLM, *see* Masked language modeling (MLM)
MLOps, 548
MNAR, *see* Missing Not At Random (MNAR)
MobileBERT, 409
Mode collapse, 13
Mode coverage, 75, 85
Model-agnostic meta-learning (MAML), 312
Model compression/inference optimization
  architectures, 386
  efficient architectures
    ALBERT and parameter sharing, 410
    engineering practices, 412, 413
    future research, 413, 415
    historical context, 407, 408
    MobileBERT and TinyBERT, 409, 410
    principles, 407–409
    sparse attention mechanisms, 411
    TinyML and edge deployment, 410, 411
    trade-offs, 412
  goals, 385
  importance, 386
  knowledge distillation, 399–406
  quantization and weight pruning, 387–398
  serving at scale, 415–424
  strategies, 386
Model decay, 471
Model parallelism, 259, 260

Modular adaptation, 325
Momentum, 207
Multi-agent adversarial ecosystems, 466
Multi-agent systems, 478
Multidimensional evaluation, 495, 496
Multi-head attention, 44
Multimodal fusion, 98, 586
  applications, 589, 590
  case studies, 596, 597
  ethical and social risks, 590, 591
  implications, 588
  integration, 587
  technical dimensions, 588, 589
  transformers, 587
  trends, 595
Multimodal models, 20, 21
Multimodal systems, 478, 497, 546, 561, 562, 571
Multi-teacher distillation, 402
Multivariate tests, 525
Music generation, 19

N

NAS, *see* Neural architecture search (NAS)
Natural language processing (NLP), 5, 98, 403
Negative loops, 469
Nesterov momentum, 208
Neural architecture search (NAS), 413
Neural networks, 124, 202, 248
Next-sentence prediction (NSP), 148
NIST AI Risk Management Framework, 578
NLP, *see* Natural language processing (NLP)
Nonequilibrium thermodynamics, 70
Normalization, 94
  algorithms, 110
  effectiveness, 229
  encoding categorical variables, 111
  methods, 226–228
  pipelines, 112, 113
  role of, 228
  techniques, 110, 111
Norm-based clipping, 224
NSP, *see* Next-sentence prediction (NSP)

O

Observability
  automated *vs*. human-in-the-loop, 569
  challenges, 564
  concept, 562, 563
  data drift and concept drift detection, 567
  ethical and legal dimensions, 569, 570
  future research, 572, 573

infrastructure, 568, 569
longitudinal and life cycle monitoring, 571
multimodal monitoring, 571
organizational and sociotechnical considerations, 572
performance and efficiency monitoring, 565
quality and alignment monitoring, 565
safety and compliance monitoring, 566
security considerations, 570
user feedback integration, 568
ONNX Runtime, 393, 415
Operator fusion, 419
Optimization
adaptive, 210–221
backpropagation and gradient descent, 201–209
challenges, 199
gradient clipping and normalization, 221–231
historical roots, 198
importance, 200, 201
learning rate schedules, 231–240
reasons, 198
themes, 199, 200
Organizational culture, 580
Overfitting
bias-variance trade-off, 163
causes
correlations, 166
data leakage, 165
dataset size and quality, 165
excessive training, 166
lack of regularization, 166
model capacity, 164
definition, 163
examples, 163

P

Pairwise comparisons, 355, 356
Parallelism
abstractions, 278
bottlenecks and challenges, 264, 265
conceptual and technical solution, 256
data, 258, 259
defined, 257
hybrid, 262, 263
model, 259, 260
pipeline, 260, 261
requirements, 257
tensor, 261, 262
Parallelizable training, 47
Parameter-efficient tuning techniques, 52
Parameter server paradigm, 276
Participatory governance, 582
Perceptrons, 202
Periodic checkpointing, 269
Per-layer clipping, 225
Perplexity, 11, 505, 506, 508, 511, 541

Personalization, 19
Pipeline parallelism, 260, 261
Polynomial decay, 235
Popularity bias, 472
Positive feedback loops, 468
Post-training quantization, 390
PPO, *see* Proximal policy optimization (PPO)
Predator-prey dynamics, 469
Preference metrics, 520
Preprocessing, 101, 103
Pretraining, 168, 170
Privacy, 576
Probabilistic learning, 429, 438
Probabilistic models, 124
Probability theory, 6–8
Professional liability frameworks, 576
Progressive distillation, 80
Progressive Growing of GANs, 58, 60
Proprietary models, 583
Provisioning, 253, 254
Proximal policy optimization (PPO), 343
  algorithms, 369
  challenges, 369, 370
  concept, 364
  conceptual principle, 371
  engineering practices, 367, 368
  policy gradient methods, 365
  pre-trained models, 367
  principles, 366, 367
  reasons, 364, 365
  research, 370
  simplification, 366
  summarization and dialogue alignment, 368
Pruning
  engineering practices, 396, 397
  historical development, 395
  illustrations, 397
  implications, 397, 398
  motivation, 388, 389
  principles, 393
  trade-offs, 395, 396
  types, 394
PyTorch ecosystem, 280

Q

Quantile transformation, 111
Quantization
  engineering practices, 396, 397
  hardware support, 391–393
  illustrations, 397
  implications, 397, 398
  motivation, 388, 389
  principles, 389
  trade-offs, 395, 396
  types, 389, 391
Quantization-aware training, 390

R

RAG, *see* Retrieval-augmented generation (RAG)
Randomization, 520

Reasoning hallucinations, 433
Recall-Oriented Understudy for Gisting Evaluation (ROUGE), 10, 503, 504
Recognition network, 63
Reconstruction losses, 155, 156
Reconstruction term, 64
Recurrent neural networks (RNNs), 25, 34, 222
Red teaming
  case studies, 459–462
  continuous monitoring and adaptation, 459
  cybersecurity practices, 457
  framework, 455, 456
  future research, 464
  implications, 463
  multi-agent ecosystems, 466
  organizational structures, 458, 459
  regulatory integration, 465
  stress test systems, 450
  techniques, 456, 457
Reference-based metrics, 491
Reference-free evaluation paradigms, 496, 497
Regression model, 489
Regression testing, 555
Regularization, 167
Regularization term, 64
Reinforcement learning, 221
Reinforcement learning with human feedback (RLHF), 11, 474
  bias mitigation, 445
  components, 344
  feedback collection, 372–379
  hallucination reduction, 436
  human annotation pipelines
    challenges and trade-offs, 351
    emerging to traditional pipelines, 351, 352
    InstructGPT, 350
    interfaces and workflow design, 348
    pairwise ranking and direct scoring, 346, 347
    quality control and inter-annotator agreement, 349
    recruitment and training, 347, 348
    role of, 345
    scalability, 349, 350
  importance, 344
  PPO, 364–372
  reward modeling, 352–363
  training loop, 343
Reinforcing loops, 469
Reparameterization trick, 64
Representation bias, 132
Resource intensity, 152
Retrieval-augmented generation (RAG), 435
Reward hacking, 358
Reward modeling
  AI-assisted feedback, 362
  challenges, 361

Reward modeling (*cont.*)
- concept, 354, 355
- constitutional signals, 362
- definition, 352
- dialogue, 359
- future research, 362, 363
- generalization, 359
- human judgments to functions, 355
- instruction following, 360
- iterative refinement, 360, 361
- overfitting and hacking, 357, 358
- pairwise comparisons and Bradley–Terry model, 355, 356
- preference distillation, 362
- reasons, 353, 354
- scaling, 356, 357
- summarization, 359

Ring-allreduce, 276
RLHF, *see* Reinforcement learning with human feedback (RLHF)
RMSNorm, *see* RMS normalization (RMSNorm)
RMS normalization (RMSNorm), 228
RMSProp, 211, 213
ROUGE, *see* Recall-Oriented Understudy for Gisting Evaluation (ROUGE)
Robust scaling, 111
Rollback systems, 556
Rule-based simulations, 123

S

Safety checks
- automated *vs.* human-in-the-loop, 453, 455
- categories, 452, 453
- cultural and ethical dimensions, 465
- future research, 464
- historical analogies, 464
- implications, 463
- life cycle, 450
- rationale, 451, 452
- trade-offs, 462, 463

Safety metrics, 520
Sampling bias, 131
Scaling law, 48
Scaling training
- checkpointing and fault tolerance, 265–274
- data and model parallelism, 256–265
- distributed training, 274–282
- GPU/TPU usage and resource provisioning, 248–256
- problem, 247
- themes, 248

Score-based generative models (SGMs), 71, 80
Second-order methods, 208
Sector-specific compliance, 579
Self-attention, 44
Self-distillation, 402
Self-evaluating models, 489

Self-healing models, 479
Self-supervised learning (SSL), 145
  advantages, 151
  applications, 149, 151
  augmented versions, 147
  challenges, 152
  concept, 146
  convergence criteria, 182–184
  emerging techniques
    automated alerts and intervention, 188
    drift tracking, 188
    large-scale experiment tracking, 189
    sharpness-aware monitoring, 188
    training loss, 189
  generalization, 167–169
  loss function design (*see* Loss functions)
  monitoring, 187
    metric myopia, 187
    noisy loss curves, 187
    overfitting to validation, 186
    resource blindness, 187
  monitoring convergence and training progress, 173–175
  overfitting, 163–166
  perspectives, 169–172
  practical monitoring tools, 184–186
  pretext tasks, 148, 149
  training diagnostics, 175–182
  variant, 147
Semantic diversity, 506
Semistructured data, 94
  hybrid zone, 99, 100
  storage, 101
Semi-supervised learning, 145
Semi-supervised VAE, 66
Sequence modeling, 88
Sequence-to-sequence distillation, 402
Serving at scale
  batching, 418
  challenges, 417
  compressed model, 415
  elastic scaling, 421
  fault tolerance and reliability, 422
  frameworks, 421
  future research, 423, 424
  hardware specialization and acceleration, 420
  memory management, 419
  operator fusion and graph optimization, 419
  optimized runtimes, 416
  quantized models, 416
  runtimes and inference engines, 415, 417
  trade-offs, 422, 423
SGD, *see* Stochastic gradient descent (SGD)
SGMs, *see* Score-based generative models (SGMs)
Sharded checkpointing, 269

Sharpness-aware minimization, 188
Shortcut learning, 152
Simulation testing, 474
Single-device training, 247
Social equity, 577
Spectral distortion, 490
Speech recognition, 153, 404
Speech synthesis, 19, 490
SSL, *see* Self-supervised learning (SSL)
Static provisioning, 253
Statistical testing, 134, 518
Step decay, 234
Stereotypical bias, 440
Stochastic gradient descent (SGD), 205, 210
Structured data, 94
  implications, 104
  nature of, 96, 97
  preprocessing, 101
  storage, 100
Structured pruning, 394
StyleGAN, 58, 60
Super-convergence, 237
Supervised generative modeling, 3
Supervised learning, 31, 103, 144
Synthetic data, 584
Synthetic data generation, 94
  categories, 120
  challenges, 129
  definition, 120
  developments, 129
  evaluation, 127, 128
  practice, 128
  reasons
    balanced class representation, 122
    controlled experimentation, 122
    data augmentation and generalization, 122
    data scarcity, 121
    privacy preservation, 121
  techniques, 123–126
Systemic bias, 441
Systemic vulnerability, 483
Systems theory, 468

T

Task-based performance assessments, 488
  adaptive and personalization, 538
  authenticity, 533
  case study
    code generation, 535
    education, 535
  challenges, 535
  complexity and cognitive load, 538
  concept, 530
  continuous, 538
  domain-specific examples, 532
  education and psychology, 537
  ethical and cultural dimensions, 539

evaluation, 536
formative *vs.* summative, 536
foundations, 531
future research, 540
governance, 539
historical roots, 532
human–AI collaboration, 534
human-in-the-loop, 536
intrinsic benchmarks *vs.* real-world integration, 534
intrinsic *vs.* extrinsic, 531, 532
outcome measures, 533
reductionism, 539
safety and fairness, 537
scaling, 537
Task-based proxies, 491
Task-specific distillation, 401
Tech regulation, 574, 575
Telecommunications technologies, 574
TensorBoard, 184
TensorFlow distributed strategies, 280
Tensor parallelism, 261, 262
Tensor processing units (TPUs)
advantages, 252
characteristics, 252
evolution, 250
workloads, 253
Text encoding techniques, 115
Text generation, 10, 16, 17
Text-to-speech (TTS) systems, 19
TinyBERT, 409
TinyML, 410, 411
Tokenization, 510
Topical drift, 471
Toxic bias, 441
Toxicity, 438
See also Bias
TPUs, *see* Tensor processing units (TPUs)
Trade secret law, 583
Training diagnostics
collapse and representation, 179, 180
data and pipeline, 180, 181
generalization, 177
integrated workflows, 181
loss and objective, 176, 177
numerical and statistical health, 179
optimization and learning rate, 178
toolkit, 175
Training dynamics
batch size and gradient noise, 174
learning rate, 174
loss curves and plateaus, 175
oscillations and instabilities, 175
Transfer learning, 168
assumptions, 291
benefits, 295
concept, 290
feature extraction to pre-trained transformers, 292, 293

Transfer learning (*cont.*)
pre-trained model utilization, 297
risks and challenges, 295, 297
significance, 298, 299
strategies, 293, 294
Transformer-based fusion, 588
Transformers, 25
advantages, 35
cross-domain adaptability, 47, 48
flexible conditioning, 49, 50
long-range dependency modeling, 46, 47
parallelizable training, 47
scalability, 48, 49
synergy, 50–52
applications, 54, 55
architecture components, 34
challenges, 36
comparative perspective, 55
configurations, 45
vs. diffusion models, 88, 89
drawbacks, 52
enterprise applications, 35
evaluation considerations, 53
factors, 44
future research, 55, 56
generative settings, 35
hybrid usage, 35
origins and design, 34
origins and principles, 44, 45
training paradigms and enhancements, 52
Transparency, 578
TRPO, *see* Trust region policy optimization (TRPO)
Trust metrics, 520
Trust region policy optimization (TRPO), 365

U

Unstructured data, 94
annotation and labeling, 103, 104
emergence of, 97, 99
implications, 105
preprocessing, 102
storage, 100
Unstructured pruning, 394
Unsupervised generative modeling, 3
Unsupervised learning, 145
Utilization challenges, 254, 255

V

Value-based clipping, 224
Variational autoencoders (VAEs), 4, 6, 125, 156
advantages, 39, 64
applications, 67
architecture and probabilistic foundation, 63, 64
architecture components, 38
challenges, 39
definition, 63

vs. diffusion models, 86, 87
drawbacks, 65
enterprise applications, 38
ethical considerations, 68
future research, 69
vs. generative architectures, 67
hybrid usage, 39
multimodal and hybrid architectures, 68
origins and objectives, 38
reparameterization trick, 64
usage in generation, 38
variants and extensions, 66
Vector-quantized VAE (VQ-VAE), 66
VAEs, *see* Variational autoencoders (VAEs)
Vision and image generation, 17, 18
Vision transformers (ViTs), 88
Visual inspection, 8
ViTs, *see* Vision transformers (ViTs)
Voice cloning, 19
VQ-VAE, *see* Vector-quantized VAE (VQ-VAE)

W

Warmup schedules, 236
Wasserstein distance, 7
Wasserstein GANs (WGANs), 59
Wasserstein loss, 157
Wav2vec, 149
Weight normalization, 228
Weight pruning, *see* Pruning
Weights & Biases (W&B), 185
WGANs, *see* Wasserstein GANs (WGANs)

X, Y

XLA compilers, 393

Z

Zero-shot learning
advantages, 318
applications, 318
case-style illustrations, 321, 322
challenges and risks, 319, 320
explainability and uncertainty, 320, 321
foundations, 311
mathematical intuition, 316, 317
meta-learning to in-context learning, 312, 313
research, 322, 323
strategies, 314–316
workflows, 317, 318
Z-score standardization, 110